T0138362

How Our Days
Became Numbered

How Our Days Became Numbered

Risk and the Rise of the
Statistical Individual

DAN BOUK

The University of Chicago Press
Chicago and London

The University of Chicago Press, Chicago 60637
The University of Chicago Press, Ltd., London
© 2015 by The University of Chicago
All rights reserved. Published 2015.
Paperback edition 2018
Printed in the United States of America

27 26 25 24 23 22 21 20 19 18   2 3 4 5 6

ISBN-13: 978-0-226-25917-8 (cloth)
ISBN-13: 978-0-226-56486-9 (paper)
ISBN-13: 978-0-226-25920-8 (e-book)
DOI: 10.7208/chicago/9780226259208.001.0001

Library of Congress Cataloging-in-Publication Data
Bouk, Daniel B., 1980– author.
  How our days became numbered : risk and the rise of the statistical
individual / Dan Bouk.
      pages cm
  Includes bibliographical references and index.
  ISBN 978-0-226-25917-8 (cloth : alkaline paper)—ISBN 0-226-25917-X
(cloth : alkaline paper)—ISBN 978-0-226-25920-8 (e-book)—
ISBN 0-226-25920-X (e-book)   1. Life insurance—United States—History.
2. Insurance—Statistical methods—History. 3. Insurance companies—Social
aspects—United States.   I. Title.
  HG8531.B68 2015
  368.3200973—dc23

                                                    2014039799

♾ This paper meets the requirements of ANSI/NISO Z39.48–1992 (Permanence
of Paper).

For my parents,
Gail and Ted Bouk

For all our days are passed away in thy wrath: we spend our years as a tale that is told. / The days of our years are threescore years and ten; and if by reason of strength they be fourscore years, yet is their strength labour and sorrow; for it is soon cut off, and we fly away. / Who knoweth the power of thine anger? even according to thy fear, so is thy wrath. / So teach us to number our days, that we may apply our hearts unto wisdom.

PSALM 90:9–12, *King James Bible*

Do you suppose, he can be estimated by his weight in pounds, or, that he is contained in his skin,—this reaching, radiating, jaculating fellow?

RALPH WALDO EMERSON, "Fate," from *Conduct of Life*

CONTENTS

Strange Books

In 1936, James Cain's *Double Indemnity* drew popular attention to a peculiarly modern substantiation of fate. At a crucial moment in the novel, Keyes, the hardened claims adjuster, explains to his younger boss, Norton, how closely life insurers have circumscribed the known world. Keyes does not yet realize that the novel's protagonist—an agent—murdered a company policyholder. But he *knows* that the policyholder did not commit suicide by jumping from a train, despite all the evidence pointing to that conclusion. Keyes shows his boss a book full of tables and says, "Here's what the actuaries have to say about suicide." He goes on: "Here's suicide by race, by color, by occupation, by sex, by locality, by seasons of the year, by time of day when committed." He continues: "And here—here, Mr. Norton— are leaps subdivided by leaps from high places, under wheels of moving trains, under wheels of trucks, under the feet of horses, from steamboats. *But there's not one case here out of all these millions of cases of a leap from the rear end of a moving train.* That's just one way they don't do it."[1]

How Our Days Became Numbered tells the story behind the books of figures that captured Cain's imagination, a story of Americans seeking certainty and security in an unsettled, industrializing nation and becoming statistical subjects in the process. It tells a story of those who became the sometimes grudging objects of powerful corporations' efforts to forecast their deaths in the years after the Panic of 1873, of those who more happily accepted corporate help to lengthen their lives in the early twentieth century, and of those who submitted to a new system created by the United States government during the Great Depression that would predict the course of

---

1. James M. Cain, "Double Indemnity," in *Cain x 3* (New York: Alfred A. Knopf, 1969), 417.

their wage-earning lives. It tells the story of statistical systems built around promises of protection from financial ruin that developed into systems for strengthening and reforming bodies too. Such systems made Americans into risks and in the process numbered their days.

Three years before Cain's book appeared, in 1933, Metropolitan Life Insurance Company statisticians Louis Dublin and Bessie Bunzel published *To Be or Not to Be: A Study of Suicide*. Page 59 featured a table of "suicides from jumping," and while it failed to distinguish between leaps under trucks and leaps under horses' feet, Dublin and Bunzel did compare jumps from skyscrapers to jumps from atop a barn.[2] And the book made age, sex, occupation, location, and race crucial factors in explaining and understanding suicide. Cain exaggerated the details, but not the spirit of the enterprise.

Dublin and Bunzel's book belonged to a long tradition in and around life insurers' offices of writing that, as Cain imagined, provided special insights into unknown pasts and uncertain futures. "Methods used by those who commit suicide follow fairly well defined lines," explained Dublin and Bunzel.[3] When Americans killed themselves in the years approaching 1930, more than a third chose shooting (among white men, the figure jumped to over 40 percent), followed by hanging, poisoning, and asphyxiation.[4] "Colored women" preferred poisoning, then firearms.[5] Statistical knowledge became, in the proper hands, evidence of regularities, even *laws* of (human) nature. With such knowledge, one could predict the future in broad strokes; one could read groups' fates; one could number a nation's days.

Yet Cain's story also reveals an impulse that reached past the usual goals of statistics and probability, past merely gaining certain knowledge about populations, to the more difficult puzzle of securing insights into the lives and fates of individuals. Cain solved the problem like a novelist—he invented a situation that statistics showed to be impossible, a situation that involved no chance. But actual life insurers struggled to square probabilistic, statistical methods with a business that contracted with individuals—indeed, a business that often treated individuals differently, demanding of one applicant more payment, rewarding to another's bereaved family a smaller claim. Statistical and probabilistic methods could not discriminate

2. Louis I. Dublin and Bessie Bunzel, *To Be or Not to Be: A Study of Suicide* (New York: Harrison Smith & Robert Haas, 1933), 59.

3. Dublin and Bunzel, *To Be or Not to Be*, 56.

4. Dublin and Bunzel, *To Be or Not to Be*, 56–57, 59.

5. Dublin and Bunzel, *To Be or Not to Be*, 61. Dublin had previously made the argument that "suicide in the United States is almost altogether limited to white people." Louis I. Dublin, "To Be or Not to Be?" *Harper's Monthly Magazine* 161, no. 964 (1930): 486–494 at 487.

so finely: they did not deal with individuals. So life insurers brought in lawyers and investigators, and most importantly, doctors—alongside countless clerks and tabulating machines—to process, evaluate, and forecast individual lives.

When we think of science and life insurance or science and risk, we usually think about actuaries (as did Cain, incorrectly)—in no small part because life insurers touted their reliance on accomplished mathematicians capable of revealing statistical laws of nature.[6] But in practice, life insurers relied just as squarely on the expertise of doctors intent on determining what particularity or difference marked each person. Indeed, when Louis Dublin landed his first job in life insurance, it wasn't in an actuarial position. Instead, he joined a medical department at the heart of a company's bureaucratic assembly line for manufacturing individualized "risks," the sort of line that would have actually been responsible for Cain's imagined catalog of fates.

Insurers hedged their bets by roping the probabilistic to the particular, the actuary to the doctor. They at once proclaimed the surety of statistical laws, but acted as if they could beat the averages—claiming that they could pick only the better lives and avoid those doomed to die too soon. Such hedging came at a cost. Cain's never-recorded suicide from the back of a slow-moving train neatly avoided the troubling tension between smoothing away particularities, as the actuaries advocated, and classing all experience into tiny boxes, as doctors preferred. That tension bedeviled life insurers as their networks expanded and their coffers filled with cash. The question that Cain avoided asking was voiced over and over in the late nineteenth century, in private corner offices and in the people's legislative halls: how should

---

6. Life insurers had a reason to boast. No other form of insurance adopted probabilistic methods so early. For a discussion of the reasons that life insurance made this jump, while marine and fire insurance did not, see Lorraine Daston, *Classical Probability in the Enlightenment* (Princeton, NJ: Princeton University Press, 1988), 131–133. In the 1840s, life insurance succeeded games of chance as the paradigm case for probabilizing, a shift stemming partly from discomfort with sordid origins in gambling, but even more so from a new commitment to a "frequentist" belief that probabilities were real things in the world that could be determined from past experience. See Theodore M. Porter, *The Rise of Statistical Thinking 1820–1900* (Princeton, NJ: Princeton University Press, 1986), 81–88. On the place of actuaries in the larger scientific community in Britain, see Timothy Alborn, "A Calculating Profession: Victorian Actuaries among the Statisticians," *Science in Context* 7, no. 3 (1994): 433–468; and Timothy Alborn, *Regulated Lives: Life Insurance and British Society 1800–1914* (Toronto: University of Toronto Press, 2009), 102–135. On American actuaries, see E. J. Moorhead, *Our Yesterdays: The History of the Actuarial Profession in North America 1809–1979* (Schaumburg, IL: Society of Actuaries, 1989).

these powerful corporations, their fancy mathematics tensely tied to their surveying ambitions, be allowed to relate to Americans in groups or alone?

A new possible answer came in the early twentieth century in a development that the novelist Cain failed to grasp. For while Cain explored the powerful dream of using statistical methods to see into a life, he missed a more radical, and increasingly influential, vision—one that Dublin, Bunzel, and their peers championed at the same moment. By the 1930s, these life insurance statisticians had new ambitions for all the figures in their books. Unsatisfied with predicting the future, they set out to change it. Where most life insurers remained satisfied with discovering the clues that could point to an early death, that could nose out a potential suicide before he bought a policy, Dublin and Bunzel—with Metropolitan's support and, more importantly, the company's data—aimed to prevent suicide.[7] *How Our Days Became Numbered* explores this fundamental transition from nineteenth-century dreams of reading fate to twentieth-century efforts to master fate, from foretelling death to grasping after (and controlling) life. This transition mattered to insurers, but also to doctors, corporate managers, bureaucrats, and legislators, and it shaped the lives of ordinary Americans (those who read Dublin and Bunzel's book, or the many more who read Cain's noir, and those who read nothing at all). It promised a new beginning, an opportunity for life insurers' tools to migrate into more settings, touch more lives, and foretell more fates than purely mortal ones. Yet even as it allowed our days to be more fully numbered, the transition from prediction to control still failed to solve life insurers' persistent puzzle; it failed to finally reconcile the individual to the statistical.

* * *

I began writing this book fascinated by strange books lurking in lonely corners of university libraries. Books titled *The Money Value of a Man*, or *How to Live*, or *Race Traits and Tendencies of the American Negro*, or *Sources of Longevity*, or *Social Security in America*. These books—concerned with the lives and deaths of men and races, intent on forecasting or extending Americans' lives, determined to make dollars stand in for people, and all hailing in one way or another from America's life insurance industry—begged an explanation. Their origins in an industry that business historians have proven to be central to modern finance and corporate capitalism started me thinking.[8]

7. Dublin and Bunzel, *To Be or Not to Be*, ix.

8. See Sharon Ann Murphy, *Investing in Life: Insurance in Antebellum America* (Baltimore, MD: Johns Hopkins University Press, 2010); J. Owen Stalson, *Marketing Life Insurance: Its History in*

Had I stumbled upon evidence of financial power transmuting into cultural power? American intellectual and cultural historians had up to that time largely neglected life insurance,[9] but the work of other scholars studying the histories of quantification, statistics, and the human sciences made clear that such transmutations were possible, even probable.[10] Looking again at those strange books, I decided that I had indeed found evidence of the cultural power of corporate capital; but more than that, I realized that before me lay a key to understanding the peculiarly powerful way that American capitalism translates people into numbers.

For help in learning how to interpret these books, I looked to historians' efforts over the last few decades to integrate numbers and statistics—as the stuff of culture—into our understanding of modern American life. Their writings taught me how numbers spread across modern societies: how a "quantifying spirit" and an "avalanche of printed numbers" engendered an

---

America (Cambridge, MA: Harvard University Press, 1942); R. Carlyle Buley, *The American Life Convention 1906–1952: A Study in the History of Life Insurance* (New York: Appleton-Century-Crofts, 1953); Morton Keller, *The Life Insurance Enterprise: A Study in the Limits of Corporate Power* (Cambridge, MA: Belknap Press of the Harvard University Press, 1963); Marquis James, *Metropolitan Life: A Study in Business Growth* (New York: Viking Press, 1947); Shepard Bancroft Clough, *A Century of American Life Insurance: A History of the Mutual Life Insurance Company of New York, 1843–1943* (New York: Columbia University Press, 1946); and Douglass North, "Life Insurance and Investment Banking," *Journal of Economic History* 14, no. 3 (1954): 209–227.

9. Thankfully, that is changing. Jonathan Levy's example suggests just how much we have been missing by this neglect. See *Freaks of Fortune: The Emerging World of Capitalism and Risk in America* (Cambridge, MA: Harvard University Press, 2012); Viviana A. Rotman Zelizer's earlier works of historical sociology are another important exception. See Zelizer, *Morals and Markets: The Development of Life Insurance in the United States* (New York: Columbia University Press, 1979); Zelizer, *Pricing the Priceless Child: The Changing Social Value of Children* (Princeton, NJ: Princeton University Press, 1994). Scholars in the "Insurance and Society" field make the general case for the cultural embeddedness of insurance. For an introduction to that field, see Geoffrey Clark, Gregory Anderson, Christian Thomann, and J.-Matthias Graf von der Schulenburg, eds., *The Appeal of Insurance* (Toronto: University of Toronto Press, 2010); or Richard V. Ericson and Aaron Doyle, eds., *Risk and Morality* (Toronto: University of Toronto Press, 2003). American intellectual and cultural historians have paid more attention to banking or credit reporting. See, for instance, Stephen Mihm, *A Nation of Counterfeiters: Capitalists, Con Men, and the Making of the United States* (Cambridge, MA: Harvard University Press, 2009); or Scott A. Sandage, *Born Losers: A History of Failure in America* (Cambridge, MA: Harvard University Press, 2005).

10. Key works in this literature include Gerd Gigerenzer, Zeno Swijtink, Theodore Porter, Lorraine Daston, John Beatty, and Lorenz Krüger, *The Empire of Chance: How Probability Changed Science and Everyday Life* (Cambridge: Cambridge University Press, 1989); Daston, *Classical Probability in the Enlightenment*; Porter, *Rise of Statistical Thinking*; and Theodore M. Porter, *Trust in Numbers: The Pursuit of Objectivity in Science and Public Life* (Princeton, NJ: Princeton University Press, 1995); Alborn, "Calculating Profession"; and Alborn, *Regulated Lives*; Ian Hacking, *The Taming of Chance* (New York: Cambridge University Press, 1990); and Audrey B. Davis, "Life Insurance and the Physical Examination: A Chapter in the Rise of American Medical Technology," *Bulletin of the History of Medicine* 55, no. 3 (1981): 392–406.

"empire of chance" to which the United States and its peers paid tribute.[11] And they taught me to think about the ways that Americans could and did live their lives through and with numbers. Numbers, I learned, marked off space and time. In the early republic, numbers named buildings on busy streets, described property lines, facilitated the comparison and exchange of commodities—and even the exchange of enslaved human lives.[12] At the same time, numbers offered a media for self-expression (in diaries and ledgers), social imaginings (in newspaper columns), and political struggle (on and off legislative floors).[13] By the eve of the Civil War, Americans had learned in schools and in their marketplaces to reckon; they regularly encountered statistics in the press, and they mobilized numbers in some of the most important public debates of the day, such as the causes of steamboat explosions and the future of slavery.[14] They had become, in Patricia Cline Cohen's words, "a calculating people."[15] Numbers, as much as words or wood-cuts, pulsed through American culture.

As I read, I came away especially impressed by the power historians attributed to numbers as tools of aggregation. Numbers, they demonstrated,

11. For a formal overview of the varied uses of numbers and quantification in modern societies, see Wendy Nelson Espeland and Mitchell L. Stevens, "A Sociology of Quantification," *European Journal of Sociology* 49, no. 3 (2008): 401–436. See also: Tore Frängsmyr, J. L. Heilbron, and Robin E. Rider, eds. *The Quantifying Spirit in the 18th Century* (Berkeley: University of California Press, 1990); Ian Hacking, "Biopower and the Avalanche of Printed Numbers," *Humanities in Society* 5, nos. 3 and 4 (1982): 279–295; Gigerenzer et al., *Empire of Chance*.

12. For numbers at their worst, used as tools for valuing and pricing the enslaved, see Walter Johnson, *Soul by Soul: Life inside the Antebellum Slave Market* (Cambridge, MA: Harvard University Press, 1999), 135–161.

13. See especially recent work on the history of accounting, including Michael Zakim, "Bookkeeping as Ideology: Capitalist Knowledge in Nineteenth-Century America," *Common-place* 6, no. 3 (2006), http://www.common-place.org/vol-06/no-03/zakim; Caitlin Rosenthal, "Storybook-keepers: Narratives and Numbers in Nineteenth Century America," *Common-place* 12, no. 3 (2012), http://www.common-place.org/vol-12/no-03/rosenthal; Bruce G. Carruthers and Wendy Nelson Espeland, "Accounting for Rationality: Double-Entry Bookkeeping and the Rhetoric of Economic Rationality," *American Journal of Sociology* 97, no. 1 (1991): 31–69; and Naomi R. Lamoreaux, "Rethinking the Transition to Capitalism in the Early American Northeast," *Journal of American History* 90, no. 2 (2003): 437–461. Laurel Thatcher Ulrich explores an exemplary diary suffused with numbers in *The Life of Martha Ballard, Based on Her Diary, 1785–1812* (New York: Vintage, 1991).

14. Patricia Cline Cohen, *A Calculating People: The Spread of Numeracy in Early America* (Chicago: University of Chicago Press, 1982); Arwen P. Mohun, "On the Frontier of *The Empire of Chance*: Statistics, Accidents, and Risk in Industrializing America," *Science in Context* 18, no. 3 (2005): 337–357.

15. Cohen, *Calculating People*. Herman Melville used the phrase much earlier to characterize Nantucket capitalists. See Melville, *Moby-Dick, or The White Whale* (Boston: St. Botolph Society, 1892 [1851]), 117.

made masses of individuals into wholes throughout American history. At the nation's start, Americans introduced to the world the idea of a free people counting itself regularly through the decennial census.[16] Over time the census gave substance to the "imagined community" that Benedict Anderson explains must exist at the root of that crucial, modern abstraction: the nation.[17] After the Civil War, in the years of American industrial expansion and corporate consolidation commonly called the Gilded Age and Progressive Era, different numbers gave substance to that other fundamental abstraction of modernity, "society." Historians' standard narrative of quantification culminates in the years between the world wars when social scientists, market researchers, and government bureaucrats wrote books and released reports aggregating Americans' economic activity, agricultural products, political opinions, and even their sexual proclivities.[18] Out of apparently thin, disconnected numbers, such books and reports created thick, organic representations of Americans and America in the mid-twentieth century.[19]

But books like *The Money Value of a Man* or *How to Live*, though published during those peak years of aggregation, pointed to a problem with this standard narrative. Here were cultural artifacts rooted in statistics and quantification, but they focused on assigning dollar values to one man at a time and on helping each person learn to live more healthfully—out of statistical materials, these books drew individualized conclusions. I read individuation in their pages as much as aggregation.

I should not have been surprised. The ties between the quantified part and the statistical whole were clear enough to nineteenth-century thinkers.

16. The US Constitution overcame the local opposition to censuses that had relegated earlier censuses to colonies and other less powerful places around the world by tying congressional representation to enumeration. But it also tied taxation to the census count, lest power-hungry states encourage overzealous counting. On resistance to censuses before the American example, see Cohen, *Calculating People*, 34–40, 47–80, 231n50. For the constitutional reasoning, see Article I, Section 2 and James Madison's explanation of that section in "Federalist 54" (1788).

17. Margo Anderson, *The American Census: A Social History* (New Haven, CT: Yale University Press, 1988); on censuses and nationalism generally, see Benedict Anderson, *Imagined Communities: Reflections on the Origin and Spread of Nationalism*, rev. ed. (New York: Verso, 1991), 163–170.

18. See Sarah E. Igo, *The Averaged American: Surveys, Citizens, and the Making of a Mass Public* (Cambridge, MA: Harvard University Press, 2008); Emmanuel Didier, "Cunning Observation: US Agricultural Statistics in the Time of Laissez-Faire," *History of Political Economy* 44, suppl. 1 (2012): 27–45; Emmanuel Didier, *En quoi consiste l'Amérique? Les statistiques, le New Deal et la Démocratie* (Paris: La Découverte, 2009); Thomas A. Stapleford, *The Cost of Living in America: A Political History of Economic Statistics, 1880–2000* (New York: Cambridge University Press, 2009); and Daniel J. Boorstin, *The Americans: The Democratic Experience* (New York: Vintage, 1973).

19. Numbers frequently appear thin even as what they are used to create is anything but. See Theodore M. Porter, "Thin Description: Surface and Depth in Science and Science Studies," *Osiris* 27, no. 1 (2012): 209–226.

The philosopher John Venn recognized in the 1860s that the basis for probabilistic laws lay in the combining of "individual irregularity with aggregate regularity"—each defining the other.[20] Quite often, Americans worried that the lone man or woman would be lost in the statistical aggregate.[21] But some found the individual-aggregate tension provocative or inspiring. Emerson, for example, echoed statistical thinking through his idea of the "Over-Soul," which emerged out of the mass of men's souls, even as it formed those souls—the one and the many defining each other.[22] Yet in telling the story of creating America's statistical communities, historians have too often allowed fraught stories behind the making of statistical individuals to fall to the wayside.

The exceptions I encountered in reading other historians' works made this state of affairs seem all the more lamentable. Studies of credit reporting in the mid-nineteenth century made clear that reducing an individual to a rating could have a profound impact on his future potential in matters economic or social.[23] Scholars looking at mental testing in the early to mid-twentieth century revealed that tools for sorting people according to statistical methods not only opened or closed doors, but affected individuals' subjectivities as well.[24] The power of statistical studies to inform ordinary people's understandings of themselves becomes all the more evident when one looks at social surveying, a category encompassing pollsters, sociologists, and sex researchers who helped generate new ways of thinking about personhood in interwar America.[25]

I came to see the potential of my odd books to help me investigate what it meant to be made into a statistical individual in modern America. Not

20. Quoted in Hacking, *Taming of Chance*, 126.

21. Thus, the genius father and son Peirce partnership (the astronomer Benjamin and his son, the logician Charles Sanders) ruffled many feathers when they presented statistical evidence in courtroom arguments to prove the certainty of a forged signature precisely because their contemporaries saw individual freedom imperiled by the very notion of statistical law, as Louis Menand has explained. Menand, *The Metaphysical Club: A Story of Ideas in America* (New York: Farrar, Straus & Giroux, 2001), 163–176.

22. Ralph Waldo Emerson, "The Over-Soul," in *Essays* (Boston: Riverside Press Cambridge, 1903 [1865]).

23. Josh Lauer, "From Rumor to Written Record: Credit Reporting and the Invention of Financial Identity in Nineteenth-Century America," *Technology and Culture* 49, no. 2 (2008): 301–324; Sandage, *Born Losers*.

24. Two recent and influential accounts are John Carson, *The Measure of Merit: Talents, Intelligence, and Inequality in the French and American Republics, 1750–1940* (Princeton, NJ: Princeton University Press, 2007); and Nicholas Lemann, *The Big Test: The Secret History of the American Meritocracy* (New York: Farrar, Straus & Giroux, 1999).

25. Igo, *Averaged American*.

only did life insurers think statistically about more ordinary people than any other set of institutions during the period when America became an industrial, corporate nation—a period mainly overlooked by the works I had encountered—but life insurers' peculiar transition from prediction to control during that period helped me realize that making people statistical had implications for bodies as well as for wallets.[26] Moreover, I saw the potential for my sources to reveal how uncomfortably Americans engaged with the entire project of being made statistical, how they often resisted the process, and yet somehow ended up being made statistical anyway.

It finally dawned on me that those strange and fascinating books I had first browsed in library stacks barely traced the edges of life insurers' cultural power. I realized that they could best be understood as *residues*: they were material remnants of a mode of statistical production operating at a massive scale that turned individuals into what life insurers called "risks." Such books had been written to ward off those resisting statistical processing, to justify risk makers' expansions, or to find new methods for thinking statistically about Americans. Those books were the tailings of specialized mills for making statistical individuals.

* * *

And books were not the only traces, the only detritus, left behind by this industrial process. I saw soon enough that my investigation had to go beyond a handful of peculiar books—however interesting and even quite popular they may have been. After all, in 1936, while presses rolled Cain's novel, bit by bit, onto magazine paper, the United States government began printing serial numbers attached to individual Americans' names on little paper cards, the first fruits of the Social Security Act. Those cards and the numbers they carried—especially as they found their ways into wallets and, through a strange 1930s vogue, onto some Americans' bodies in ink—deserved a place in this story as well.

Books, cards, and even tattoos were all ultimately by-products, the ephemeral residues left behind when an individual was made into a "risk."

26. In the period covered by this book, life insurers' methods did not have much impact on people's senses of self. So our story will only in a few places become what we might call, borrowing from Ian Hacking, a "making up people" story. Hacking's work has been particularly influential in pointing out the capacity of statistical, medical, and social scientific categories to change the way people see themselves and therefore live their lives. See Ian Hacking, "Making Up People," *London Review of Books*, 17 August 2006, 7–11; Hacking, "The Looping Effects of Human Kinds," in *Causal Cognition: An Interdisciplinary Approach*, ed. D. Sperber, D. Premack, and A. Premack (Oxford: Oxford University Press, 1995), 351–383.

But just because they were by-products does not mean they lacked value or do not deserve sustained attention. On the contrary, as this book will show, the by-products of risk making—which circulated widely in American culture and through American institutions—came to shape lives in profound and lasting ways. And through the analysis of those by-products, we can come to a deeper appreciation of the larger risk-making systems that spawned them.

It is worth pausing at this point over the idea of making risks. I do not mean this phrase to imply that life insurers created new hazards, that they made existence more dangerous—although they and those around them did sometimes worry that insuring lives made people live more dangerously.[27] To understand what I mean by making risks, we must first think differently about what a "risk" was and is. Societies have always had ways of thinking about the dangers and hazards that face them and their constituents—we sometimes talk about those dangers as risks.[28] But with the spread of insurance, risk took on a more specific definition: a risk became a kind of commodity.[29]

The making of risks from human lives and the events of human lives resembled the other crucial commodification processes that came with America's commercial and industrial development. The commodified

27. Concerns that insurance could encourage immoral behavior—murder or suicide, chief among them, and gambling too—led to prohibitions against life insurance lasting into the eighteenth century. After life insurance became common, insurers still worried about "moral hazard," the potential for insurance to actually make the insured live more dangerous lives. More recently, scholars have discussed other ways that life insurance changes behavior. Deborah Stone finds "moral opportunity" in the incentives life insurers create. And Jonathan Levy notes that reliance on life insurance and other financial risk management tools often created new financial risks that individuals had to bear. Deborah Stone, "Beyond Moral Hazard: Insurance as Moral Opportunity," in *Embracing Risk: The Changing Culture of Insurance and Responsibility*, ed. Tom Baker and Jonathan Simon (Chicago: University of Chicago Press, 2002); Levy, *Freaks of Fortune*. On moral and social concerns inhibiting life insurance, see Zelizer, *Morals and Markets*, 32–39.

28. Mary Douglas and Aaron Wildavsky offer a general argument for the cultural and social origins (and therefore historical specificity) of risk conceptions in *Risk as Culture: An Essay on the Selection of Technological and Environmental Dangers* (Berkeley: University of California Press, 1983). For a model example of a historian considering the multiple construction of risk conceptions in the United States, see Arwen P. Mohun's *Risk: Negotiating Safety in American Society* (Baltimore, MD: Johns Hopkins University Press, 2013).

29. See Levy, *Freaks of Fortune*. For an attempt to trace out deeper roots of risk thinking and commodification, see Peter Bernstein, *Against the Gods: The Remarkable Story of Risk* (New York: Wiley, 1998); Daston, *Classical Probability in the Enlightenment*; and Ian Hacking, "Risk and Dirt," in *Risk and Morality*, ed. Richard V. Ericson and Aaron Doyle (Toronto: University of Toronto Press, 2003).

objects differed, but the process (and the by-products) of making some natural thing into a medium of exchange, a financial creature, did not differ in kind. When nineteenth-century farmers sent trainloads of grain to Chicago to be stored and sold from that city's grain elevators, they incidentally set quite a bit of paper in motion: elevator receipts recorded the amount and quality (class) of the wheat deposited; "to arrive" contracts made possible a futures market for gambling on the value of that slip.[30] These were the material manifestations of commodifying grain.

Those same farmers shipping grain—and to an even greater extent the merchants who received their grain shipments—could and often did have their life risks (the risk of death in particular) commodified too through the vehicle of a life insurance policy. Life insurance contracts charged an individual a regular fee (called a premium), while requiring the insurer to pay a set sum (a claim) in the event of that individual's death during an agreed-upon period of protection. That, at least, was the simplest form of a policy. But by the last quarter of the nineteenth century, where this book begins, life insurance policies came in a dizzying variety of forms and commodified a corresponding variety of life risks, for example: the risk of financial embarrassment took on commodity form through policies that accumulated value over time, but could lose that value if a policyholder could not pay his premiums; risks of the early death of a loved one or a business partner came to be commodified through policies that paid out in the form of annuities to a beneficiary.

And while Chicago accepted the lion's share of America's grain, New York in the final decades of the nineteenth century commodified the most life risks, by far. The so-called Big 3 companies of the era—Mutual Life of New York (MONY), New-York Life, and Equitable Assurance Society of the United States—all hailed from New York where they sold "ordinary" policies to those well-off enough to pay yearly premiums, and each in 1890 had extended a quantity of insurance more than double that of their closest rivals.[31] They were key innovators who used their market power (and their close relationships with state legislatures and regulators) to set industry standards. New York's commodifying reach extended further with the rise of "industrial" life insurers like Metropolitan Life and Prudential

---

30. William Cronon explains this process in *Nature's Metropolis: Chicago and the Great West* (New York: Norton, 1991), 120–125. See also Jonathan Ira Levy, "Contemplating Delivery: Futures Trading and the Problem of Commodity Exchange in the United States, 1875–1905," *American Historical Review* 111, no. 2 (2006): 307–335.

31. Stalson, *Marketing Life Insurance*, 798.

Insurance Company of America (in nearby Newark), which sold small policies (mainly useful to pay for burials) with weekly premiums to workers and their families. These two newcomers joined the earlier three to make the "Big 5," which together dominated the industry well into the twentieth century. Because of their size, power, and influence, these five companies shaped the way that most Americans' life risks came to be commodified and so play central roles in this book's narrative.

The commodification of life risks through life insurance had only gained significant purchase in the United States in the 1830s and 1840s and succeeded most in reaching male heads of households in or near urban, commercial environments, especially men whose work generated an income: clerks or merchants, lawyers, doctors, or even clergy. Throughout the mid-nineteenth century, life insurers' local agents sold applicants on the need to replace themselves, at least in a monetary sense, upon their deaths, to ensure that their wives (striving to maintain the new middle-class ideal of separate gender spheres) never had to work and that their children would not become orphaned wards of the state.[32] Life insurers promised to make this alchemical transformation possible: they could turn a life into money. Using a thermodynamic metaphor appropriate to an industrializing age, one contemporary insurance proponent explained this transformation: "Life insurance comes in as a financial invention, by which capital in the shape of a productive life—a life controlling and directing some branch of the wealth-begetting or wealth-distributing machinery of the age—can perpetuate itself, or convert a part of its productive energy into a contingent fund, that will be immediately available in case of death."[33]

But to qualify for a life insurance policy, the applicant had to submit himself to a thorough interrogation by the agent about his age, his habits, and his economic situation (sometimes verified with a credit report), had to agree to set a value on his life (the value of his eventual policy), and had to see a doctor, who would weigh him, record his pulse, and make a slew of other measurements. In the bargain for economic security, insurance applicants allowed themselves to be made into commodified risks. About eight hundred thousand Americans made that bargain and stood as risks on insurers' books in 1872. With the industry's embrace after the Panic of 1873 of southerners, westerners, workers, women, children, and people of color,

---

32. On the prime targets for life insurance and the role of life insurance vis-à-vis gender roles, see Murphy, *Investing in Life*, chapter 5.

33. Elizur Wright, *Massachusetts Reports on Life Insurance: 1859–1865* (Boston: Wright & Potter, 1865), 303.

the number of those in insurers' corporate files reached over ten million (or one-ninth of the population) by the turn of the century.[34]

Books, cards, medical examination forms, policy applications, credit reports, and sheaves of other paperwork were the equivalents for people of grain slips and futures contracts for wheat. That commodifying life risks generated so much more paper than did commodifying grain—and that life risks resisted so neat and full a commodification as grain—only makes sense: lives were more complicated than grain and the discriminations and classifications that went into making life risks demanded explanation, since after all, an adverse classification meant much more to a person than to the wheat. With mortgages often tied to a lender having life insurance and with few alternatives to life insurance for saving for old age, getting a policy at a reasonable rate mattered especially to those barely holding on to a place in the boom-bust economy of the Gilded Age.[35] Once risk makers began pitching their techniques as means for altering fates and not just predicting them—and as their techniques became tools employed outside of life insurance by doctors, public health officials, and the state—the commodification of life risks came to exert power over individuals' bodies and lifestyles as well. The stakes could be very high indeed.

34. More than a million Americans had been enumerated by the thirty largest life insurers in 1872. See Levi W. Meech, *System and Tables of Life Insurance: A Treatise Developed from the Experience and Records of Thirty American Life Offices, under the Direction of a Committee of Actuaries* (New York: Spectator Company, 1898), 4–5. In 1900, over fourteen million policies were in force. While some significant portion may have been policies on the same life, those should be easily countered in our estimates by policies recently lapsed or terminated on an insured's death. See aggregate statistics prepared by Frederick L. Hoffman in United States Department of Commerce and Labor, *Statistical Abstract of the United States*, no. 33 (Washington, DC: Government Printing Office, 1911), 33, 581. Ten million insured would have amounted to about one-ninth of those counted by the US Census that same year.

35. Edward Bellamy offered a compelling metaphor for late nineteenth-century insecurity when he imagined a "prodigious coach which the masses of humanity were harnessed to" drawing the lucky, comfortable few who made much of pitying the many even as they worried about "slipping out" of their seats, or worse "of a general overturn in which all would lose their seats." See Bellamy, *Looking Backward 2000–1887* (Boston: Ticknor, 1888), 11–14. The historiography on late nineteenth-century responses to American industrialization and corporate growth is immense. Of particular relevance to this book and my thinking have been works interested in legal and social policy responses to the particular insecurities that riddled corporate capitalism, especially Daniel T. Rodgers, *Atlantic Crossings: Social Politics in a Progressive Age* (Cambridge, MA: Belknap Press of Harvard University Press, 1998); John Fabian Witt, *The Accidental Republic: Crippled Workingmen, Destitute Widows, and the Remaking of American Law* (Cambridge, MA: Harvard University Press, 2004); Levy, *Freaks of Fortune*; and Barbara Young Welke, *Recasting American Liberty: Gender, Race, Law, and the Railroad Revolution, 1865–1920* (Cambridge: Cambridge University Press, 2001).

It was never a foregone conclusion that individuals would submit to being measured and commodified. In the mid-nineteenth century, those few Americans who were teaching themselves to number their neighbors' days—*not* their own, as the psalmist prayed long before—often lamented how little data they possessed to serve that end. As nearly all observers admitted, the state—from which data seekers expected quite a bit in an age featuring governments that printed reams of statistical ephemera—knew much about American commerce, but significantly less about Americans themselves.[36] Expressing a sentiment so common as to be almost clichéd, one statistician lamented: "The minds of the leading men in this country, as well as of the masses, are so much absorbed in commerce and the arts, agriculture, and politics, that it is almost an impossibility to turn them aside, and secure their attention and action upon a subject, whose results, however good and necessary for the progress of the race, are so remote, as are those of mortality statistics; they deem it better to transmit to posterity an account of stocks and merchandise than the births of their children, more important to record the destruction of their ships than the deaths of their fellow men."[37] The state knew far too little, they complained, about the life span.

Life insurance corporations did better because they turned Americans' concern with commercial affairs—a supposed hindrance to statistical collecting—into a boon for gathering numbers about individual lives. In the process of commodifying the life risks of their applicants, life insurers accidentally made the most commercially minded men into some of the best quantified Americans of the mid-nineteenth century. In the following years, many more Americans would (often unwittingly) jockey for that status at the impulse of risk makers' expansive, innovative ambitions.

Some readers may still think it more natural to talk about "calculating" risks than making risks. But the stories that follow of risk makers wrestling with the challenges of mass producing a nation's worth of risks should demonstrate how incomplete a story attuned solely to calculation would be. It is certainly true that risk makers borrowed from gamblers, astronomers, and probabilistic mathematicians the tools to smooth, to search for stable, average, graphable truths amidst the frenetic blooming buzz of life. Yet making

36. For an example of more successful state statistical activities, see Michael Zakim, "Inventing Industrial Statistics," *Theoretical Inquiries in Law* 11 (January 2010): 283–318.

37. G. S. Palmer, "Partial Report upon a Uniform System of Registration of Births, Marriages, and Deaths, and the Causes of Death," *Transactions of the American Medical Association* 9 (1856): 775–777 at 775.

risks did not stop with arithmetic or mathematics. Risk makers also looked to salesmen, jurists, and economists for help in assigning prices to lives. And they turned to doctors, credit reporters, and statists to learn how to categorize and class. Then using those categories, they devised bureaucracies that could write and reduce a life to an index card.[38]

Moreover, risks did not simply exist in the world, waiting to be calculated—they had first to be defined, created. The richness of risk making—its multiplicity of tools, its interdisciplinarity, its blending of cultural practices[39]—helps explain how it provoked concern and critique and also why so many outside of life insurers' offices eventually built their own risk assembly lines or trafficked in the books, cards, and other by-products (such as height-weight tables, medical impairment studies, and race or gender valuations) that have shaped so many lives.

* * *

The first half of this book follows life insurers as they fumbled, improvised, debated, and fought their way toward national systems for processing lives and producing risks in the last quarter of the nineteenth century. Each chapter in this half takes its name from one of the handful of controversial activities that life insurers used to commodify life risks and make statistical individuals. Chapter 1 eavesdrops on industry insiders and state regulators (in a public courtroom) taking shots at *classing*—at paying attention to individual or group particularities—and asserting the primacy of the broad average that smooths away particularities. Closer examination, however, reveals

---

38. This book returns repeatedly to points of exchange between fields and institutions, especially in those spaces where commerce, science, and the state come together. Previous studies on the interstices of such fields include Geoffrey C. Bowker and Susan Leigh Star, *Sorting Things Out: Classification and Its Consequences* (Cambridge, MA: MIT Press, 1999); Mary Poovey, *A History of the Modern Fact: Problems of Knowledge in the Sciences of Wealth and Society* (Chicago: University of Chicago Press, 1998); Jacob Soll, "From Note-Taking to Data Banks: Personal and Institutional Information Management in Early Modern Europe," *Intellectual History Review* 20, no. 3 (2010): 355–375; and Porter, *Trust in Numbers*. On the ways that scholars of capitalism, science, and the state have already learned a great deal from one another, see Jeffery Sklansky, "The Elusive Sovereign: New Intellectual and Social Histories of Capitalism," *Modern Intellectual History* 9, no. 1 (2012): 233–248.

39. Throughout this book, I will explore risk making by examining "practices"—shared modes of doing that in our story go by names like smoothing, classing, writing, fatalizing, and valuing. In this emphasis, I continue along a path now well trod in the history of science, which is quite far along in its practical turn. More recently, Joel Isaac has proved the utility of bringing practice to American intellectual history too, in Isaac, *Working Knowledge: Making the Human Sciences from Parsons to Kuhn* (Cambridge, MA: Harvard University Press, 2012), 6–7.

that classing and smoothing could not be so easily opposed, that in making statistical individuals they were locked in a fundamental tension. Without really resolving that tension, the Panic of 1873 spurred traditional "ordinary" life insurers to expand into new regional (and international) markets and made space for new "industrial insurers" offering small burial policies to workers and their families.

Soon enough, expansion revealed new problems in insurers' constructions of risks. Chapter 2 shows African American activists leading a revolt against life insurers' application of probabilistic methods to Americans, especially their *fatalizing* assumption that past data could be used to predict a group's future. In that revolt, those made into risks challenged the risk makers. Chapter 3 reveals that consequential challenges also emerged from within companies, where complaints about the cost and trouble of maintaining extended *writing* networks had the unexpected effect of pushing risk makers to become even more committed to thinking statistically about individuals. All three chapters reveal questions unresolved, systematic troubles, and discontent simmering (with the occasional eruption) beneath insurers' expansion of risk making. Internal tensions met a broad public distrust of corporate power bred by the first efflorescence of corporations in the late nineteenth century. The result was a shadow cast over risk making, even as more and more risks got made.

When, in 1905, the heir to the founder of the giant Equitable company threw an ill-conceived costume ball (with Gilded Age worthies dressed as courtiers from the days of Louis XVI, no less)[40] and thereby drew attention from financial vultures looking to control Equitable's hundreds of millions in assets, the industry's sunny outlook turned suddenly cloudy. New York State launched a special inquiry into the Big 5 companies—the Armstrong investigation—revealing webs of power and wealth that Americans could hardly imagine: life insurance executives earned large salaries, manipulated the press, bought legislators, and freely distributed sinecures to friends and

---

40. On the ball and its aftermath, in lively style, see Patricia Beard, *After the Ball: Gilded Age Secrets, Boardroom Betrayals, and the Party That Ignited the Great Wall Street Scandal of 1905* (New York: HarperCollins, 2003). The ball follows in the tradition of Gilded Age re-creations of French court life. Sven Beckert details one such ball from 1897, featuring fifty attendees dressed as Marie Antoinette! Sven Beckert, *Monied Metropolis: New York City and the Consolidation of the American Bourgeoisie, 1850–1896* (New York: Cambridge University Press, 2003), 1. The tradition continues into the twenty-first century with tech entrepreneur David Sacks's similarly themed "Let Him Eat Cake" fortieth birthday party in 2012. See Owen Thomas, "The Guy Who Might Sell His Company to Microsoft for $1B Says, 'Let Him Eat Cake,' " *Business Insider*, 14 June 2012.

family, while their backroom deals with big investment banks fueled the great merger movement, the metastasization of trusts, and the ascendance of corporate octopi. Concerns about life insurers' enormous capital reserves and the power that came with those reserves had instigated the investigation, but once it began, all sorts of other details of life insurers' operations came under scrutiny—including their methods of making risks. In this setting, the state challenged risk makers' commitment to applying statistical methods to individuals. Inquiries focused on insurers' mathematical methods for *smoothing* away the vagaries of life and capitalism—and on their assertions of the deep reality of the risks they made—raising eyebrows and drawing scorn from many in the process, as chapter 4—the book's hinge—explores. By the time the investigations ended and the nation endured another panic, life insurers emerged chastened, more heavily regulated, and more fundamentally conservative—at least, that has been the traditional interpretation.[41]

Yet if we care about corporations' power to shape Americans' lives as much as those corporations' raw economic power, we must reconsider the impact of the Armstrong embarrassment. This is the burden of the book's second half. Even as life insurers backed away from (some) overt political scheming and broke down (many) interlocking directorates, clever and ambitious intellectuals (discussed in chapter 5) found space within disturbed corporate ecosystems to advocate increasing life insurers' influence in new ways. Most importantly for this story, they cultivated a new kind of power (a species of what Foucault called "bio-power") derived from reappropriating risk-making techniques—given extra force by still-enormous corporate resources—that aimed to lengthen individual lives, combat some forms of racism, and argue for increased social cooperation.[42] One influential reformer imagined, quite plausibly, turning his company (Metropolitan) into the model for "A New Socialism," a socialism built on medical exams as much or more than pension plans, that led not just to welfare capitalism, but to the medicalization of society. These reformers succeeded to a great

41. Morton Keller made this case most forcibly in 1963, using the Armstrong investigations and reforms to explain how brash Gilded Age corporations became model corporate citizens with limited power. See Keller, *Life Insurance Enterprise.*

42. In its blending of medical and actuarial methods, risk making as a force for change touched both poles of Foucault's bio-power. It disciplined individuals in a form of "anatomo-politics of the human body" with the tools originally intended for the "biopolitics of the population." Michel Foucault, *History of Sexuality, Volume 1*, trans. Robert Hurley (New York: Vintage, 1990), 139–141.

extent in winning support for public health measures and in convincing Americans to submit themselves to personal risk assessments—on the bathroom scale and in the doctor's office—successes easily overshadowed by reformers' more spectacular failures, as in their battle for national health insurance.

Their greatest success, moreover, came in packaging risk making to appeal to broader audiences and work in new settings: early twentieth-century reformers had stumbled upon a new answer to that old question, which had long challenged insurers—how should corporations be allowed to use statistical knowledge to intervene in individual lives? Their answer resonated with the larger American public, as chapter 6 demonstrates, and facilitated the migration of risk making into new contexts. Risk-making by-products, in an important instance, seeped into debates over immigration restriction, as chapter 7 explains, and became tools that could be used (simultaneously) to legitimate and contest discrimination against African Americans. Corporate risk makers predicting individual fates remained controversial, but using the same methods and tools to improve lives won broad cultural purchase.

Then, as the conclusion explains, risk makers expanded their reach once again in response to the Depression, reaching out to forecast Americans' economic lives from a new perch in the American welfare state. Out of the ashes of the Armstrong conflagration a new species of risk-making machinery had arisen, developed by corporations, but embraced by doctors, the state, and many ordinary Americans even now.

* * *

How did our days become numbered? For the most part, I offer here an answer to that question grounded in the ordinary and mundane, in stories about science and medicine in corporate hands, about living people interpreted statistically, about the making of modern American life. But I must admit an apocalyptic streak too. I admit wondering about the implications of having so widely turned over the numbering of our days to risk makers and of placing so much faith in their methods. I imagine the psalmist would not approve, and I am not sure that I do either.

I understand the "our" in the above question to be simultaneously expansive and parochial. Expansively, it applies to all moderns. Life insurance took on its present form in Britain in the mid-eighteenth century and became a signature feature of modernity around the world in the nineteenth century. American risk makers looked eagerly across the Atlantic for models, tools, and ideas from the start. By the end of the nineteenth century, world audiences had to reckon with Americans' risk-making methods in return.[43] For a few decades, American life insurers even reigned as the largest corporate risk makers in Europe and Latin America.[44]

Yet, for the most part, the subjects of American life insurers' quantifications lived in North America, under the (sometimes) watchful eyes of American regulators. Life insurers, for their part, operated under the constraints of state laws and courts that often also empowered them. As they numbered Americans' days, they contended with local politics, with the peculiarities of the American statistical apparatus. Risk makers crafted statistical individuals out of transnational ideas and local particulars, out of American methods and global phenomena. This book tells the national story—of American corporations numbering the days of Americans—that emerged out of these complex interactions.

It dwells especially on the risk-making systems that such interactions produced. My curiosity for old books led me to their authors—the risk makers—and to their novel statistical assembly lines. I discovered nation-spanning networks of observers funneling intelligence to central offices filled with clerks guided by algorithms for determining individual probabilities of mortality. These human treadmills, as one insurer conceived of them, churned out people on paper that could be bought or sold, shared or kept.

Throughout those networks, I encountered surprising and engaging characters: the composer Charles Ives, the biomathematician Alfred J. Lotka, the statistician Frederick L. Hoffman, the economist Irving Fisher, and the African American novelist Richard Wright. Life insurers housed at once the socially comfortable—Yale alumni like Ives looking for a steady income and time for their art—and the marginal (like Bunzel, Dublin, and Lotka): the immigrants, Jews, and occasionally women who, when the academy would not afford them space to do science, found it in corporate skyscrapers. I enlisted these characters to tell this story, sometimes because they profoundly shaped how risks were made, sometimes because of the window their fame opened into the risk-making process, and always because they reminded me that these systems and networks grew out of human actions, choices, and contingencies.

---

43. Timothy Alborn and Sharon Ann Murphy, eds., *Anglo-American Life Insurance, 1800–1914* (London: Pickering & Chatto, 2013); Martin Lengwiler, "Double Standards: The History of Standardizing Humans in Modern Life Insurance," in *Standards and Their Stories: How Quantifying, Classifying, and Formalizing Practices Shape Everyday Life*, ed. Martha Lampland and Susan Leigh Star (Ithaca, NY: Cornell University Press, 2009).

44. Keller, *Life Insurance Enterprise*, 81–123.

Some of these characters expressed enormous faith in the power of risk-making tools to improve the world. But those closest to the risk mills and the numbers they generated just as often worried, doubted, and in some cases retreated when their by-products escaped their grasps.[45] Their concerns deserve greater scrutiny in this age of "Big Data," the blueprints for which we can discern in the processes of the Gilded Age life insurance corporations investigated here. Risk makers have found niches reducing individuals to market types within our vast social networks, predicting terrorist activity for our prying governments, and rating minor leaguers among our "moneyball"-style baseball enthusiasts, sometimes in service of professional teams.[46] Today, there is no question that Americans' lives will be translated into tables of numbers and curves on graph paper. But important questions remain to be asked more often as ordinary people are more frequently made into data: who controls the tables, who controls the curves, what are their limitations, and how will they be used? Many late nineteenth-century Americans would have found such questions distressingly familiar.

It is, in fact, only fitting that this story now begins with a fight over life insurers' tables and curves.

45. On why, in some cases, numbers come to be themselves determinative, displacing the reasoned judgments of experts, while in other cases, numbers serve only as tools (not to be trusted too far) in making decisions, see Porter, *Trust in Numbers*.

46. "Big Data" is everywhere these days. See, for instance, Nate Silver, *Signal and the Noise: Why So Many Predictions Fail—but Some Don't* (New York: Penguin Press, 2012); or Viktor Mayer-Schönberger and Kenneth Cukier, *Big Data: A Revolution That Will Transform How We Live, Work, and Think* (Boston: Houghton Mifflin Harcourt, 2013). Michael Lewis, *Moneyball: The Art of Winning an Unfair Game* (New York: Norton, 2003).

ONE

Classing

Thomas Scott Lambert delivered a lecture on 2 January 1878. He was by then well known as a speaker and writer on "popular" medical topics. His *Popular Anatomy and Physiology, Adapted to the Use of Students and General Readers* had, by the mid-nineteenth century, become an oft-used teaching text, joining prior works like his *Popular Treatise on Bathing*. He lectured widely and had held teaching positions at Young Ladies' Institutes in Brooklyn and Pittsfield.[1]

On this winter afternoon, his large audience heard a very different talk in extraordinary circumstances. Packed into the Court of Oyer and Terminer in New York City, the crowd had come to hear Dr. Lambert sentenced for perjury in connection with the failure of his company, American Popular Life.[2] He seized the moment to mount a final, public defense of his behavior, of his company, and of his pet medical theory: "biometry," a full-body phrenology encompassing the secrets to predicting any person's life span (and much more too).

The court clerk, having instructed Lambert to rise to receive his sentence, asked the convicted perjurer "what have you to say why judgment should not now be pronounced upon you, according to law?" At that point, according to a reporter for the *New York Times*, continuing its extended coverage of the sensational trial: "The prisoner's tall form bent somewhat, and he

---

1. Lambert, *Popular Anatomy and Physiology, Adapted to the Use of Students and General Readers* (Portland, ME: Sanborn & Carter, 1851). To that volume, Lambert added a shorter, more practice-focused work, *Practical Anatomy, Physiology, and Pathology; Hygiene and Therapeutics* (Portland, ME: Sanborn & Carter, 1851). For more, see Lambert's obituary: "Death List of a Day," *New York Times*, 23 March 1897.

2. "Dr. Lambert Sentenced," *New York Times*, 3 January 1878.

reached down on the table beside him for a thick roll of manuscript. He straightened himself up, and for one hour and fifty minutes addressed the court, talking rapidly, and, from an elocutionary or rhetorical standpoint, very well."[3]

Lambert denied the charges against him. He had not knowingly signed a false financial statement. He had, if anything, he insisted, been duped by his underlings.[4] Moreover, he deemed it suspicious that of sixty or seventy failed life insurers with sixty or seventy failed presidents at their head, only he and one other had been brought before the law, and both were the poorest—the least adept, he intimated, at bilking their companies for cash on the way down.[5] New York's superintendent had singled him out unfairly, he persisted, while allowing other life insurers to "swindle" the public for millions of dollars without punishment. Lambert further argued that the superintendent felt threatened, like Lambert's competitors, by his innovative approach to life insurance. He read testimonials from *Popular Science Monthly* to support his claims to biometry's scientific status and practicality. Finally, Lambert begged for mercy so that his family would not endure further hardships atop his financial embarrassment.[6]

Lambert's case and the story of American Popular Life's collapse packed courtrooms and fascinated readers who, in the 1870s, were anxious to see someone brought to justice for the cascade of life-insurer failures brought on by the Panic of 1873. A flood of Americans had bought life insurance in the prior decade. Life insurers had advertised themselves in glossy sales pamphlets proffered by eager agents as sources of security and stability, a security and stability made manifest in elegant, lavishly decorated offices and magnificent new edifices housing growing operations (as exemplified by Mutual Life of New York's gorgeous building at the corner of Broadway and Liberty, built in 1864).[7] They offered protection from death's economic

---

3. "Dr. Lambert Sentenced," *New York Times*, 3 January 1878.

4. The judge subsequently deemed such ignorance and negligence a crime against the public in its own right: "You were placed as a sentinel there to protect the public. If you were led hither and thither by those whom you employed, it is no excuse." "Dr. Lambert Sentenced," *New York Times*, 3 January 1878.

5. It is unclear if Lambert's claim to have been among those most wiped out by the failure of his company is true. He had deemed himself wealthy enough before the collapse to abandon his own life insurance policy as a protection he no longer needed. See "Dr. Lambert on the Witness Stand," *New York Times*, 15 December 1877. It certainly is true that 75 companies, out of 129 at the peak of the life insurance bubble, failed between 1871 and 1877. Stalson, *Marketing Life Insurance*, 434, 751.

6. "Dr. Lambert Sentenced," *New York Times*, 3 January 1878.

7. Mutual Life Insurance Company of New York, *Annual Report* (1865): 8.

consequences and the certainty of a high rate of return on policyhold-ers' premiums. Many Americans in the commercializing, industrializing North—very much aware of their own fragile mortality after the devasta-tions of the Civil War and also increasingly at the mercy of a boom and bust economy—had bought the promises of security and opportunity that life insurers sold. Life insurance had become, in historian Sharon Murphy's words, "a pillar of modern middle-class life."[8] But by 1878 experience had exposed the hollow cores of many such promises. Corporations supposed to ensure stability had become causes of distress. Someone had to pay.

Prosecutors in New York offered up Lambert. They levied a charge of per-jury. But that was to some extent a matter of convenience—it was a criminal charge they thought would stick. Even as the prosecution set out to prove that Lambert knowingly lied about his company's financial status, the pros-ecution made their case into a broad demonstration of Lambert's misman-agement, his abuse of policyholders' trust. In prosecuting Lambert, they put a face on the broken promises of life insurance.

Lambert, even as his attorneys sought out every technical or formal av-enue to win his freedom, directed his energies toward a defense of his ap-proach to the prediction of individual fates and the making of risks. Thanks to Lambert, spectators eager to hear about corporate malfeasance were treated (whether they liked it or not) to an unexpected glimpse into an on-going and unresolved debate taking place among life insurers—with state insurance commissioners and the courts involved too—over the propriety of using probabilistic, statistical methods to make risks. Lambert occupied an extreme position in such debates, as he would himself admit. He thought actuarial and statistical methods should be superseded by medical judg-ment in the making of risks, and he wore his unorthodoxy proudly. He believed, perhaps correctly, that his unorthodoxy played a part in his being brought to trial while so many other companies' executives suffered their failures in private. His critics looked at Lambert's eccentricities and saw only further evidence of either incompetence or hucksterism.

Lambert's extreme position made a certain sense. He was not (entirely) a crazy outlier—he had simply gone further than most of his fellow life insur-ers along a course that many of them were also following. Lambert's biom-etry served the more general purpose of individualizing life insurance—and individualizing life insurance was the goal of most life insurers at the time. So long as life insurers made contracts with individuals, life insurance could

8. Murphy, *Investing in Life*, 300.

not help but be individualized to some degree. A risk, by definition, derived from an individual. But the extent to which individual characteristics determined the way a life became a risk remained always up for negotiation. To say that Lambert and his peers individualized life insurance is to say that they cared more and more about the factors that made people distinct and less about those that bound them together into masses.

Life insurers' drive to individualize manifested itself in a growing volume of "equity" talk. Equity, to a risk maker, meant fitting each life insurance contract as best as possible to each individual's projected life span. Life insurers pursued equity in deference to ideals of individual justice (so that the morally good and mortally blessed should not subsidize intemperate weaklings), but also because caring about equity allowed them to better compete for prime applicants: it allowed life insurers to attract sound lives with insurance offered at a lower price. In a packed marketplace, equity made commercial sense even as it appealed to a liberal sensibility. Equity could be achieved by changing the way that individuals, in the process of being made into risks, were grouped together and offered policy terms—by "classing." Leading life insurers often constructed classes based on the regions policyholders came from (the South having been a particularly disfavored area in the years before the Civil War), the types of policies they held, the years they bought those policies, their sexes, and most fundamentally, their ages.

Equity could also be achieved by changing the cost of insurance for already existing groups in accordance with each group of risks' statistical record of survivals and deaths—by "smoothing." Thus a company's actuary could mathematically review the mortality of the company's policyholders and suggest to his president and trustees an "equitable" cost structure for their insurance that charged more to the short-lived and less to the long-lived. Lambert joined many of his peers in looking for better roads to equity, but he left the pack in his conviction that classing should dominate, with little room left for smoothing in the making of risks. Lambert's approach to equity thus resisted—or, more precisely, *claimed* to resist—making either statistical communities or statistical individuals.

This chapter aims neither to prosecute nor to defend Lambert. Nor does it attempt to adjudicate the proper roles of classing and smoothing in life insurance. Rather, it uses Lambert's company, its failure, and his subsequent trial to investigate the tensions that existed between life insurance offices over the best way to individualize life insurance. As life insurers built larger and more diverse statistical communities within a competitive environment, they struggled to reconcile their commitment to broad groupings with their preference—shared by many in the shadow of what Emerson called the "age of the first person singular"—for exalting individuality.[9] Most life insurers refused to make a choice: they hewed to statistical laws while simultaneously classing their risks. Out of the play between smoothing and classing, they made statistical individuals from a growing number of Americans.

As this chapter will show, the panic ultimately damned Lambert's approach to classing risks and his condemnation of the statistical individual. But the tension that his unorthodoxy exposed between preserving individuality and equity through classing while finding overarching regularities through smoothing found no easy resolution—that is why it deserves our attention. After the panic, that tension only became more pronounced. Those companies that survived—and many did not—learned any number of lessons about, for instance, the value of maintaining a large capital reserve or of keeping in good standing with regulators. But one lesson of particular importance to the study of making Americans into risks came in the form of a widespread realization by life insurers of the hazard they faced in selling policies to only the northern, male, middling classes. As a result, many life insurers decided to look toward other regional markets, workers, and even women and children, all in a quest to sell more policies and make new risks. Life insurers thus expanded the reach of their classing and smoothing systems, without settling on a right way to smooth, a right way to class, or resolving tensions between smoothing and classing. As more Americans fell subject to risk makers' efforts, debates over the proper way (if there was one) for a corporation to make a statistical individual and predict that individual's fate would only become more heated, and as subsequent chapters will show, more public.

* * *

Lambert founded American Popular Life in 1866, alongside twenty other companies that opened their doors the same year in the United States.[10] With the Civil War finally ended and the shock of demobilization fizzling out, the American economy began a spectacular boom. The life insurance industry did too. War's carnage reminded the American middle classes of their own mortality and spurred them to consider purchasing policies.[11]

---

9. See Louis P. Masur, "'Age of the First Person Singular': The Vocabulary of the Self in New England, 1780–1850," *Journal of American Studies* 25, no. 2 (1991): 189–211 at 205–206.

10. Massachusetts Insurance Department, *Annual Report of the Commissioner of Insurance of the Commonwealth of Massachusetts* 16 (1866): 2:c. On other upstart companies, see Stalson, *Marketing Life Insurance*, appendix 9, table A, 751.

11. On life insurers during and immediately following the Civil War, see Murphy, *Investing in Life*, especially 273–274.

Increased commercial and industrial activity also made it seem more important to many striving Americans—whose wealth came from their hands or heads, rather than from land or property—to protect their families (or secure their debts) in the event of their death. It did not hurt that life insurers also sold their policies as an easy way to invest (indirectly) in the raging economy.[12]

"Whole life" insurance policies offered their purchasers protection at a fixed rate (usually with an annual premium payment) for their entire lives—the premium primarily dependent on the age of the applicant and inversely proportioned to the applicant's expectation of life.[13] In the oldest nod to equity in life insurance, older applicants paid more, younger less (in proportions dictated by a "life table" or "mortality table" like that in figure 1.1). Most whole life policies came with "dividends" announced every few years, or sometimes yearly, as a share of the company's profits handed back to policyholders, which added value to policies or could be used to pay upcoming premiums. When a policyholder died, the insurer promised to pay the entire face value of the policy—usually around $1,000–$5,000, but sometimes much more—to a beneficiary.

Whole life policies drove industry growth as the amount of insurance in force increased by a factor of five from 1865 to 1870 and the number of policyholders grew into the hundreds of thousands by the mid-1870s.[14] At

---

12. This trend began, like many life insurance innovations of this period, in Britain. Moses Knapp quoted the eminent British actuary De Morgan: " 'Probably,' says De Morgan, 'if the following question were put to all those whose lives are now insured, 'What is the *advantage* you derive from investing your surplus income in an Insurance office?' more than half would reply: 'The *certainty* of my executors receiving a sum at my death, were that to take place to-morrow.' This is but half an answer; for not only does the office undertake the equalization of life, but also the *return of the sums invested, with compound interest.*" As Knapp put it: "No person forms a correct idea of a Life Insurance Office who does not regard it as a savings bank, yielding compound interest for life, without interruption." Moses Knapp, *Lectures on the Science of Life Insurance Addressed to Families, Societies, Trades, Professions—Considerate Persons of All Classes,* 2nd ed. (Philadelphia: E. S. Jones, 1853), 38, 121. For more on British life insurance as an investment, see Timothy Alborn, "The First Fund Managers: Life Insurance Bonuses in Victorian Britain," *Victorian Studies* 45, no. 1 (2002): 65–92. On the close ties between insurers and antebellum mortgages, see Tamara Plakins Thornton, " 'A Great Machine' or a 'Beast of Prey': A Boston Corporation and Its Rural Debtors in an Age of Capitalist Transformation," *Journal of the Early Republic* 27, no. 4 (2007): 567–597.

13. Eventually, whole life would be subsumed under the category "ordinary" insurance, a distinction created to set apart large policies paid annually from "industrial" life insurance paid with pennies, nickels, and dimes weekly with small benefits.

14. There were over a million policies recorded in a thirty-company report containing data up through 1874. Because individuals could hold more than one policy, the data does not allow us to determine precise numbers of lives insured. But that uncertainty can be countered by the

SCHEDULE.

Table of Mortality based on American Experience.

Age.	Numbers living.	Numbers dying.	Expectation of life.	Age.	Numbers living.	Numbers dying.	Expectation of life.
10	100,000	749	48.72	53	66,797	1,091	18.79
11	99,251	746	48.08	54	65,706	1,143	18.09
12	98,505	743	47.44	55	64,563	1,199	17.40
13	97,762	740	46.82	56	63,364	1,260	16.72
14	97,022	737	46.16	57	62,104	1,325	16.05
15	96,285	735	45.50	58	60,779	1,394	15.39
16	95,550	732	44.85	59	59,385	1,468	14.74
17	94,818	729	44.19	60	57,917	1,546	14.09
18	94,089	727	43.53	61	56,371	1,628	13.47
19	93,362	725	42.87	62	54,743	1,713	12.86
20	92,637	723	42.20	63	53,030	1,800	12.26
21	91,914	722	41.53	64	51,230	1,889	11.68
22	91,192	721	40.85	65	49,341	1,980	11.10
23	90,471	720	40.17	66	47,361	2,070	10.54
24	89,751	719	39.49	67	45,291	2,158	10.00
25	89,032	718	38.81	68	43,133	2,243	9.48
26	88,314	718	38.11	69	40,890	2,321	8.98
27	87,596	718	37.43	70	38,569	2,391	8.48
28	86,878	718	36.73	71	36,178	2,448	8 00
29	86,160	719	36.03	72	33,730	2,487	7.54
30	85,441	720	35.33	73	31,243	2,505	7.10
31	84,721	721	34.62	74	28,738	2,501	6.68
32	84,000	723	33.92	75	26,237	2,476	6.28
33	83,277	726	33.21	76	23,761	2,431	5.88
34	82,551	729	32.50	77	21,330	2,369	5.48
35	81,822	732	31.78	78	18,961	2,291	5.10
36	81,090	737	31.07	79	16,670	2,196	4.74
37	80,353	742	30.35	80	14,474	2,091	4.38
38	79,611	749	29.62	81	12,383	1,964	4.04
39	78,862	756	28.90	82	10,419	1,816	3.71
40	78,106	765	28.18	83	8,603	1,648	3.39
41	77,341	774	27.45	84	6,955	1,470	3.08
42	76,567	785	26.72	85	5,485	1,292	2.77
43	75,782	797	25.99	86	4,193	1,114	2.47
44	74,985	812	25.27	87	3,079	933	2.19
45	74,173	828	24.54	88	2,146	744	1.93
46	73,345	848	23.80	89	1,402	555	1.69
47	72,497	870	23.08	90	847	385	1.42
48	71,627	896	22.36	91	462	246	1.19
49	70,731	927	21.63	92	216	137	98
50	69,804	962	20.91	93	79	58	80
51	68,842	1,001	20.20	94	21	18	64
52	67,841	1,044	19.49	95	3	3	50

1.1. The American experience table, a life table originally authored by actuary Sheppard Homans for the Mutual Life of New York (MONY), here presented in 1868 as New York's legal standard for policy valuations, in chapter 623, Ninety-First Session in New York Legislature, *Laws of the State of New York*, 2:1315–1317 at 1317.

the same time, new insurers proliferated, eager to meet rising demand and probably even more anxious to grab premiums and turn them into investments in a rebuilding, westward-looking nation's vast market for mortgages and railroad debt. The number of companies tripled from about 40 before the war to a peak of 129 in 1870.[15] Of those 129, only 6 boasted at least $100 million of insurance in force—with Mutual Life of New York (MONY) in a class by itself at $242 million—while 14 firms had between $30 and $70 million in force, with the remaining mass of companies mainly doing business on a much smaller scale.[16] In that mass of small upstarts, American Popular Life had plenty of company.

But even in such a crowd, American Popular Life stood out. One of its directors later admitted "that considerable money had been invested in advertising under the peculiar style of the President, by literary documents and circulars setting forth his theories."[17] In one such book, *Sources of Longevity*, the reader encountered two essays by Lambert on the principles of "biometry" and its application to life insurance, two medical essays having won a hefty $500 prize, and only in the final pages a short advertisement for the company itself.[18] American Popular Life sold more than insurance; it sold a way of thinking about lives.

Biometry complemented phrenology. While phrenology promised that scalp measurements could provide insights into the nature of individual minds, biometry promised to use a wide range of medical measurements and inquiries to reveal the mysteries of individual bodies and lives.[19] Skilled

---

fact that many other companies' records were not included. It is safe to say that the number fell at least in the high hundred thousands. See Meech, *System and Tables of Life Insurance*, 5.

15. On insurance in force, see S. C. Chandler, Jr., *A Comparative Atlas and Graphical History of American Life Insurance Embracing a Period of Twenty Years Previous to January 1, 1880* (New York: Spectator Company, 1880). On the number of companies, see Stalson, *Marketing Life Insurance*, 751.

16. Stalson, *Marketing Life Insurance*, 798.

17. "The Life Insurance Troubles—The Officers of the American Popular Life Still Denouncing Superintendent Smyth," *New York Times*, 25 April 1877.

18. American Popular Life Insurance Company, *Sources of Longevity, Its Indications and Practical Applications* (New York: Wm. Wood, 1869).

19. Far from a mere "pseudoscience," phrenological assumptions so permeated polite society that the respected doctor and occasional life insurance statistician James Wynne would print in *Harper's* his judgment that Harvard's Louis Agassiz possessed "unquestionably the finest head and the most strikingly intellectual countenance" of America's scientific luminaries See James Wynne, "Louis Agassiz," *Harper's New Monthly Magazine* 25, no. 146 (July 1862): 199–200. For more on phrenology, its uses, and its place in American scientific and intellectual culture, see Sandage, *Born Losers*, 112–121; Ann Fabian, *The Skull Collectors: Race, Science, and America's Unburied Dead* (Chicago: University of Chicago Press, 2010); and John D. Davies, *Phrenology: Fad and Science* (New Haven, CT: Yale University Press, 1955).

practitioners of biometry could, Lambert promised, help with "the selection of vocations, of education, of partners, either in business or for life—in fact in all the most momentous concerns of life."[20] Biometry, like phrenology, depended on "a harmony, a perfect relation between all parts" of the body, whether internal or external. Lambert argued that "a difference in constitutions . . . *should be observable at the surface of the body*," and thus that careful observation and measurement of superficial characteristics could speak to inner vitality.[21]

The correspondence of the superficial to the fundamental allowed Lambert to overcome a major problem he faced in predicting longevities. He considered ancestry the most important determinant of life span. "Blood," rather than "climate" or hygiene—factors more amenable to medical thinkers with a statistical bent[22]—made for long lives.[23] Lambert classified whole peoples by their longevities: Laplanders, Esquimaux, and citizens of Zurich suffered notably short existences, he asserted, while Jews, Welsh, Scots, Italians, and New Englanders lived very long indeed.[24] Each individual's family history mattered even more. But therein lay the problem: "Most persons do not know the longevities of their ancestors, nor the diseases of which they died," complained Lambert.[25] So, lacking oral family histories, Lambert resorted to reading family longevity off of individual bodies.

Upon seeing the ideal body, Lambert argued, "There need not be any hesitation in concluding that the ancestry on both sides were naturally

20. American Popular Life, *Sources of Longevity*, preface: ii.

21. American Popular Life, *Sources of Longevity*, section 1, part 3:2.

22. When MONY commissioned the doctor and statistician James Wynne in the 1850s to write a report on American vital statistics (that would justify the penalties they levied against southerners at the time), the company expected and received a statistical treatise that attributed much determinative power to climate and geography. See James Wynne, *Report on the Vital Statistics of the United States, Made to the Mutual Life Insurance Company of New York* (New York: H. Bailliere, 1857). Wynne, like the man he most closely emulated—William Farr, Britain's eminent statistician—drew on an older tradition of "political arithmetic" hailing from eighteenth-century England, as well as more recently founded Continental traditions being advocated by Louis René Villermé in France and Adolphe Quetelet in Belgium. On English and Continental precursors, see Julian Hoppit, "Political Arithmetic in Eighteenth-Century England," *Economic History Review* 49, no. 3 (1996): 516–540; Libby Schweber, *Disciplining Statistics: Demography and Vital Statistics in France and England, 1830–1885* (Durham, NC: Duke University Press, 2006); and Alain Desrosières, *The Politics of Large Numbers: A History of Statistical Reasoning* (Cambridge, MA: Harvard University Press, 1998), 16–44,67–91. On Farr's contributions and his appropriation of actuarial methods, see John M. Eyler, *Victorian Social Medicine: The Ideas and Methods of William Farr* (Baltimore, MD: Johns Hopkins University Press, 1979).

23. American Popular Life, *Sources of Longevity*, section 1, part 1:35.

24. American Popular Life, *Sources of Longevity*, section 1, part 1:7–8.

25. American Popular Life, *Sources of Longevity*, preface: ii.

long-lived." He sketched out such a body for his readers: "The members of the body are well proportioned, the trunk relatively long, the orifice of the ear decidedly below the eyebrow and distinct from it" and so on in minute detail. He favored large noses, "true hazel" eyes, lips and gums "of a good color," "well colored" skin, sound teeth, and a broad head.[26] One of the essays Lambert published cited the soul of quantification himself, Alexander von Humboldt, as the model individual from the perspective of longevity—with his small stature, "remarkable length of body, mounted on equally remarkable short, small, lower limbs."[27] Another essay by a prominent New York doctor, John H. Griscom, offered an even more finely drawn "Portraiture of a Man destined to Longevity."[28]

Though Lambert trusted external observations, he also accepted the ideas of those who favored new instruments that could measure the body's hidden interior. Stethoscopes told the story of the heart better than outwardly evident musculature, argued one essay published by Lambert, just as a spirometer measured lung capacity, and could warn of tubercular tendencies, more accurately than external chest measurements.[29] In adopting these new instruments, as in many other aspects of biometry, Lambert drew on techniques and ideas developed in and around insurance companies by doctors hired to keep the sick or disabled from purchasing life insurance.[30]

Lambert not only depended on life insurance medical practices; he considered life insurance to be a crucial proving ground for biometry. Examining doctors employed by insurers could demonstrate their capacity—to their employers and simultaneously to the world—to precisely assess the vitality

26. American Popular Life, *Sources of Longevity*, section 1, part 1:32.

27. J. V. C. Smith, "Physical Indications of Longevity in Man," in American Popular Life, *Sources of Longevity*, section 1, part 2:6.

28. Griscom had, in 1848, chaired the American Medical Association's first committee to petition state and local governments to develop systems for collecting birth, death, and medical registration data. John H. Griscom, "To the United States National Medical Association," *Transactions of the American Medical Association* 1 (1848): 339–340. Griscom's essay had little to do with ancestry, but Lambert's conviction that bodies displayed constitutions allowed him to still adopt Griscom's ideas of bodily assessment as his own. John H. Griscom, "Physical Indications of Longevity in Man," in American Popular Life, *Sources of Longevity*, section 1, part 2:8–9.

29. Smith, "Physical Indications of Longevity in Man," 10–12.

30. Doctors had found a secure home in life insurance companies in the first half of the nineteenth century because the scale of operations made it nearly impossible for boards of directors to examine each new insurance applicant directly, as had been the eighteenth-century norm. By the 1860s, insurance medicine enjoyed its own specialist literature. Lambert refers readers interested in biometry to works on insurance medicine including J. Adams Allen, *Medical Examinations for Life Insurance* (Chicago: Clarke, 1866); and D. S. Gloninger, *The Medical Examiner's Manual* (Philadelphia: Claxton, Remson, & Haffelfinger, 1869).

of any individual and forecast his or her life span. Lambert felt certain that if American Popular Life survived and thrived, his theory's future would be assured. Many of his contemporaries, in a similar but far less charitable vein, also thought of the company "more in the light of an experiment than as legitimate business."[31]

Relying on biometry, Lambert and his company's doctors built a body of policyholders consisting, so they claimed, of "Lives *thoroughly* (no other does this) classed and sub-classed according to the risk and to the kind of premiums paid."[32] Lambert called on doctors employed as medical examiners to submit "pen-portraitures of the applicants" which he and the expert doctors in his office would read for signs of longevity.[33] Those judged to have a long life ahead of them paid less. Those with a short life paid more. (Simple chronological age, alone, mattered much less than it did in other companies.) "Thorough classing," by which Lambert meant categorizing each policyholder according to his anticipated life span, "therefore gives a fair chance to the long-lived to pay less and have more."[34] Such was the soul of "equity," of charging each according to his individual risk.

Biometry promised equity for everyone. American Popular Life sold life insurance as a man's responsibility to his family, one that his family should "scrupulously, even religiously, care for and sustain . . . as its choicest blessing."[35] It extended the opportunity to fulfill that responsibility to all by "thoroughly classing" applicants using "biometrical science."[36] No one needed to be rejected for poor health, although the price for the sick or disabled would be much higher, perhaps prohibitively so.[37] But American Popular Life promised family protection on the cheap to healthy applicants whom it pursued with advertisements like that shown in figure 1.2. "The best Company for the 'Best' lives," exclaimed another ad.[38]

Lambert and his colleagues' medical evaluations can be read to support the company's boast that it attracted a disproportionate number of good lives. In its first year, the company sold 295 policies, of which 255 applicants

31. "Life Insurance Topics; The American Popular Failure," *New York Times*, 24 April 1877.
32. American Popular Life, *Sources of Longevity*, back cover.
33. American Popular Life, *Sources of Longevity*, section 1, part 3:2.
34. American Popular Life, *Sources of Longevity*, section 2: iv.
35. American Popular Life, *Sources of Longevity*, section 1: 12–13.
36. This variant appears in Massachusetts Insurance Department, "Classification of Lives," in *Annual Report of the Commissioner of Insurance of the Commonwealth of Massachusetts* 18 (1872) 2:xlviii–1 at xlix.
37. American Popular Life, *Sources of Longevity*, section 1, part 3:3.
38. American Popular Life, *Sources of Longevity*, back cover.

INTERESTING QUESTIONS.

—◆◆◆—

No First-Class Life can afford to Insure in any other Company.

Is not Long Life usually inherited ? (From Parents, Grandparents, &c.

Do not some sound persons have Constitutions that show greater promise of living to Old Age than do those of others ?

Will not Good Habits tend to prolong life?

Is every place of Residence equally healthy?

Does every Vocation equally endanger life?

Will the Ignorant, on an average, live as long as the Intelligent, other things being equal?!!

Are then all SOUND PERSONS *of the same age* EQUAL RISKS? Or, are Risks that pay different kinds of Premiums equal?

Is it scientific, fair, or honest, to insure them as equal?

Ought they not then to be insured in classes according to their probabilities or risk, as is done in Marine and Fire Insurance? Can this be done? Certainly. Why not?

UNDER ITS SYSTEM

This Company can give to any person who can be put in the best class, ten per cent. more Life Assurance for the same premium than he can obtain in any Mutual or other company, and will in the policy guarantee to him such proportionate dividends (so called) upon the whole, as such other company will pay upon a part of it, and to do whatsoever such company does, writing its policy verbatim, if required.

This Company can give any kind of Whole Life Insurance to persons who can be put in the best class; and guarantee in the policy that they will not be required to pay in yearly or other premiums more than four to six tenths of its whole amount or face.

Four to six hundred dollars will thus be the total outgo, in securing a one thousand Dollar Whole Life Policy.

This Company can give these advantages to the long-lived, and do a profitable business.

1.2. An American Popular Life advertisement, one of a series appearing on the back cover of American Popular Life, *Sources of Longevity* (1869).

were deemed in the view of the examiners to be fit to live longer than the actuarial tables predicted for their actual age, while only 11 applicants were charged premiums greater than those required by the tables.[39] There are two possible explanations for this fact. Either Lambert and his colleagues erred on the side of applicants, judging a vast majority to be longer lived than average, or only those applicants deemed to have a long life, and thus granted

39. Massachusetts Insurance Department, *Annual Report of the Commissioner of Insurance of the Commonwealth of Massachusetts* 12 (1866): 2:c.

a cheaper policy, chose to insure with American Popular Life instead of one of its many, many competitors.

Policyholder death statistics, on their surface, give more weight to the latter explanation. In 1868, the company had the lowest ratio of death claims to policyholders among companies operating in Massachusetts.[40] After the company had survived three years with lower than expected rates of mortality, Lambert crowed: "A financial, competitive, not to be disputed value, is now yielded to the teachings of Biometry; which, as an unapplied science, might have been regarded as wholly speculative, and been viewed with doubt, if not with decided unbelief."[41] Yet, later revelations disrupt so triumphant a story. The company never kept formal "loss books" recording the deaths held against it.[42] It frequently contested and litigated any death claims, thereby delaying deaths from appearing in regulators' records.[43] And on at least one occasion it reported a "death" as a "surrender" (where the policyholder gave up the policy in exchange for cash) to keep its loss ratio artificially low.[44] The company admitted applicants that it predicted would have long lives and then it refused to let them die. They became statistically immortal.

"Is not classing risks the pith of Insurance?" asked an American Popular Life ad, assuming an affirmative: "Is not then that Company best that classes best?"[45] This was a subversive question, one meant to further distinguish American Popular Life from its contemporaries. Other companies certainly did class their risks. Among policyholders, age classes formed the basis of modern life insurance. Ever since the British mathematician James Dodson's appropriation of the life table (a happier name for the same old mortality table) for insurance purposes in the early 1760s, policyholders had paid premiums corresponding to their age, a proxy for their risk, with the reasoning that all individuals of a given age contributed to insure one another.[46]

40. American Popular Life, *Sources of Longevity*, section 2: preface: ii.

41. American Popular Life, *Sources of Longevity*, section 2: preface: i.

42. "The American Popular Life," *New York Times*, 25 October 1877; "The American Popular Frauds," *New York Times*, 26 April 1877.

43. It may have been that Lambert's faith in his system encouraged the company to be more than usually litigious. If he thought he could accurately forecast a life span, any death that occurred before he predicted it might be considered suspect of fraud. Or, he may have just been trying to save biometry's face. "The American Popular Frauds," *New York Times*, 26 April 1877.

44. "Miscellaneous City News: The American Popular Life," *New York Times*, 7 December 1877.

45. American Popular Life, *Sources of Longevity*, back cover.

46. For more on Dodson, his influences, his table, and his plans for a new "assurance" society, see C. G. Lewin, *Pensions and Insurance before 1800: A Social History* (East Linton, UK:

Most companies further separated policyholders into classes based on sex or on region and climate. Alongside American Popular Life arose upstart companies that classed separately those who abstained from alcohol or who practiced homeopathic medicine and charged them lower rates based, in part, on specially constructed life tables.[47] On the most basic level, all insurers—usually relying on the advice of their medical examiners—classed individuals into two groups: those who met the basic standard set for insurance and those who did not. To make a risk from a life, insurers *all* classed that life in one way or another.

Yet most companies and most insurance promoters did not, in fact, tout classing risks as the "pith" of their business. When contemporaries spoke of the "science of life insurance," as they often did, they usually had statistical laws in mind more than medical classification.[48] As the Massachusetts commissioner of insurance put it in 1872, expressing a widely held belief: "In

---

Tuckwell, 2003), 377–381; Maurice Edward Ogborn, *Equitable Assurances: The Story of Life Assurance in the Experience of the Equitable Life Assurance Society 1762–1962* (London: George Allen & Unwin, 1962), 22–33. Jacob L. Greene, the secretary of the Connecticut Mutual Life, told a typical story about the making of a risk community in 1871, propounding the bookkeeping fiction of homogenous groups insuring one another: "[Life insurance] brings together a multitude of men aged thirty, and makes them agree with each other that when one dies, the survivors will pay to his family the cash value of his life; which they accomplish, after ascertaining the rate of mortality and assuming a safe rate of interest, by paying into the common fund a fixed sum each year of their lives." Connecticut Mutual Life Insurance Company, *Annual Report* (1871): 20–22.

47. The actuary Elizur Wright advocated temperance rates and authored "newly calculated homeopathic mortuary tables" for Hahnemann of Cleveland, which, along with Atlantic Mutual of Albany and eight other boom companies, discriminated in favor of homeopaths. Lawrence B. Goodheart, *Abolitionist, Actuary, Atheist: Elizur Wright and the Reform Impulse* (Kent, OH: Kent State University Press, 1990), 144–145. Asher S. Mills, "Hahnemann Life Insurance Company," *Ohio Medical and Surgical Reporter* 1, no. 5 (September 1867): 151–152; "Circular of the Atlantic Mutual Life Insurance Company," *North American Journal of Homeopathy* 15 (November 1866): 291–295. See also Massachusetts Insurance Department, *Annual Report of the Commissioner of Insurance of the Commonwealth of Massachusetts* 12 (1866): 2:c. On the number of homeopathic companies, see John S. Haller, Jr., *The History of American Homeopathy: The Academic Years, 1820–1935* (New York: Pharmaceutical Products Press, 2005), 117. Homeopathic companies hoped to sell their methods as well as insurance, as did American Popular Life, but they put more faith in statistics than Lambert and his colleagues. "Each life is carefully recorded;" explained Hahnemann's publicists, "each journey of any importance, and changes of residence from one climate to another, noted; and all items, which may affect the person's health or physical condition, carefully gathered. It is, in fact, as nearly as possible, the medical history of each person's life from the time of its acceptance until its close." Quoted from "Hahnemann Life Insurance Co.," *Hahnemannian Monthly* 2, no. 11 (June 1867): 506–508 at 507.

48. For example: "The science of Life Insurance is greatly indebted for the valuable tables which have been constructed, for showing the law of average duration of human life; these tables are the groundwork of the calculations, giving the value of annuities, and the tariffs of premiums on Life Insurance." Mutual Life Insurance Company of New York, *Annual Report* (1845): 15.

the tables of mortality upon which the fabric of life insurance is founded no distinction is made between persons, and the chances of living and dying are considered as being the same for all."[49] He had his own idea about what constituted "the essence of insurance": the "principle of average," the smoothing away of the messiness of life (most commonly by drawing in pencil a smooth line through data plotted on graph paper) to reveal ordered, durable laws.[50]

Even American Popular Life averaged and smoothed, argued the commissioner. Once the company assessed the risk of an applicant, it classified him according to his life expectation, essentially giving him an ideal age, usually different from his real age. But each applicant still had to be assigned to some group according to his (ideal) age in a process perfectly parallel to that followed by all other companies. In the end "the [mortality] probabilities of all belonging to the same class are treated as being equal—a practice really based upon the same principle of average" employed in other insurers.[51] By charging the same premium to all individuals of the same rated (ideal) age, Lambert admitted the principle of average even if his company put less effort than most into averaging and smoothing.

The argument over the "pith" or "essence" of insurance had everything to do with emphasis—with the methods that dominated in a company. Classing and smoothing, in combination, made every insurance office run, but they nearly always worked in tension with one another. Despite Lambert's intimations to the contrary, classing had just as strong a claim to statistical practice as did smoothing. Lives had to be classed before their mortality probabilities could be smoothed and averaged, but classed lives did insurers no good without smoothing. This tension worked itself out in insurers' or commissioners' offices, and also in the realm of advertising, in the ways insurers sold themselves to potential customers. In that context, the ideas at stake—big questions about the possibility of certain knowledge, the responsibility of individuals to society, and the nature of equity or fairness—bubbled up for public consideration. In selling insurance, companies simultaneously sold competing epistemological, social, and ethical positions.

49. Massachusetts Commissioner of Insurance, "Classification of Lives," *Report of the Insurance Commissioner* 18 (1872): 2: xlviii–1 at xlviii.

50. Massachusetts Commissioner of Insurance, "Classification of Lives," *Report of the Insurance Commissioner* 18 (1872): 2: xlviii–1 at xlix.

51. Massachusetts Commissioner of Insurance, "Classification of Lives," *Report of the Insurance Commissioner* 18 (1872) 2: xlviii–1 at xlix.

Whether classing or smoothing, risk makers claimed some certainty to be possible, but the sorts of certainty that could be grasped depended on which activity the risk maker favored. Lambert claimed that through biometry he could accurately and precisely predict an individual's longevity, that he could approach certainty in classing. Most other companies began with the *probabilistic* certainty that made smoothing possible: that over a large number of individuals, a consistent number would die at any given age.[52] Actuaries inscribed their faith in the certainty of smoothing into the life table, which began with a fixed number of (ideal, abstract, statistical) living individuals in one column (100,000 in figure 1.1) and then reported the number still living with each successive year of age. From the columns of a life table, explained most life insurers, one could read quantified fates (for a *group*) as certain as anything in the world.

For smoothers and averagers, however, uncertainty reigned at the level of *individuals*. (Not coincidentally, this was also the level at which classing mattered most.) Consider one of the first annual reports released by the largest and most influential life insurer of the time, Mutual Life of New York (MONY), which presented ten stories of death, including a Dutchess County farmer who paid one $21.70 premium on a $1,000 policy before being struck down by apoplexy; a young Rochester widow receiving $1,000 after her husband died of a short illness having paid one $23.10 premium; a New York merchant who died after two payments of $152.50, leaving $5,000 for his mourning widow.[53] Each had been classed well and admitted to the company on standard terms, with each hoping for and expecting a long life. Then each died suddenly. These were narratives of tragedy, of the hopelessness of forecasting for an individual life.

---

52. The so-called law of large numbers, coined by the mathematician Poisson in 1835, played a crucial role here. Porter defines it simply as "the observed regularity of statistical aggregates," or more formally "the proposition that the frequencies of events must, over the long run, conform to the mean of their probabilities when those probabilities fluctuate randomly around some fixed, underlying value." See Porter, *Rise of Statistical Thinking*, 12–13. Chapter 4 will discuss life insurers' smoothing practices in greater detail. For now, it is enough to recognize that actuaries understood mortality to behave according to a quantifiable, regular, statistical "law," akin to the "laws" that Newton discovered for motion and gravity and the laws of plant distribution that Alexander von Humboldt graphed and mapped. Actuaries were "Humboldtians," in the sense defined by Susan Faye Cannon, "Humboldtian Science," in *Science in Culture: The Early Victorian Period* (New York: Dawson and Science History, 1978).

53. Other stories also show the prudence of life insurance taken out on a wife whose inheritance a man depended on, on a debtor by his creditor, or on one's own life in the amount of one's debt. Mutual Life Insurance Company of New York, *Annual Report* (1845): 8–9.

They were also narratives of a windfall, of unexpected death accompanied by extraordinary gain, the uncertainty of life eased by the certainty of payment. These stories shifted the emphasis away from equity, from assuring that the long-lived got what they deserved—the value that Lambert made so important to his sales pitches—and toward charity, toward assuring that the unfortunate were provided for. Such narratives also hinted at the gambles inherent to life insurance: some people would get (monetarily) much more than they deserved.

Beliefs in statistical certainty, limited faith in classing lives too narrowly, and an ethical stance favoring charity often came bound together with notions of social responsibility. Where Lambert emphasized the individual's responsibility to family alone, contemporary commentators often justified life insurance as a social good too. They raised, for instance, the specter of orphans left without protection: their fathers dead without insurance, they would become street children in roaming gangs, one man's failure to insure now a cause of social disorder.[54] Life insurance proponents told stories of vital and financial disaster; stories of the uncertainty and instability of commercializing, industrializing American society; stories that life insurance could prevent from coming true. As one put it in 1859: "The handle of a [railroad] switch pointing a few degrees in the wrong direction, may not only extinguish a constellation of the lights of society, but precipitate a score of families from affluence to destitution."[55] With such social consequences in mind, one life insurance agent recommended that his colleagues become Sunday school teachers evangelizing the moral value and social responsibility of insurance.[56] The lecturer Moses Knapp, having himself been led to proselytizing life insurance by his minister, even went so far as to equate modern life insurance with the community of early Christians who sold their private possessions and "laid them down at the Apostles' feet; and distribution was made unto every man according as he had need," only now operating on a much larger scale made possible by probabilistic science.[57]

54. Knapp, *Lectures on the Science of Life Insurance*, lecture 1.

55. Elizur Wright, January 1859 report in *Massachusetts Reports on Life Insurance: 1859–1865* (Boston: Wright & Potter, 1865), 2. This passage reflected contemporary fears, but it also perpetuated them. Insurers must be understood as purveyors of risk consciousness, as the tellers of stories about modernity's dangers.

56. Henry Clay Fish, *The Agent's Manual of Life Assurance* (New York: Equitable Life Assurance Society, 1867), 86.

57. Knapp quoted here from the book of Acts, chapter 4. Knapp, *Lectures on the Science of Life Insurance*, 26, 36.

The contrasts between American Popular Life and other insurers should not be overdrawn, however. In the 1860s and 1870s, few or no (probably no) companies really acted like the apostles or understood themselves primarily as smoothers redistributing wealth from the lucky to the unlucky in order to stabilize society. We should, again, think in terms of emphases here. And even those companies (including the older, larger firms like MONY) that emphasized smoothing were edging toward equity, toward individualist conceptions of responsibility, toward more classing.

To understand why many life insurers amplified their equity talk and edged toward classing, we must first understand their increasing emphasis on the value of their product as a form of *investment*. Life insurance looked like an investment thanks to the mechanism of the dividend or bonus— payments to policyholders (or additions to their policies) made regularly on whole life policies, meant as a kind of return of company profits. By the 1870s, MONY had replaced its short, descriptive lists of windfalls awarded with long tables of policyholder deaths, announcing both the amount insured and the amount that the deceased's beneficiaries would gain through the accrual of dividends. When Charles Noble died in 1875, for instance, his widow received the $3,000 face value of his policy, but also the added consolation of $1,151.29 in returns on her husband's "investment" in the company.[58] Such rolls reminded readers, implicitly, of a death to come (who knew when?) and of economic embarrassment that threatened (all too often).[59] Insurers sold their product as a solution to the (physical and financial) uncertainty of life by promising certain protection and, increasingly, certain investments. But the dividend did more than allow life insurers to emphasize investment; it also drew attention to individual equity.

To ensure equity, life insurers classed. Acting prospectively, companies classed individuals together to create groups composed of individuals exposed to more or less the same risk of death. Equity talk had roots at the beginning of modern life insurance. The Equitable Assurance Society—the first modern life insurer—chose its name to emphasize its innovation in classing individuals by age and charging them different rates accordingly. Actuaries afterward defined "equity" as dependent upon each policyholder paying only for his or her own mortality risk and not for the excessive

---

58. Mutual Life Insurance Company of New York, *Annual Report* (1875): 64–88 at 64.

59. Henry David Thoreau's *Walden*, in 1854, claimed that ninety-seven out of one hundred merchants failed at some point in their lives. This unsubstantiated statistic originated in the 1830s and lived on for nearly a century. See Sandage, *Born Losers*, 7–8.

mortality probabilities of others. Dividends acted after the fact—as a *fly-wheel or governor,"* according to a MONY actuary, that could correct "any injustice" done by faulty prospective classing or out-of-date mortality tables.[60] Actuaries recommended a dividend be paid back to policyholders when the company enjoyed higher interest rates than expected, experienced less mortality in some class than projected, or spent less on administration than budgeted.[61] Where Lambert focused on improving classing up front to improve equity, MONY and its peer companies (the older, larger ones, as well as most upstarts) relied on post hoc actuarial analyses and the mechanism of the dividend to do the same thing.

The drift toward individualist equity and classing took on a peculiar, but extraordinarily important, form in a new insurance offering called a tontine or deferred dividend policy. Policyholders purchasing such a policy agreed to forgo receiving dividends for up to twenty years. If they died (or lapsed their policy) before that time, they and their beneficiaries lost all claim to their dividends. If they survived, they received their own dividends as well as those of their less fortunate peers.[62] The policies thus compensated individuals (after the fact) for their long lives: they retrospectively classed policyholders into groups of longer- and shorter-lived people. Speaking of the system as "exceedingly equitable," one company proposed that giving extra dividends to the beneficiaries of those who died late made sense since the beneficiaries of those who died young already received a monetary windfall.[63] Americans in the second half of the nineteenth century bought this argument, or, if not the argument, they bought deferred dividend policies—in droves.[64]

60. Sheppard Homans, *Report Exhibiting the Experience of the Mutual Life Insurance Company of New-York for Fifteen Years Ending February First, 1858* (New York, November 1859), 33. Emphasis in the original.

61. Sheppard Homans, "On the Equitable Distribution of Surplus," *Journal of the Institute of Actuaries* 11 (October 1863): 121–129.

62. Henry Hyde's Equitable Life Assurance Society of the United States first rolled out its "Tontine-Dividend System" in 1870. Others followed soon after. Equitable Life Assurance Society of the United States, *Annual Report* (1870): 33–34.

63. Equitable Life Assurance Society of the United States, *Annual Report* (1870): 33–34.

64. The appeal of the policy as a gamble may very well have mattered more than its appeal to equity. Contemporary critics certainly thought so. Elizur Wright called it a particularly unjust gamble too: "As a game of long purses against short ones," Wright wrote, "it can hardly be said to be fair." He worried particularly about those who lost their dividends when their policies lapsed. Wright, *Politics and Mysteries of Life Insurance* (Boston: Lee & Shepard, 1873), 207.

Metropolitan (known today as MetLife)—one of the approximately one hundred postwar upstart companies, and thus one of American Popular Life's close, but more conventional, peers—illustrates the popularity of deferred dividends. Metropolitan had emerged as an ordinary life insurance company in 1868, presumably planning to capitalize on the boom market. To compete, it tried a variety of strategies: it employed a very popular trick that allowed policyholders to pay part of their premiums with "notes" (IOUs that were supposed to be paid off by future dividends); it also toyed with insuring the laboring classes via a temporary alliance with a new friendly society targeting German emigrants (called the Hildise Bund); and it even dipped its toes in southern and Canadian markets—all with limited success.[65] Metropolitan's most rapid and sustained growth came only when, in 1872 (as the life insurance market stood on the brink of collapse— although no one realized that yet), it began selling a new kind of policy on the deferred dividend plan. Riding the success of its tontine business, Metropolitan nearly doubled its insurance in force in three years.[66] Equity and classing proved powerful as competitive tools.

Lambert's subversive question had something to it after all. Perhaps the company that classed best really was best. But who could say what the best way to class was? The fundamental difference between American Popular Life and companies like MONY or Metropolitan was not whether or not they classed. Everyone classed. The fundamental difference lay in whether a company put more power and trust in classing or in smoothing. In fact, maybe the true essence of insurance was the *tension*, the *conflict*, the *dialectic* between classing and smoothing.

But as we are about to see, the difference Lambert cared about most turned on questions of power and methods. Who should control classing? Doctors or actuaries? And how could an individual's fate be foretold? By medical expertise or statistical probability?[67] Lambert sided—emphatically—with the weaker group and with disfavored methods, decisions fated to undo him.

---

65. Stalson, *Marketing Life Insurance*, 340, 431; James, *Metropolitan Life*, 43–44, 48, 53.

66. See the chart of Metropolitan's insurance in force in Chandler, Jr., *Comparative Atlas*.

67. Medical expertise and statistical probability did not need to be at odds with one another. Doctors had subscribed to statistical and probabilistic methods before and many continued to do so. But Lambert represented a medical tradition that looked suspiciously at statistics. On antebellum American physicians' commitment to statistical methods and the displacement of statistical thinking in much medical practice in the second half of the nineteenth century, see James Cassedy, *American Medicine and Statistical Thinking, 1800–1860* (Cambridge, MA: Harvard University Press, 1984). As we will see in future chapters, life insurance offices often served as crucial sites for the preservation of the ties between medical and statistical practice.

* * *

Only two upstart companies, aside from American Popular Life, used health as a factor in setting rates and both of them only "rated up"—they accepted unhealthy lives at higher rates.[68] American Popular Life's decision to rate up *and down* had no precedent. It stood alone in empowering doctors to such a degree. Likewise, it alone picked a fight with actuaries, their statistical methods, and their life tables. In the long run, the fight turned out badly.

Lambert and his medical allies disdained "mere mathematicians, dignifying themselves with the name of actuaries."[69] Lambert thought the public ("no longer inclined to worship a veiled Mokanna") might share his feelings. "Great mistrust of these 'mortality' and 'expectation' tables is felt by many persons," he claimed. He deemed it suspicious that these so-called laws and "calculations based upon them . . . carried to many decimal places, as if the tables were exquisitely exact" were "never used, nor could they be with safety, as a perfect measure in insuring." Companies started with a life table, then computed premiums assuming "a lower rate of interest . . . than that which it is expected will be realized" and then topped them off with "a larger addition" to cover expenses or unexpected contingencies.[70] Artificially low interest rates and a "loading" addition that usually amounted to 30 percent of the premium indicated to Lambert that even the actuaries did not really trust their tables. (And, in that judgment, he was right).[71]

Others had attacked actuaries before. The abolitionist and early advocate for state insurance regulation Elizur Wright—having trained himself in actuarial methods—prefigured Lambert's critique of actuaries' antidemocratic, secretive, or even dogmatic expertise: "The hieroglyphic veil which concealed from the common herd the learning of the ancient Egyptian priesthood was thin; and that which renders a priesthood of professional

68. On the various practices for dealing with so-called underaverage or substandard or subprime lives developed by British companies, see Alborn, *Regulated Lives*, 273–278; on the earliest American up-raters, see Stalson, *Marketing Life Insurance*, 325–326.

69. Lambert didn't say this, but printed the comments approvingly in Smith, "Physical Indications of Longevity in Man," 24. Smith also wrote: "Usually a few mathematical tables and a general average of the length of life has been considered as the important substratum of a life insurance organization. As well might a dry skeleton be considered as the living man."

70. American Popular Life, *Sources of Longevity*, section 2:13–14.

71. Actuaries' conservative mistrust for their tables had business benefits. All such assumptions (low projected interest rates, conservative tables, 30 percent loading) created space ostensibly for unexpected disaster and epidemic or long-term economic downturns, but in practice they mainly served to pay for company expenses or inflated company reserves and thereby increased the pots of money that company finance committees could invest, or became dividends handed back to policyholders (profits in non-mutual companies.)

actuaries necessary for the safe conduct of modern Life Insurance is not thick."[72] But Wright did not share Lambert's cynicism about mathematical tables. In fact, he believed life tables should replace actuaries entirely, that mere clerks with accurate tables could do all that a life insurer needed. They could not. Tables and actuaries, it turned out, went hand in hand.

When Lambert inveighed against life tables in 1869, he probably had one particular table foremost in his mind: the "American experience table," so named by a New York State law of 1868.[73] The handiwork of Sheppard Homans (America's most prominent actuary, next to Elizur Wright), the American experience table began its life as an analysis of MONY's first fifteen years of "experience"—of the company's records of what happened to those it insured over its first fifteen years (i.e., for any given age: how many died, how many let their policies lapse, how many survived to pay another premium).[74] MONY adopted the table for the purpose of setting dividends in 1861, and then made it the basis for setting premiums in 1867.[75] In both cases, the new table allowed MONY to lower the cost of insurance to middle-aged American men, who lived longer according to the table than those described by the British tables otherwise in use at the time.[76] New York legislators and regulators, who listened to MONY *very* carefully,

72. "Life Insurance—Wright's Tables," *Hunt's Merchants' Magazine and Commercial Review* 31, no. 6 (December 1854): 734–736 at 734.

73. On the establishment of the American experience table as the state standard, see NYS chapter 623, 91st session, 1868 in New York Legislature, *Laws of the State of New York, Passed at the Ninety-First Session of the Legislature, Begun January Seventh, and Ended May Sixth, 1868, in the City of Albany* (Albany: Van Benthuysen & Sons, 1868), 2:1315–1317.

74. Joffe links the curve to the fifteen-year experience in S. A. Joffe, "Concerning the American Experience Table of Mortality," *Transactions of the Actuarial Society of America* 14, no. 49 (1913): 24–37. How Homans first constructed the curve that became the table is evident from a piece of graph paper sewn into the Minutes of the Insurance Committee (Book 1), MONY Papers. See figure 4.1.

75. "For the Private Use of the Trustees of Mutual Life Insurance Company of New-York," bound into Minutes of the Insurance Committee (Book 1): 379–380, MONY Papers. Clough, *Century of American Life Insurance*, 61.

76. Homans claimed that Americans lived longer than Europeans on the whole and enjoyed a *"vitality* at middle ages, say from 30 to 60," that was "undoubtedly greater among Americans than among persons of similar ages and circumstances in any other country." Higher rates of mortality in early adulthood and at ages over sixty tempered the advantage in the middle ages. Sheppard Homans, "Life Insurance: Sheppard Homans on Life Insurance. Paper read before the American Social Science Association. Revised by the Author," *Insurance Times* 2 (1869): 798. High mortality among young adults, Homans had speculated in 1858, sprung from America's entrepreneurial fixation. "The excessive mortality at younger ages," he wrote, "is perhaps owing to the fact that in this country young men are induced, by a spirit of enterprise, to enter upon the anxieties and cares of active business life too early." The twenty-seven-year-old Homans could speak from experience. Homans, *Report*, 24.

subsequently made Homans's table the state standard for valuing policies. That meant that regardless of the table a company used to set its premiums, it had to show the state that it had sufficient reserves to cover its liabilities as computed using MONY's table. The decision elicited significant controversy from those who supported the established standard put forth by the transnational actuarial community and from those who did not enjoy seeing New York State attempt to singlehandedly impose a national industry standard based on one company's limited experience.[77]

American Popular Life, incorporated in New York State and subject to its insurance laws, could argue with actuarial tables but it could not ignore them. In its own calculations, American Popular Life used its rated (ideal) ages to determine its future liabilities and it assumed a 7 percent return on its investments (which may not have been entirely unreasonable during the boom). New York, however, mandated valuations based on policyholders' actual ages and a 4.5 percent interest assumption. Both differences caused the state to demand that American Popular Life keep much more money in reserve for future payments than the company itself calculated to be necessary. In the long run, Lambert believed such accounting differences would be the foundation for his company's profit. But in the meantime, he had to keep the state's actuaries happy. To do that, Lambert and his close colleagues on the company's board, along with early investors, turned to "conditional certificates." These certificates entitled their bearers to a future share of anticipated profits, but also to a future obligation to contribute cash to the reserve fund should the company need it. American Popular Life counted this unpaid capital, these glorified IOUs, as an asset in the reserve fund and went on with its business. Lambert and his fellow officers also enjoyed the privilege of drawing a monthly advance of $125 on future profits from these certificates.[78] So long as no one looked too closely—and with so many new upstart companies running, who had time to look closely?—the company squared biometrical with probabilistic classing.

* * *

77. Elizur Wright, "National vs. State Supervision," *Insurance Times* 1 (July 1868): 263–265; Jon Sanford, "Fourteenth Annual Report of the Insurance Commissioner of the Commonwealth of Massachusetts," reprinted in *Insurance Times* 2 (September 1869): 652–665 at 662–663; Charles F. McCay, "American Tables of Mortality," *Journal of the Institute of Actuaries* 16 (October 1870): 20–33.

78. The scheme appears in outline in coverage of Deputy Insurance Superintendent John A. McCall, Jr.'s testimony against the company and in Lambert's defense of his actions. "The American Popular Frauds," *New York Times*, 26 April 1877; "Dr. Lambert on the Witness Stand," *New York Times*, 15 December 1877.

In May 1873, a Viennese housing bubble popped, sending shock waves through credit systems propping up similar bubbles in American railroad building and western land development. The key event in the US panic came in September when Jay Cooke's bank failed to find a market abroad for its Northern Pacific railroad securities and subsequently unwound in a most spectacular and disastrous fashion. The rest of the nation, and much of the world, followed.[79] Most life insurers' growth stalled in 1871, ahead of the full-on panic, but the industry did not feel the entire impact of the depression until the mid-1870s, when Americans lost the capacity to keep up their premium payments. Insurance-in-force plummeted in 1876. And well over half of all existing companies at the end of 1870 (75 out of 129 companies at the peak of the bubble) failed between 1871 and 1877.[80] Suddenly, state regulators had the urge to look closely at companies' books.

John A. McCall, Jr., deputy superintendent of the New York insurance department, investigated American Popular Life in 1877 and discovered a disaster. More than a third of the company's 2,905 policies appeared bogus—kept on the books only to disguise policyholders' flight from the company. McCall gave no credence to holding "checks" or "certificates" as legitimate assets, and he uncovered a number of dubious transfers and loans between directors and the company, apparently to allow the company to temporarily meet state asset standards. Out of cash, the company lied about its bank balances and bond holdings and overvalued its mortgage portfolio (which had been hit badly by the panic). Most distressing of all, McCall realized that American Popular Life had convinced many policyholders to exchange their older policies for new ones—ostensibly of a greater value, because sharing in company profits—that substantially decreased the reserve held to pay up those policies in the future. Hemorrhaging policyholders and assets, American Popular Life had struggled to keep up appearances and hold down its reserve requirements.[81] It failed.

Lambert still thought biometry might save him. He talked about the policy swap fraud alleged by McCall as if it were an act of resistance against injustice. The old policies carried a large "fictitious reserve," by which he

79. Charles P. Kindleberger, *Manias, Panics, and Crashes: A History of Financial Crisis* (New York: Wiley, 2000), 8, 27–29, 37, 39, 49, 99, 111, 120, 131–132; Nouriel Roubini and Stephen Mihm, *Crisis Economics: A Crash Course in the Future of Finance* (New York: Penguin, 2011), 22.

80. Stalson, *Marketing Life Insurance*, 434, 751.

81. "Dr. T. S. Lambert on Trial," *New York Times*, 6 December 1877; "American Popular Life Failure," *New York Times*, 24 April 1877; "American Popular Life. Testimony as to False Reports of the Company," *New York Times*, 29 April 1882.

meant an actuarially defined reserve, and the company organized the swap to free itself from that fiction.[82] Its apparent trouble had everything to do with the flaws in the probabilistic prediction of life spans, thought Lambert, his risk making with biometry assuring the company's long-term health. Hoping to hold off the insurance superintendent's denunciation, Lambert appeared in Superintendent John F. Smyth's office with calipers in hand. He measured Smyth's head, presumably offering phrenological or biometric advice. He apparently hoped to convince Smyth that his methods obviated actuarial standards—that biometrical classing should free him from the rules of statistical classing. He failed. Smyth reportedly told Lambert that he "better get a pair of calipers in that cash-book."[83] On 23 April 1877, Smyth called for the company to be dissolved.[84]

Lambert had once declared "life insurance" to be a misnomer. He preferred "family assurance," since it had to be paid for out of family funds for the benefit and protection of the family.[85] Yet Lambert's company let many families down, including Mrs. M. Leverty of Bridgeport, Connecticut, and her three children. Leverty's husband died on 2 October 1874. On 18 April 1877, when she filed a complaint with Smyth's office, she had still never seen a cent of the claim on his life, leaving her "trusting and hoping in a kind Providence."[86] Wasn't insurance supposed to be the agent of Providence?

Industry observers almost universally interpreted failures as a judgment rendered by wrathful economic gods against companies designed less to pool and distribute policyholders' money than to support the extravagant lifestyles of ill-prepared company officers.[87] Lambert's company

---

82. "The American Popular Life," *New York Times*, 23 October 1877.

83. "Dr. T. S. Lambert on Trial," *New York Times*, 6 December 1877.

84. "American Popular Failure," *New York Times*, 24 April 1877.

85. American Popular Life, *Sources of Longevity*, section 1, part 1:11–13. In fact, life insurance did much more than protect families. Creditors often took out life insurance policies on their debtors as a precondition for a loan or mortgage, and business partners insured one another against the loss of an invaluable set of hands.

86. Quoted in "The American Popular Frauds," *New York Times*, 26 April 1877. The nearly two thousand legitimate policyholders remaining when the company failed were left without insurance and would have to wait for years until the company could be unwound and each receive some small portion of the $100,000 reserve held in trust by New York State. On the distribution of that fund, see "American Popular Life," *New York Times*, 29 April 1882.

87. See, for instance, "Life Insurance," *New York Times*, 26 August 1875, 3; John A. McCall, *A Review of Life Insurance from the Date of the First National Convention of Insurance Officials, 1871–1897* (Milwaukee, 1898); "An Insurance Imbroglio," *Chicago Tribune*, 14 January 1871, 2. "Financial Affairs," *New York Times*, 6 February 1873, 3; "Life Insurance Management," *New York Times*, 29 October 1876.

fit the paradigm—better, it was the paradigm. Yet other, more orthodox companies failed too. Emory McClintock, who would go on to become the most respected actuary in America, started out his career with a postwar upstart called Asbury Life, another victim of the panic. He could not believe "that, without exception, the officials of at least thirty companies were incompetent or reckless or worse."[88] Instead, he blamed high competition costs paired with too-strenuous regulations (including changing standards for valuing assets and especially the introduction of "non-forfeiture" and tougher reserve standards) that small companies could not abide in a downturn.[89] Of course, given the intensity of the panic, failure may have been inevitable for many companies regardless of the competence of their management or the severity of regulatory laws.

Still, those companies that survived the panic offered—implicitly, in their actions—yet another interpretation: failure had been encouraged by the narrowness of life insurers' markets. Some companies, as a result, looked to diversify regionally. In 1877, for instance, after a year of mild decline, MONY's president, Frederick Winston, went to California to gather data that could justify speeding up its already intended westward expansion and lay the groundwork for the networks of agents and doctors that the company would eventually need throughout the West. If that trip went like previous, better-documented trips, he likely spent his time talking to actual and potential agents, while quizzing doctors about local conditions.[90] On this trip, he met a doctor in Colorado keeping statistics on tuberculosis and arranged to have the company pay for their tabulation.[91]

When MONY's condition had only worsened a year later, Winston headed south as well. Returning home in May 1878 armed with "researches and observations as to the climate and other facts relating to several of the States at the South Western and South Eastern portions of our Country," Winston prevailed upon his board to commit to significant southern expansion.[92]

---

88. New York Legislature, *Testimony Taken before the Joint Committee of the Senate and Assembly of the State of New York to Investigate and Examine into the Business and Affairs of Life Insurance Companies Doing Business in the State of New York* (New York: J. B. Lyon, 1906), 2:1703–1704.

89. New York Legislature, *Testimony*, 2:1698–1704. "Non-forfeiture" laws required companies to either pay a surrender value or offer a paid-up insurance policy in the event that a whole life insurance policy lapsed. On Elizur Wright's campaign for such a law in Massachusetts, see Goodheart, *Abolitionist, Actuary, Atheist*, part 4; Levy, *Freaks of Fortune*, 97–103.

90. For instance, 17–18 May 1869 meeting, Minutes of the Agency Committee (Book 1): 205, MONY Papers.

91. 25 May 1877 meeting, Minutes of the Mortuary Committee (Book B), MONY Papers.

92. 17 May 1878 meeting, Minutes of the Insurance Committee (Book 4): 51, MONY Papers.

MONY reopened agencies (closed because of the war) in Arkansas, Texas, and South Carolina in 1878. In the mid-1880s, the company removed all regional climate penalties and by the end of that decade agents sold policies in every southern state.[93] Whatever Winston found out on his fact-finding trips, it could not have been all that encouraging, especially in the South. Other reports from this time period show that rates of mortality among the insured throughout the South still ranged from 11–60 percent higher than those predicted by Homans's American experience table, while the rest of the United States had rates safely below.[94] Expanding west and south had less to do with those regions becoming more welcoming and much more to do with Winston's judgment that the North could no longer sustain the company.[95]

Metropolitan came to a parallel conclusion—that competing for northern, well-off, male risks could not sustain it—much more dramatically. It very nearly joined its brethren upstarts in failure in 1877 when it lost one-third of its business. By the start of 1880, Metropolitan had less than half its peak level of insurance in force.[96] Metropolitan's president, Joseph Knapp, set sail for London in 1879 to reinvent the company on the model of the British Prudential Assurance, a company famed for its ability to insure the working classes.[97] Knapp eventually shipped back two thousand experienced industrial insurance agents, skilled in selling (and collecting premiums weekly on) what amounted to small burial policies favoring working-class men, women, and children.[98] They helped him create an

---

93. Clough, *Century of American Life Insurance*, 158–159, 170.

94. Meech, *System and Tables of Life Insurance*, 167–179, 199–201.

95. Life insurers' national expansion has been largely overshadowed in contemporary and historical accounts by the spectacular overseas escapades of a few. The Big 3—Equitable, New-York Life, and MONY—all started selling policies in Latin America, Europe, and Asia in the 1880s and 1890s. They built fancy offices, invested heavily in foreign securities, and wrestled with protectionist foreign officials. It is and was a glamorous story and an exciting one. These were among America's first corporate multinationals. The excitement ended as quickly as it began with the onset of World War I. In the long run, national expansion mattered much more. For a compelling account of life insurers as multi-nationals, see Keller, *Life Insurance Enterprise*, chapters 6 and 7.

96. See the chart of Metropolitan's insurance in force in Chandler, Jr., *Comparative Atlas*.

97. James, *Metropolitan Life*, 61, 71–72. As early as 1875 the company experimented with selling to workers, but it struggled with the details. Other companies, including New-York Life, made similar halfhearted, homegrown efforts. "Industrial Insurance," *New York Times*, 2 August 1879.

98. Louis I. Dublin, *A Family of Thirty Million: The Story of the Metropolitan Life Insurance Company* (New York: Metropolitan Life, 1943), 123–125.

organization capable of handling the huge volume upon which industrial insurers depended.[99] It took the threat of destruction, brought on by the panic, for Knapp to risk so drastic a transformation.

Is not that company best that classes best? Perhaps. But the Panic of 1873 revealed vulnerabilities inherent to companies based on statistical laws just like those inherent to the company based on Lambert's biometric classing. Surviving companies acted as if they had failed to class well in the broadest sense. Now they would turn to prospects who were not northern white men in the professional and commercial classes.

The panic—and Lambert's apparent frauds—roundly discredited biometric risk making. It also set risk makers in motion, spurring them to create national and international networks for selling insurance and measuring lives, while inspiring the creation of industrial insurance corporations that made risks from male and female workers and their children. It did not, however, settle the deeper questions raised by Lambert's challenge. Risk makers expanded without having agreed upon a proper balance between classing and smoothing, without having convinced the broad public of the wisdom in their approach to making statistical individuals, as new challenges would very soon make clear.

* * *

When T. S. Lambert came to trial, classing had nothing to do with the indictment. He stood accused of perjury, of having sworn in a financial statement that assets existed which, in fact, did not. During the proceedings, he proved himself an able operator. He sobbed for the fate of his family at all the right moments, and when that did not seem to be working he arranged for associates to find jury members and attempt to sway them.[100] One of the attorneys assigned to dissolve American Popular Life declared Lambert "'the toughest customer' he ever had to manage."[101] His incredible final plea only confirmed his audacity.

Judge Brady, having listened with remarkable patience for nearly two hours' worth of peroration, saw no sign that conviction had taught Lambert anything and decided that five years of hard labor might do the trick better.

99. The scale of industrial insurance struck ordinary insurers dumb. The British Prudential processed as many policies in one year as did all American ordinary companies combined. "Novelty in Insurance," *New York Times*, 25 August 1871.

100. Lambert's attorney admitted to attempted jury tampering in his closing remarks. "The Trial of Dr. Lambert," *New York Times*, 20 December 1877.

101. "The American Popular Life," *New York Times*, 25 October 1877.

Lambert returned to New York City's house of detention, the Tombs, awaiting transfer to the state prison, while his attorney drafted his appeal. After a
year in jail he won his freedom on a technicality.[102]

Lambert's obituary neglected to mention biometry or his career in life
insurance.[103]

102. Lambert contended that the notary who observed his signature had been a resident
of New Jersey at the time and therefore ineligible to witness acts in New York. The court of appeals agreed and set him free on bail. In 1880, the prosecutor, with Judge Brady's permission,
decided against a retrial. "The Case of Dr. Lambert," New York Times, 27 December 1877; "The
American Popular Life," New York Times, 16 December 1879; "Dr. Lambert At Last Set Free," New
York Times, 1 July 1880.

103. "Death List of a Day," New York Times, 23 March 1897.

TWO

Fatalizing

In an 1860 essay entitled "Fate," Ralph Waldo Emerson pointed to phrenology and statistics as twin proofs of fate's reality and of its calculability—measuring skulls accessed the body's history to see its future, while tallying deaths uncovered the race's past to foretell its fate.[1] Lambert's challenge to life insurance rested on just such an understanding of how head- or body-reading resembled statistical calculation in some ways while differing from it in others. Phrenology and statistics shared one crucial similarity: both imagined the roots of the future to lay in an accessible past, in a past one could glean from bodies or tabulate from ledgers. But they each looked toward different futures. Statistics spoke to a population's future, phrenology (or biometry) to an individual's. In that difference lay the source of the tension between classing and smoothing that characterized life insurers' making of risks.

But Lambert's challenge—built on the critique that individuals ought not to be interpreted through the lens of statistics—was hardly the most important thorn in the side of late nineteenth-century risk makers as they looked out of windows or read newspapers from their corporate life insurance offices. No doubt even lowly clerks realized that outside company walls a much more troubling challenge was brewing in the guise of increasing skepticism about the assumption that statistics shared with phrenology: that the past could be used to predict the future. Doubts that the past could be relied upon to forecast the future were, of course, hardly new,[2] but those

---

1. Emerson, *Conduct of Life* (Boston: Ticknor & Fields, 1860), chapter 1.
2. Moments of rupture abound. One important such rupture between past and present, for instance, came with the idea of Christian conversion and second birth.

doubts took on a new shape as a growing portion of the American population came into contact with life insurers' fatalizing—the insistence that past statistics guide their prognostications of future life spans and that company experience be used to set life insurance premiums. Life insurers played a central role in bringing the fundamental premises of statistical probability to the American public, only to discover that many people did not like seeing a powerful corporation yielding stern, fatalistic laws. The greatest challenge that risk makers faced lay not in the tension between classing and smoothing, but in the opposition of fate to hope, of prediction to change.

Life insurers' desperate post-panic expansions set the stage for this challenge. In the 1880s, life insurers responded to the previous decade's panic and depression by directing their risk-making energies toward a wide variety of communities they had neglected in the past. They stood poised to bring much of the American population within their statistical community and thus to make many more Americans into statistical individuals. With the arrival of the first American "industrial" life insurers—Prudential, Metropolitan, and John Hancock, being the pioneers—wage workers and their children, regardless of sex or race, could finally afford to purchase life insurance, now offered to them by roaming agents who dropped by at their tenement doors to collect weekly nickel premiums. Life insurance had reached the urban masses. Life insurers began to make the masses into risks.

But even as American workers and their families bought life insurance, they resisted the risk makers. Many workers avoided being made into risks at all by eschewing corporate life insurance in favor of insurance sold by the so-called cooperative movement. Among those who did allow themselves to be made into risks, the validity of using their statistical pasts to predict individual and group futures—the validity of fatalizing—became a point of sustained contention, a point debated in newspapers and considered in legislatures. This chapter focuses on the challenge to risk makers' fatalizing as it played out most visibly and most consequentially: in African Americans' backlash against life insurers' risk-justified schemes for racial discrimination.

African American activists fought life insurers over the relevance of history to their race's future, and thus also to the individual futures of America's black population. They challenged the risk makers' fatalizing approach and in the process made their marks on broader debates over racial futures in the post-Reconstruction United States. Where life insurers initially assumed continuities between the slave-era past and a free future, African Americans and their allies interpreted the Civil War as a moment of rupture, severing the past from the present and future. In legislatures, this argument won the day and its success inspired other groups facing discrimination, especially women angered by paying more than men for life insurance.

But in an ironic twist of fate—the sad kind of irony—African Americans' argument eventually convinced insurers too. At least, it convinced them that they could not win by justifying fatalism. Instead, they had to attack hope. Many life insurers began talking of a fundamental rupture separating the days of slavery from those after; they imagined a future for African Americans in which the past held no sway, just as their opponents had done. However, where African Americans had argued for a future of equality, most life insurers adopted an old argument given new life by Prudential's recently hired statistician, Frederick L. Hoffman. They began prophesying (with statistics) the black race's inevitable extinction. Life insurers set aside fatalizing—as they had done before and would do again—and embraced a narrative of rupture, of change. But in their hands, rupture and change brought doom, not hope.

* * *

The risk makers failed to foresee African Americans' revolt against fatalizing. Indeed, they do not even appear to have foreseen insuring African Americans. When Prudential Insurance Company of America began doing business as an industrial life insurer in 1875 and when Metropolitan joined it in 1880, neither thought much about insuring African Americans.[3] That changed when thousands of African Americans applied for admittance into industrial insurers' risk communities.[4] In its first year of operation, the

3. African Americans had been among the more heavily insured antebellum populations because of the odious practice of masters insuring slaves, particularly slaves working in urban, industrial settings. Outside of slavery, blacks could seldom afford ordinary insurance policies requiring large, annual premium payments and may have been rejected outright because of racism. Murphy, *Investing in Life*, 184–206.

4. By 1891, at least thirty-five thousand African Americans held industrial policies in the state of New York alone. See William H. Johnson, "Insurance Legislation: The Lawmaking Power Invoked to Prevent Discrimination on Account of Color," *New York Age*, 31 January 1891, letter to the editor. The initial number of black applicants and policyholders is uncertain. Frederick Hoffman reports about fifty-four hundred deaths among African American policyholders between 1891 and 1898. It's notoriously hard to work backward from death figures, but it's worth noting that more African Americans died in seven years with Prudential than were insured by Penn Mutual over almost 60 years. Also, since lapse rates among the African American industrial insured were quite high, we can guess that for each death there was at least one policyholder who lapsed, and probably more. See Frederick L. Hoffman, *History of the Prudential Insurance Company of America* (Newark, NJ: Prudential Press, 1900), 166–167, 301–302, 310; Harry Toulmin, "The American Negro as an Insurance Risk," *Abstract of the Proceedings of the Association of Life Insurance Medical Directors of America* 24 (1913): 152–179 at 153–154.

influx of black applicants frightened Metropolitan; the company decided in December 1880 to decline all future African American applicants until it could figure out the place of race in risk. Before the suspension, Metropolitan awarded identical policies to blacks and to all other industrial policyholders by default. In 1881, Prudential's actuaries completed a review of the mortality experience of their African American industrial policyholders from the company's founding in 1875 through 1881. In the grand scheme of things, six years' experience on a small sample population did not amount to much, but Prudential's actuaries went looking for racial difference and managed to find it. Their experience exhibited excessive levels of mortality among African Americans and promised to justify the company's decision to impose the same sort of penalty on the basis of race that the industry had begun to phase out with regard to region.[5] While insurers' decisions to remove regional penalties (against southerners mainly, despite continued mortality differences) proved politically astute, their flight from African Americans (justified by mortality differences) would prove the opposite.

Industrial insurers operated a high-volume business; so to simplify sales they charged the same nickel to everyone. The home office then calculated benefits according to actuarially defensible discriminations, by age initially and then by race. In November 1881, Metropolitan decided to mimic Prudential, allowing policies to be sold to African Americans once again, but with the understanding that black policyholders' survivors only received two-thirds of the standard benefit. Their discrimination, they would insist for years to come, wasn't personal—it was only business. They did

---

5. Insurers' rationale for discrimination has been well studied, even as antidiscrimination arguments have generally been explored on only a superficial level. Benjamin Alan Wiggins has recently lamented how little we know about the origins of antidiscrimination legislation, noting a gap that this chapter fills. Wiggins, "Managing Risk, Managing Race: Racialized Actuarial Science in the United States, 1881–1948" (PhD diss., University of Minnesota, 2013), 60. The first work to investigate the intellectual origins of the decision to discriminate was John S. Haller, Jr., "Race, Mortality, and Life Insurance: Negro Statistics in the Late Nineteenth Century," *Journal of the History of Medicine and Allied Sciences* 25, no. 3 (1970): 247–261. Megan J. Wolff expanded this story by explaining how insurers' discriminatory science became central to the "Negro question" generally, but still undervalued the power of the antidiscrimination case. Wolff, "The Myth of the Actuary: Life Insurance and Frederick L. Hoffman's *Race Traits and Tendencies of the American Negro*," *Public Health Chronicles* 121 (January–February 2006): 84–91. The legal historian Mary L. Heen has written the fullest survey to date of life insurers' race-based practices in the United States in "Ending Jim Crow Life Insurance Rates," *Northwestern Journal of Law and Social Policy* 4, no. 2 (2009): 360–399. Michael Ralph considers insurers' discrimination in the context of disability studies and social movement politics in " 'Life . . . in the midst of death': Notes on the Relationship between Slave Insurance, Life Insurance, and Disability," *Disability Studies Quarterly* 32, no. 3 (2012) http://dsq-sds.org/article/view/3267/3100.

not discriminate against a race. They discriminated, as insurers' fatalizing practices dictated, against high rates of mortality. Race happened to be a cost-effective marker of high mortality.

Because discrimination occurred behind the scenes in insurers' offices, many African Americans may not have realized that they were charged the same as their white neighbors but granted smaller claims.[6] Those that did notice faced the company's "ultimatum . . . 'Take what we offer or nothing.'"[7] Some refused to stand the insult. The stage was set for organized resistance against life insurance corporations and the statistical foundations of risk making.

* * *

To understand the possibilities that existed for resistance to risk makers and to statistical fatalism in late nineteenth-century America, one must first understand the evolving place of probability and statistics. Midcentury life insurers had traded on the growing cultural capital of probability and statistics, and on the astronomical methods from which much probability theory developed. As early as 1845, MONY bragged: "It is now a well settled principle, that the system of Life Insurance does not rest upon uncertain theories or vague speculation, but that it is reduced to an exact science."[8] Life insurers were unique among insurers in relying on probabilistic mathematics; marine and fire insurers priced risk without recourse to fatalizing. Other insurers lacked a crucial advantage that life insurers possessed: the certainty that the thing they insured against would eventually come to pass. All men would die. But, one hoped, not all ships would sink and not all buildings would burn.[9]

Even as life insurers relied on probabilistic methods and past statistics to calculate the odds of future events, they provided one of the most often-cited proofs that such activities were possible, that science could make fates legible.[10] A London reviewer, having read Emerson's essay, exalted mortality

6. "The Legislature," *Springfield Republican*, 1 April 1884, 8.

7. T. McCants Stewart, "Insurance Bill Signed," *New York Age*, 11 April 1891, 2.

8. Mutual Life Insurance Company of New York, *Annual Report* (1845): 17.

9. Lorraine Daston argues that mortality became an early subject of probability because of a tendency, absent in shipwrecks or fires, toward simple and complete quantification. Counting deaths would eventually catch all people. Counting shipwrecks would not catch all ships. Similarly, death came with a long association to a quantitative variable: age. The risks posed to ships did not admit such ready quantification. Given quantifiable data on mortality, early practitioners of probability created regularities, the same regularities on which insurers later capitalized. See Daston, *Classical Probability in the Enlightenment*, 131–133.

10. Porter, *Rise of Statistical Thinking*, 81–88.

probabilities as the best evidence that there was even such a thing as human fate: "Thus far we may readily admit the existence of fate: statistics will tell us how many deaths shall occur, in a given period, among a million of men; and the most extraordinary chances may be foretold by science, in the annals of human life, as well as the re-appearance of a comet or the occurrence of an eclipse; these are fated and foreseen."[11] Life insurance and statistical fatalism depended upon one another.

The example of life insurers' fatalizing helped convince leading thinkers in the last decades of the nineteenth century of the enormous potential for statistical methods to solve even the most daunting problems of the day. Looking back at the mid-1890s in his celebrated early twentieth-century memoir, Henry Adams remembered the promise of "statistics" for him and his peers, a promise that lay precisely in that field's capacity to render fate scientifically certain. "Even the Government volunteered unlimited statistics, endless columns of figures, bottomless averages merely for the asking," Adams enthused.[12] Once one had past data, one could move to scientific prediction, Adams explained, as "one's averages projected themselves as laws into the future."[13] Adams sought after some "sort of satisfactory answer to the constructive doctrines of Adam Smith, or to the destructive criticisms of Karl Marx or to the anarchistic imprecations of Élisée Reclus."[14] Some nineteenth-century evangelists of life insurance imagined that corporate organization and the statistical methods undergirding risk making provided just such an answer to socialism.[15]

Adams admitted his ultimate disappointment with statistics and especially with statisticians: "They should have reached certainty," he exclaimed, "but they talked like other men who knew less." Adams's lamentations might be read in part as disappointment that statistics had failed not only to solve social problems, but also to silence those "other men who knew less." The statistician's "method did not result in faith," Adams continued,

---

11. Review of *The Conduct of Life* by Emerson in *Ladies' Companion* 19 (1861): 105–108 at 106.

12. Adams wrote with too little understanding of the very real limits on government figures. He should also have noted that private insurers filled in some, but far from all, of the gaps that government left, especially when it came to data on Americans' lives.

13. Henry Adams, *The Education of Henry Adams* (New York: Modern Library, 1999 [1918]), 351.

14. Adams, *Education of Henry Adams*, 351.

15. See D. R. Jaques's vision of "society united on the basis of mutual insurance," quoted in Tom Baker, "Containing the Promise of Insurance: Adverse Selection and Risk Classification," *Connecticut Insurance Law Journal* 9, no. 2 (2012): 371–396 at 371–372.

and so left space for other faiths to flourish.[16] Indeed, historians of chance and prediction in the United States have stressed the hardy persistence of alternatives to statistical prediction in American culture, especially outside of the elite chambers in which Adams socialized and theorized. Competing modes of thinking about and predicting the future refused to be beaten back by the arguments of science or bureaucracy. "Weather prophets," to take one important instance, made long-term forecasts, even as government officials railed against their methods as little more than superstitions.[17] More generally, Americans stubbornly continued to read their fates from almanacs, Bibles (opened at random, by spiritual inspiration), astrological charts, dice, dream books, and numerological treatises.[18] Statisticians and risk makers enjoyed no monopoly on the interpretation of fate.

Close investigation reveals that even in life insurance corporations, officers could and did choose to ignore their own fatalizing dogmas. To take one significant example, industrial life insurers did just that when they began selling insurance despite having little to no reliable data on mortality experience to go on—a necessary prerequisite for fatalizing. Metropolitan had to guess at what mortality table to use until, more than a decade after the company began selling industrial policies, it could analyze its corporate mortality experience and derive a sound mortality table.[19] In another instance of choosing not to worry about fatalizing principles, ordinary insurers in the 1890s succumbed to competitive and political pressures (the latter in the form of a climate favoring postwar reunion/reconciliation after many years of Reconstruction) when they dropped penalties against southern risks despite studies of thirty companies' mortality experience revealing that mortality in the Atlantic-coast South exceeded that in the North

16. Adams, *Education of Henry Adams*, 351.

17. Jamie L. Pietruska shows how such weather prophets also served as the foils for those engaged in professionalizing the US weather bureau. See Pietruska, "US Weather Bureau Chief Willis Moore and the Reimagination of Uncertainty in Long-Range Forecasting," *Environment and History* 17, no. 1 (2011): 79–105.

18. Adams's memoir appeared not long after books like Luo Clement, *The Ancient Science of Numbers: The Practical Application of Its Principles in the Attainment of Health, Success, and Happiness* (New York: Roger Brothers, 1908). On alternatives to chance tamed by statistical probability, see Jackson Lears, *Something for Nothing: Luck in America* (New York: Viking, 2003).

19. When Metropolitan began selling industrial insurance in 1880, its actuary recommended charging workers premiums based on the American experience table (see figure 1.1), but with a *very high* loading percentage appended to cover excesses in mortality and expense. Dublin, *Family of Thirty Million*, 381. On the first "investigation into the mortality of Industrial policyholders" at Metropolitan, see Metropolitan Life Insurance Company, *An Epoch in Life Insurance: A Third of a Century of Achievement* (New York: Metropolitan Life, 1924), x.

and West by 15–20 percent, while the Gulf States showed nearly 70 percent higher mortality rates.[20]

Life insurers outside of corporations proved even more willing to forgo statistical fatalism. Cooperative insurers, who offered an alternative form of life insurance that had become very popular in the decades following the Panic of 1873, looked warily at fatalizing from the start. They had risen in popularity on the strength of public distrust of corporate insurers and the failure of a fatalizing "science of life insurance" to save so many companies from insolvency.[21] They saw the taint of corporate capital in the making of risks. "In the thought of these apostles," of cooperative insurance, wrote Charles O. Hardy in 1923, "mortality tables, compound interest calculations, legal reserves, were implements of darkness, designed to separate the trusting citizen from his money and place it at the disposal of the stock manipulators and trust magnates of the East."[22] Cooperatives did not need fatalizing techniques; they did not need forecasting: if a fraternal brother died, his peers chipped in their contributions to cover his family's needs at that time.[23] Nor did they want to discriminate among members, the other reason that insurers needed fatalizing practices. All fraternal brothers paid the same rates, regardless of age or health or anything else.[24] By 1895, such institutions held more than half of all American life insurance.[25] For the insurance-buying public, fatalizing must not have been all that important.

In short, the cultural capital of probability and statistics in the late nineteenth century should not be overestimated. When disgruntled poli-

---

20. Meech, *System and Tables of Life Insurance*, 167–179, 199–201. By 1898, only two American companies out of forty-two maintained significant penalties against southerners. A majority of companies continued to levy penalties for the first two years that a policy was in force. George L. Amrhein, "The Liberalization of the Life Insurance Contract" (PhD diss., University of Pennsylvania, 1933), 207.

21. See Levy, *Freaks of Fortune*, 199.

22. Charles O. Hardy, *Risk and Risk-Bearing* (Chicago: University of Chicago Press, 1923), 275.

23. On assessment and fraternal dues practices, see Andrew J. Hirschl, *The Law of Fraternities and Societies: With Special Reference to Their Insurance Feature* (St. Louis, MO: William H. Stevenson, 1883), 30; Stalson, *Marketing Life Insurance*, 448–450; David T. Beito, *From Mutual Aid to the Welfare State: Fraternal Societies and Social Services, 1890–1967* (Chapel Hill: University of North Carolina Press, 2003), 131–132.

24. This democratic egalitarian strain within fraternals came alongside highly exclusionary membership policies, which made each fraternal lodge remarkably homogeneous by race, religion, or ethnicity. See Brian J. Glenn, "Fraternal Rhetoric and the Development of the U.S. Welfare State," *Studies in American Political Development* 15, no. 2 (2001): 220–233 at 226. Women were also generally excluded. Levy, *Freaks of Fortune*, 202.

25. Stalson, *Marketing Life Insurance*, 806, 818; Witt, *Accidental Republic*, 71–102.

cyholders and applicants—the targets of life insurers' discriminatory fatalizing—chose to resist the risk makers in their corporate offices, they bucked the supposed spirit of the industrializing, bureaucracy-building age, and yet in doing so they had plenty of company. They battled powerful corporations and commanding rationalities, but the critics of fatalizing did not face an unbeatable foe. They just needed a potent rationality of their own and a competing vision of the future. African Americans found just such a powerful conception—not in dream books or almanacs—but rooted in the egalitarian, civil rights tradition of Reconstruction.[26] Armed with that alternate futurity, they faced off against life insurers in the 1880s and 1890s.

* * *

Industrial insurance agents came into applicants' homes on the most delicate of missions. They came to talk about death and protection from death's indignities for those surviving. Paying for a funeral or for the costs of final medical care could easily devastate a family. Even worse—in the eyes of the working poor—not being able to pay for a funeral meant a family's social embarrassment. The "honest, industrious, self-reliant Afro Americans," who insured themselves and their families in the late nineteenth century sought merely, in the words of a New York State activist, to "anticipate and protect them from the bare possibility of a pauper burial."[27] Trouble courted the white agent who arrived to sell so important a protection and implied the inferiority, through the rates he offered, of the women of color whom he met at home during the day. When that implication came from an agent delivering a much needed, but too small, claim payment, it hurt that much more.

One story in circulation (in the 1910s) attributed the first antidiscrimination legislation to the complaints of an African American woman of some prominence in her community. The unnamed wife of an unnamed messenger for the secretary of state claimed—the story went—that an agent insulted her. Her husband took offense: "[I] determined then and there to make it hot for Life Insurance Companies and force them into a better recognition of the colored race, so I appealed to our colored representative, a man

---

26. David W. Blight names this tradition "emancipationist" unionism and notes its decline vis-à-vis "reconciliation" and "white supremacy" in the post–Civil War era. See Blight, *Race and Reunion: The Civil War in American Memory* (Cambridge, MA: Belknap Press of Harvard University Press, 2001).

27. William H. Johnson, "Insurance Legislation: The Lawmaking Power Invoked to Prevent Discrimination on Account of Color," *New York Age*, 31 January 1891, 2.

named Chappell [*sic*], to bring this bill before the Legislature."[28] An African American woman with some wealth or position, like the one in this story, may have been particularly sensitive to the agent's attitudes. She was, after all, purchasing industrial insurance—a sign that she did not have sufficient wealth for an ordinary policy, or could not because of her race pass a medical exam.

Moreover, purchasing any sort of life insurance as a woman in the late nineteenth century could be fraught. At that time, being *unable* to purchase ordinary life insurance could have been taken as a marker of middle-class status. Life insurers shied away from married women outside the working classes, viewing such women as potential femmes fatales who could hide impairments from doctors beneath their corsets and whose womanly sensibilities gave them special insight into their own latent illnesses—illnesses that they then hid by being "conspicuously inexact in their assertions concerning themselves."[29] Giving those reasons, ordinary life insurers—who sold to the middle classes—avoided insuring women. An African American woman, therefore, might have felt ambivalent about her purchase even before being slighted because of her race. By gaining security through insurance (a privilege largely unavailable to white women of similar status), she potentially sacrificed her claim to the middle class.[30]

28. Frank Wells in discussion following Toulmin, "American Negro as an Insurance Risk," 170.

29. Stable sex differences had been among the most obvious conclusions drawn by early statistical investigators and had long shown women in general to live longer than men, since at least the eighteenth century. Yet among insured lives, women had long been seen, from insurer experience dating back to the mid-nineteenth century, to be more likely to die than their male counterparts. See John K. Gore, "Should Life Companies Discriminate against Women?" *Transactions of the Actuarial Society of America* 6, no. 24 (1900): 380–388 at 380–383. For life insurers' explanations of these differences, see for instance, "Women and Life Insurance," *Springfield Republican*, 6 August 1883, 5; Walter H. Barnett, "In Woman's Behalf: Women and Life Insurance," *Huntsville Gazette*, 17 September 1892, 4. Quoted in text from "Editorial Article," *New York Tribune*, July 26, 1883, 4. In the end, as industry statisticians eventually pointed out, bad experiences with insured women stemmed mainly from the volatility of small sample sizes. Not enough married, white, middle-class women had access to insurance to escape from the domain of randomness into actuarial certainty. See Louis I. Dublin, *The Insurability of Women: An Address Delivered before the Medical Section of the American Life Convention in St. Paul, Minnesota* (New York: Metropolitan Life Insurance Company, 1913).

30. The interplay of race, class, and gender here makes an interesting contrast with the way those factors played out in the contemporary case of Jim Crow railroads. There, respect for middle-class "ladies" trumped race prejudice. But in the 1890s, broader concerns with protecting white privilege throughout the South led to mandatory segregation on trains. Women's cars became white cars, and smoking cars became Jim Crow cars. In 1896, the Supreme Court upheld this new regime in *Plessy v. Ferguson*, with the explanation that separate could be equal. See Barbara Y. Welke, "When All the Women Were White, and All the Blacks Were Men: Gender,

We do not know if such a lady ever existed. But we do know that African Americans grew tired of being discriminated against, of being imputed inferior by life insurers. Their frustration became political on 12 February 1884, when Massachusetts state representative Julius C. Chappelle requested that a committee consider legislation to limit the powers of life insurers to discriminate.[31] Chappelle, an African American born in antebellum South Carolina and a janitor for the state of Massachusetts by trade, had somehow discovered that many of his black constituents suffered from insurers' discriminatory rates, sometimes without knowing it.[32] In its 31 March report, the legislative committee assigned to assess the situation in Massachusetts in 1884 sided with the companies. But Chappelle refused to accept its claim that it was "inexpedient to legislate" against racial discrimination and so offered his own bill.[33] "He did not think it was a sufficient answer that it was a matter of business," reported the *Springfield Republican*. "It was a matter of right."[34]

Chappelle argued for the dignity of his constituents and made plain the insult of overt discrimination. Some of the "colored families" affected had been residents of Massachusetts for two hundred years, he noted.[35] Chappelle himself had a good job and had married into a family with local status, one whose members were classified by a census enumerator as "mulatto." Chappelle's wife, Eugenie, was the Boston pastor Charles O. Brady's daughter. She laid claim to at least one marker of the middle class, telling an 1880 census enumerator that "keeping house" was her profession.[36] When Chappelle thought about protecting families from embarrassment, he probably had his own family foremost in mind.

Chappelle tried to turn the argument toward the future, toward possibilities. No statistics that existed actually proved that blacks in the state would not *at that moment and in the future* live as long as whites. He cited the

---

Class, Race, and the Road to *Plessy*, 1855–1914," *Law and History Review* 13, no. 2 (1995): 261–316, especially 310. See also Kenneth W. Mack, "Law, Society, Identity, and the Making of the Jim Crow South: Travel and Segregation on Tennessee Railroads, 1875–1905," *Law and Social Inquiry* 24, no. 2 (1999): 377–409.

31. *Journal of the House of Representatives of the Commonwealth of Massachusetts* (1884): 191.

32. "The Legislature." *Springfield Republican*, 1 April 1884, 8.

33. *Journal of the House of Representatives of the Commonwealth of Massachusetts* (1884): 436.

34. "The Legislature," *Springfield Republican*, 1 April 1884, 8.

35. "The Legislature," *Springfield Republican*, 1 April 1884, 8.

36. See the entries for Charles O. Brady, Martha Brady, and Eugenie Brady, 1870 US Federal Census, City of Boston, County of Suffolk, MA, 304, www.ancestry.com (27 July 2011); and the entries for Julius and Eugenie Chappelle, 1880 US Federal Census, City of Boston, County of Suffolk, MA, SD 60, ED 642, 15, www.ancestry.com (27 July 2011).

eminent statistician Carroll D. Wright to the effect that, as one reporter put it, "it was all nonsense that colored people cannot live as long as whites."[37] One colleague agreed (in an unintentional echo of T. S. Lambert), noting that a good doctor should be able to see the "seeds of disease in colored people as readily as in white people" and that such individual judgments should be the only basis for discrimination.[38]

But the opposition peered backward, armed with statistics. Frederick Homer Williams, a lawyer from Brookline, chaired the committee that opposed the bill. Williams had only graduated from Brown seven years earlier and was not even thirty years old. He was a joiner—he belonged to the Masons, the Odd Fellows, and the Grange—whose interest in business placed him on the insurance and mercantile affairs committees.[39] Williams cited statistics from around the nation showing shorter life spans for blacks, including 1870 census figures showing a 17.28 death rate for "colored people" against 14.74 for whites.[40] These numbers, Williams argued, and not any "discrimination on the ground of color" motivated insurers' rates. It was a "matter of business," and any interference, he warned ominously and presciently, "would probably cut off insurance entirely from the colored race."[41]

Chappelle's allies noted that Williams's statistics, while bleak enough, answered the wrong question. The question was not whether blacks in slavery or adjusting to freedom were poor insurance risks, or even whether southern blacks were poor risks. The question was African Americans' *potential* for equality and specifically the present and future state of Massachusetts' African Americans—about whom no statistics had been offered by either side.[42] Through the glass of postbellum hope, antidiscrimination ought to

37. "The Legislature," *Springfield Republican*, 1 April 1884, 8.

38. "The Legislature," *Springfield Republican*, 1 April 1884, 8.

39. On Williams, see his biographical entry in William T. Davis, *Bench and Bar of Massachusetts* (Boston: Boston History Company, 1895), 1:395; and A. M. Bridgman, ed., *A Souvenir of Massachusetts Legislators, 1898* (Stoughton, MA: A. M. Bridgman, 1898), 7:117.

40. It is not clear how Williams arrived at these figures. The 1870 census reported, nationally, 356,771 white deaths against a total white population of 33,589,377. That translates to 10.6 deaths per thousand. For the "colored" population, the figures are 67,461 deaths against a total of 4,880,009 persons for a death rate of 13.8 deaths per thousand. In Massachusetts, there were 20,137 deaths against 1,443,156 white persons and 341 deaths against 13,947 "colored" persons for rates of 14 and 24.4 deaths per thousand. See 1870 Federal Census table I, 3–5 and table VII, 306–307, 345. Raw death rates are generally unsatisfying however. They do not take into account the age structure of a population (older populations generate more deaths), for instance. The census of 1870 did not differentiate among American-born white ethnics.

41. "The Legislature," *Springfield Republican*, 1 April 1884, 8.

42. "The Legislature," *Springfield Republican*, 1 April 1884, 8.

proceed on the assumption of equality with a heavy burden of (statistical) proof laid on the discriminator. It was the presumption of equality that made the bill "a matter of right."[43]

Subsequent debates over the bill revealed a tangled mess of ideas concerning racial character, justice, and American destiny. Williams resumed his assault with census data and some other statistics that he claimed showed high mortality rates for African Americans in Massachusetts. But he inferred from his data more general conclusions concerning the limits of African Americans' adaptability. "The colored race in the North is not at home," he asserted. Chappelle, the "colored champion of the bill," as one reporter called him, refused Williams the right of naming African Americans' home.

In a riposte that drew "frequent laughter of the House," Chappelle wondered what African American mortality statistics would look like if blacks, like southern whites, had the privilege to "cut sticks and whistle" all day long while others cared for them. African Americans only wanted "equal terms with whites," which were now long due to them. Time would tell how African Americans did from a position of equality. Representative Clark of Boston went so far as to call the bill a "simple act of justice," while another representative foresaw the discovery from new statistics that "colored people were longer lived than whites." Advocates in the state senate pointed to statistics showing recent increases in the black population and in birth rates. Their critics responded that fertility had nothing to do with longevity. But the debate clearly had to do with much more than longevity. Either the North had no place for blacks, as Williams held, or as Chappelle posited, the entire country was on its way to becoming "colored."[44] Those looking backward saw continuities and racial inferiority, even racial segregation. Those looking forward saw a rupture: a new nation with new opportunities for African Americans and for racial unity after the pain of Civil War and Reconstruction.

In the end, the state's representatives and senators proved Chappelle's supposition correct that the state would not endorse discrimination. They passed the bill and sent it to Governor George Dexter Robinson. Frank Wells, a public health man employed simultaneously by the state and by the state's major industrial insurer, tried to get Robinson to kill the bill. As editor of the state's *Annual Report of Vital Statistics*, Wells insisted that

43. "The Legislature," *Springfield Republican*, 1 April 1884, 8.
44. "The Legislature," *Springfield Republican*, 12 April 1884; "Massachusetts Legislature," *Boston Journal*, 12 April 1884, 6. "The Legislature," *Worcester Daily Spy*, 25 April 1884, 2; "Massachusetts Legislature," *Boston Journal*, 25 April 1884, 3.

his data showed much higher rates of mortality for Massachusetts's African Americans. The state did record "race" on its birth and death certificates, so Wells's claim is plausible. Yet he never published any race mortality data.[45] As medical director for John Hancock Life—another industrial insurer— Wells insisted that his company could not insure blacks on the same terms as whites. According to Wells, Robinson agreed to the facts but rejected the politics. He, a Republican governor, could not afford to veto a civil rights bill on the eve of a national election.[46] We'd be wise to trust Wells's testimony only so far, but of one thing we can be sure: Robinson signed the bill. The state's commitment to African Americans' equal rights triumphed over the priorities of business. The state sided with hope and rupture.

The new law read: "No life insurance company organized or doing business within this Commonwealth shall make any distinction or discrimination between white persons and colored persons wholly or partially of African descent, as to the premiums or rates charged for policies upon the lives of such persons; nor shall any such company demand or require greater premiums from such colored persons than are at that time required by such company from white persons of the same age, sex, general condition of health and hope of longevity." The penalty for disregarding the law was a $100 fine directed against the corporation or its agent.[47]

As sure as Chappelle had been of the justice of his bill, the leading industrial insurers and their allies insisted on its injustice: they could not abide it. The law had the greatest effect on John Hancock Life Insurance, since it was incorporated in Massachusetts. In a sign of things to come, John Hancock halted its discriminatory pricing strategy, but also instituted new guidelines that refused the chance of equality that Chappelle fought for. The company forbade agents from soliciting blacks and refused commissions to agents who did. Any black applicants who came to the home office had to be examined by company doctors who undoubtedly exercised extra vigilance and those who did receive policies had to pay their weekly nickels at the office.[48] Bragging of the company's ingenuity, Frank Wells summarized the effect of these guidelines in 1913: "This has all resulted in my Company absolutely

---

45. Commonwealth of Massachusetts, *Forty-First Report to the Legislature of Massachusetts Relating to the Registry and Return of Births, Marriages, and Deaths in the Commonwealth* (Boston: Wright & Potter, 1883), vii, cciii.

46. See the discussion in Toulmin, "American Negro as an Insurance Risk," 170–171.

47. Chap. 235, *Acts and Resolves Passed by the General Court of Massachusetts in the Year 1884* (Boston: Wright & Potter, 1884), 194–195.

48. "Higher Premiums Charged Negroes by New York Insurance Men," *Atlanta Constitution*, 2 February 1891, 4.

not writing any colored risks."[49] African Americans were to be systematically excluded from life insurance in Massachusetts and written out of insurers' risk community.

News of the Massachusetts victory spread thanks to African American newspapers, "colored Republican" organizations, and the Afro-American League until antidiscrimination became, as the *New York Freeman* put it in 1887, "a subject of more or less complaint and agitation in most Northern and Western States."[50] Connecticut passed a copy of the Massachusetts law in 1887. Ohio, New York, Michigan, and New Jersey followed suit in 1889, 1891, 1893, and 1894 respectively.[51] With each new debate, insurers pressed for recognition that past and present differences necessitated discrimination. They brought more statistics and multiplied their arguments, apparently to no avail. Antidiscrimination forces nitpicked the opposition's statistics, but mostly conceded the past to insurers. Yet they kept winning because they made a better case for the future, a case that statistics could not touch.

Insurers and their allies continued to argue, however, that the past, as documented by statistics, really did control the future. Southern wits oozed obvious glee in seeing respectable northern businessmen and their newspaper allies assert, repeatedly, racial difference and even inferiority next to arguments for the necessity of discrimination. The *Atlanta Constitution* smirked: "According to the New York World, life insurance companies in this country charge negroes one-third higher premiums than they do other people. The World remarks that this is not controlled by the federal constitution. The negro constitution settles the matter, as is shown by vital statistics."[52] As cruel as it was clever, the quote captured perfectly the opposition between those who believed in a fixed "negro constitution," the inferiority of which could be proved by vital statistics, and those who saw no place for old statistics after the Reconstruction amendments (the "federal constitution") had opened new doors to African Americans. Debates over access to a necessary commodity spilled into challenges to the scope

49. Discussion following Toulmin, "American Negro as an Insurance Risk," 170.

50. "The Discrimination Made against Colored People," *New York Freeman*, 30 April 1887, 2. The history of northern civil rights activities during and immediately following Reconstruction is fascinating, but still receives too little attention, especially in explanations of declining northern opposition to Jim Crow. For a good overview, see David A. Gerber, "A Politics of Limited Options: Northern Black Politics and the Problem of Change and Continuity in Race Relations Historiography," *Journal of Social History* 14, no. 2 (Winter 1980): 235–255.

51. "Discrimination against Colored People by Insurance Companies," *Cleveland Gazette*, 9 April 1887, 1; *Cleveland Gazette*, 13 April 1889; "Signed by Governor Hill," *New York Age*, 11 April 1891, 2; "New Jersey Negroes Pleased," *Freeman* (Indianapolis, IN), 5 May 1894, 4.

52. "Editorial Comment," *Atlanta Constitution*, 20 December 1888, 4.

of the Reconstruction amendments, and the challengers now had powerful corporate allies.

Those corporate allies insisted on their objectivity and their dispassion—the facts, they said, compelled them to side with racial discrimination. The *Cleveland Gazette* celebrated the Connecticut bill, but noted the dark clouds in the background: "The bill further provides that, if the application is received, the company shall not require any higher premium than is paid by a white person of the same age, sex, general health, and 'hope of longevity.' This longevity clause, it is claimed by the insurance companies, makes the law practically of no effect, as the actuaries' tables discriminate against colored persons."[53] Tables discriminated, not company employees or anyone else capable of racist feeling. Fears that the "hope of longevity" clause might one day be abused would eventually prove well founded.

As the debate moved to New York and New Jersey, both sides lamented the poor state of American statistics. John F. Collins, from Prudential's New York branch, noted African Americans' poorly recorded lives, particularly the difficulty of obtaining accurate vital data or family health histories—a problem that he failed to note was shared by many white Americans too. Collins argued that this paucity of family history data offered just as strong an argument for discrimination as did Prudential's bad experience with black risks during its first few years.[54] When John B. Lunger, Prudential's actuary, published southern statistics in order to make his case that antidiscrimination laws forced "discrimination against the whites, for the increased losses would result in a general rise in rates all around," A. B. Cosey, president of the Hudson County Colored Republican General Association, rejected the underlying data.[55] "The idea of any sensible man going to the Southern States for information in regard to natural deaths of colored people," Cosey

53. "Discrimination against Colored People by Insurance Companies," *Cleveland Gazette*, 9 April 1887, 1.

54. "Higher Premiums Charged Negroes by New York Insurance Men," *Atlanta Constitution*, 2 February 1891, 4. Insurers were not alone in finding it difficult to get reliable data on African Americans. Census officials, charged in 1890 with delving into African American family histories—by direct questioning to a limited extent and mostly by reading skin tones—struggled to distinguish between black, mulatto, quadroon, and octoroon. In the end, enumerators and analysts tended to settle for either black or mulatto, precision be damned. See Martha Hodes, "Fractions and Fictions in the United States Census of 1890," in *Haunted by Empire*, ed. Ann Laura Stoler (Durham, NC: Duke University Press, 2006), 240–270 at 244–247, 262–263.

55. "Colored Persons Poor Risks; Why Gov. Werts Vetoed the No-Discrimination Insurance Bill," *New York Times*, 24 April 1893.

exclaimed. "Why did he not go to the torrid zone in search of icebergs?"[56] Insurers found a defense for discrimination in American failures to register lives and deaths, while civil rights proponents looked at similar statistical failures and drew the opposite conclusion.

For T. McCants Stewart, the statistics hardly mattered. The politics of equality offered African Americans all they needed, if they summoned the courage to grasp it. He cheered William H. Johnson and his Albany Afro-American League for championing the anti-discrimination bill. He stood with T. Thomas Fortune's *New York Age* as it fought race prejudice throughout the North. But Stewart also believed in cultivating allies across race and party lines. Those who today recognize Stewart's name already know him as an optimist. C. Vann Woodward told Stewart's story many years ago in *The Strange Career of Jim Crow*. In 1885, as Woodward recounted, Stewart went south after the Democrat Grover Cleveland entered the presidency, expecting to find the South devolving into a racial nightmare. Instead, Stewart encountered less discrimination on his southbound trains than he often experienced in the North—which may say more about the postbellum North than about the South.[57] What Woodward never said was that Stewart went on to become a leading black Democrat. "The true political policy for the Afro Americans is division," he wrote in 1891, arguing that blacks should not limit themselves to a single party.[58] The success of the New York insurance antidiscrimination bill—passed unanimously with votes from Democrats and Republicans—convinced Stewart of the rightness of his beliefs.

Stewart proclaimed the bill a "death blow to discrimination on account of color by insurance companies." He foresaw a bright future for African Americans, and for Americans generally. "In time," he wrote, "the South will come around all right, and we shall be able to sink our color in our American citizenship."[59] The unanimity he witnessed lent support to his view, although it also showed the extent to which insurers had resigned themselves to the legislation and determined to work around it.[60] Stewart's optimism, like Chappelle's, drew on the strongest traditions of equality

56. "A Colored Man's Protest: Denunciation of Gov. Wert for His Veto of an Insurance Bill," *New York Times*, 27 April 1893.

57. C. Vann Woodward, *Strange Career of Jim Crow* (New York: Oxford University Press, 2002), 38–41.

58. T. McCants Stewart, "Insurance Bill Signed," *New York Age*, 11 April 1891, 2.

59. T. McCants Stewart, "Insurance Bill Signed," *New York Age*, 11 April 1891, 2.

60. He may also have been witnessing unity in anticorporate feeling, perhaps a consequence of the economic distress roiling the state and nation since 1893.

coming out of the Civil War. Those traditions had power, even after the end of Reconstruction and even with a Democrat in the presidency. Yet it took less than five years for Stewart to abandon the Democratic Party and only two more years for him to abandon the United States. In 1907, Stewart considered the ongoing fight for civil rights in the United States from his new position as Liberia's attorney general and deemed it "a hopeless struggle."[61] Stewart's disillusionment had many sources. One might have been that he placed faith in politics and a political vision for future equality—in rupture—just as corporate insurers discovered that they could use rupture to serve opposite ends.

* * *

Stewart's high hopes and eventual disappointments contrasted sharply with the fate of his white contemporary, Frederick L. Hoffman. In 1884, Hoffman arrived in the United States from Germany at the age of nineteen, innocent of the fight over vital statistics and antidiscrimination then taking place in Massachusetts. He came penniless and began to drift in search of work, anticipating the restless peregrinations west and south that Walt Wyckoff would record in his celebrated 1897 slumming sociology, *The Workers: An Experiment in Reality*. From New York to Cleveland to St. Louis to New Orleans to Georgia to Boston to Virginia, Hoffman jumped from job to job and scheme to scheme, saw much of the nation, and fell in love with both the South and a Confederate belle. He also developed an interest in the progress (or lack thereof) of blacks since freedom. Hoffman's statistical investigations began informally when he happened upon a pamphlet by Dr. Eugene R. Corson, a southern physician writing in a tradition that assumed racial inferiority and predicted the inevitable extinction of free blacks. He began to gather statistics more seriously after he married and settled down to sell industrial insurance for Life Insurance Company of Virginia.[62]

---

61. Charles E. Wynes, "T. McCants Stewart: Peripatetic Black South Carolinian," *South Carolina Historical Magazine* 80, no. 4 (1979): 311–317 at 316. For a fuller biography of Stewart, see Robert Joseph Swan, "Thomas McCants Stewart and the Failure of the Mission of the Talented Tenth in Black America, 1880–1923," (PhD diss., New York University, 1990); and Albert S. Broussard, *African-American Odyssey: The Stewarts, 1853–1963* (Lawrence: University Press of Kansas, 1998), 1–101.

62. Pegram, Secr'y, Life Insurance Company of Virginia to FLH, 6 March 1891, Folder "1891–94 Letters to FLH," Box 27, Hoffman Papers. Beatrix Hoffman, "Scientific Racism, Insurance, and Opposition to the Welfare State: Frederick L. Hoffman's Transatlantic Journey," *Journal of the Gilded Age and Progressive Era* 2, no. 2 (2003): 150–190 at 156–159.

In Virginia, Hoffman cultivated a network of statistical informants and collecting habits that would serve him well for the rest of his career.[63] Considering himself well versed in the existing "physiological aspects" of the *"real status* of the Negro," Hoffman aimed to make his mark on the world by linking physicians' traditions to economic data. Carroll Wright, whom Hoffman apparently contacted in his DC statistical offices at the National Bureau of Labor, drew Hoffman's attention to "the economic aspect of the race question."[64] Wright's bureau had little power to gather sustained series of statistics, which were still mostly the domain of local and state offices, and so Wright focused on specific, in-depth investigations.[65] Hoffman continued to sell insurance, but in his free time he worked as if he were just one more of Wright's statistical investigators.

In 1892, Hoffman published the first fruits of his labors in Boston's influential magazine *The Arena*, under the title "Vital Statistics of the Negro." Hoffman pointed out the difficulty of his appointed task: "Vital statistics of the colored race are, perhaps, the most difficult body of facts to collect in the United States." It was a self-serving remark, but the antidiscrimination debates suggested its truth. Hoffman did muster a small collection of data from census records, state registration data (limited as it was), and data collected by the army during the Civil War. Then he leaped to a big conclusion: that the very nature of the "Negro," and not merely circumstances or environment had doomed the race, outside slavery, to "final extinction."[66] Like life insurers, Hoffman insisted on stable racial tendencies that could be understood from statistics. But unlike them, Hoffman also saw a rupture. The Civil War, in destroying slavery, had not created a new hope of equality, he argued. It had set African Americans on the path toward extinction.

The article impressed Prudential officials whose interest in the subject correlated directly with their failing efforts to win another veto of the New Jersey antidiscrimination bill. The company hired Hoffman to do statistical work in the actuarial office and gave him time to expand his article into the monograph *Race Traits and Tendencies of the American Negro*. Published by

63. For a sample of the wide variety of institutions that Hoffman corresponded with—to varying degrees of success—in order to gather data, see Folder "1891–94 Letters to FLH," Box 27, Hoffman Papers.

64. Frederick L. Hoffman to B. O. Flower, 7 September 1891, Folder "Letters by F.L.H," Box 27, Hoffman Papers. Emphasis in the original.

65. Stapleford, *Cost of Living in America*, 22–58.

66. Frederick L. Hoffman, "Vital Statistics of the Negro," *Arena* 24 (April 1892): 529–542 at 532, 541.

the American Economic Association in 1896, the book became an immediate sensation.[67]

Hoffman's book promised to overwhelm doubters with a flood of statistics ripped from reports prepared by state and local institutions. He amassed data from city auditors and state comptrollers, from boards of health and committees on prisons and charities, just to name a few typical sources.[68] Yet the volume of data disguised its limitations. Most of Hoffman's figures came from the South—and thus had little direct relevance to the insurance debates—and they did not prove all that different from the figures others had been trotting out for years. To this data, he added well-worn theories of racial deficiencies in morality and physiology (advanced in the prior three decades by a host of southern physicians) that he had encountered years before, as historian John Haller has demonstrated. Hoffman interpreted his statistical findings through theories of African Americans' inability to cope with modern life outside of slavery—theories that African Americans succumbed more easily than whites to diseases like tuberculosis, tended toward sexual licentiousness, and possessed diminished physical capacities (whether indicated by lighter brains or smaller lung capacities).[69] Altogether, Hoffman claimed to show that blacks could not compete with whites outside slavery, that white philanthropy only made the situation worse, and that the black race was a vanishing one.

What Hoffman really did, however, was add to the insurers' side of the antidiscrimination debate an alternate (non-fatalizing) vision of the future

67. Frederick L. Hoffman, "Race Traits and Tendencies of the American Negro" *Publications of the American Economic Association* 11, no. 1/3 (1896): 1–329. Reviews were mixed. William Graham Sumner, the doyen of American sociologists, saw promise in Hoffman as a scholar and praised *Race Traits* for its "good method of work." William Graham Sumner to Frederick L. Hoffman, 26 October 1900, in Folder "FLH: Misc letters 1900–1929," Box 26, Hoffman Papers. Favorable reviews included: Miles Menander Dawson, review of "Race Traits and Tendencies of the American Negro," by Frederick L. Hoffman, *Publications of the American Statistical Association* 5, no. 35/36 (1896): 142–148; W. J. McGee, review of "Race Traits and Tendencies of the American Negro," by Frederick L. Hoffman, *Science* 5, no. 106 (1897): 65–68. The book also had prominent critics. See Gary N. Calkins, review of "Race Traits and Tendencies of the American Negro," by Frederick L. Hoffman, *Political Science Quarterly* 11, no. 4 (1896): 754–757; Kelly Miller, *A Review of Hoffman's Race Traits and Tendencies of the American Negro* (Washington, DC: American Negro Academy, 1897), 3; W. E. B. Du Bois, review of "Race Traits and Tendencies of the American Negro," by Frederick L. Hoffman, *Annals of the American Academy of Political and Social Science* 9 (January 1897): 127–133; and "How to Figure the Extinction of a Race," *Nation* 64, no. 1657 (1 April 1897): 246–248.

68. Frederick L. Hoffman to B. O. Flower, 7 September 1891, Folder "Letters by F.L.H.," Box 27, Hoffman Papers. Hoffman, "Race Traits and Tendencies of the American Negro."

69. For a detailed exposition of southern physicians' theories and their impact on Hoffman, see Haller, "Race, Mortality, and Life Insurance."

to compete with Chappelle's and Stewart's racial optimisms. Insurers had always had the better, albeit still very limited, statistical case. Hoffman improved it only marginally. But Hoffman blended statistics with southern physicians' older theories of racial inferiority and by that alchemy generated a response to black activists' dreams of a color-blind or—more radical—colored future.

By the time *Race Traits* entered the public debate, there was no point in trying to justify the sort of discriminatory penalties that had generated so much controversy—and Hoffman's text did not try.[70] It did not have to make a convincing case that present conditions justified higher rates for African Americans. That battle had been lost. Antidiscrimination laws in Massachusetts, New York, and New Jersey prevented the largest industrial insurers from granting smaller benefits to blacks. The antidiscrimination side had won in northern state legislatures. Fatalizing practices could not, legally, take notice of race.

And life insurers had largely abandoned African Americans throughout the North. The John Hancock had stopped soliciting blacks in 1884—now Prudential and most of the industry followed.[71] The major exception came from Metropolitan, which by 1894 determined to actively sell policies to blacks at standard rates, but only for those applicants who passed the most strenuous medical scrutiny.[72] At the same time, new black-run fraternal societies and insurers expanded to pick up some of the slack and in the process became key aggregators of capital in African American communities.[73]

70. Previous interpretations attribute more causal power to Hoffman's book. See Wolff, "Myth of the Actuary," and Haller, "Race, Mortality, and Life Insurance." In an otherwise brilliant article, Brian Glenn goes so far as to claim Hoffman's book to be the pivotal work that taught underwriters to hide their normative claims beneath objective data. In fact, actuaries' objective data had always been bound up in normative claims and actuaries had for a long time endeavored to mute those normative views. Brian J. Glenn, "The Shifting Rhetoric of Insurance Denial," *Law and Society Review* 34, no. 3 (2000): 779–808 at 791–792. Wiggins also gives Hoffman a central place in his dissertation, "Managing Risk, Managing Race."

71. As Hoffman later explained: "Fortunately, the companies can not be compelled to solicit this class of risks, and very little business of this class is now written by Industrial companies, and practically none by The Prudential." Hoffman, *History of the Prudential*, 153. For data on mainstream companies' policies toward African American applicants, see Winfred Octavus Bryson, Jr., "Negro Life Insurance Companies: A Comparative Analysis of the Operating and Financial Experience of Negro Legal Reserve Life Insurance Companies" (PhD diss., University of Pennsylvania, 1947), 7–9.

72. James, *Metropolitan Life*, 338.

73. This story has been told often and well. See, for instance, C. G. Woodson, "Insurance Business among Negroes," *Journal of Negro History* 14, no, 2 (April 1929): 202–226 at 216–220; Walter B. Weare, *Black Business in the New South: A Social History of the North Carolina Mutual Life Insurance Company* (Durham, NC: Duke University Press, 1993); and Alexa Benson Henderson,

Hoffman's book—to serve life insurers—needed to justify abandonment, to justify abandoning fatalizing practices for African Americans, to justify writing African Americans out of a national risk community. That was exactly what it did for years to come.[74] *Race Traits* was not supposed to change insurers' behavior—life insurers did not need statistics for a race they supposed bound for extinction, for people they had abandoned. Southern segregationists, on the other hand, could use Hoffman's text (with its apparent objectivity, bearing the imprimatur of life insurers and the American Economic Association) in their campaigns to disenfranchise and disempower African Americans. Jim Crow's proponents, men like Benjamin Tillman, who read from Hoffman's book on the Senate floor, extended the insurers' case for the exclusion of blacks from the national risk community and argued for exclusion from the nation too.[75]

* * *

In the 1880s and 1890s, rupture temporarily displaced continuity in a range of discussions about insurance. Consider the case of married women outside the working classes who sought more favorable treatment from life insurers. Emily Ransom, an advocate for charging women the same rates as men for life insurance, spoke of "new women," no longer "*dependent* upon others" but "*independent* having within herself the power and ability to make money and invest it." New women, unlike old women, would be solid insurance risks. Industry statistics told a story of middle-class women living shorter lives than men (in marked contrast to the longer lives of women in the general public). Within the fatalizing paradigm, the best hope for reformers was to undermine such statistics, to refute them, as some advocates did indeed do. But questioning the statistics was never so crucial an element of reforming women's strategy as questioning the link of yesterday to tomorrow, of turning away from statistical fatalism. Middle-class

Atlanta Life Insurance Company: Guardian of Black Economic Dignity (Tuscaloosa: University of Alabama Press, 1990).

74. The president of the Association of Life Insurance Medical Directors of America, for instance, drew from Hoffman the conclusion in 1913 that "there must be a lack of vitality, or resisting-power, in the race, and all other causes are merely contributory" in claiming that African Americans necessarily made poor risks. Toulmin, "American Negro as an Insurance Risk," 165.

75. Benjamin R. Tillman, *The Race Problem: Speech of Hon. Benjamin R. Tillman of South Carolina; in the Senate of the United States: February 23–24, 1903* (Washington, DC, 1903), 19–24.

women deserved insurance, reformers argued, because they had become a new breed of women—and they would live longer, even when insured.[76]

Statistical fatalism did not fare well at the end of the nineteenth century. Black civil rights advocates had attacked it in fighting life insurers' discriminatory rates. And they had won. They had even convinced insurers to question their own commitment to statistical fatalism. By turning to Frederick Hoffman, Prudential and its peers put limits on their fatalizing power. For African Americans, they agreed, an impassible chasm lay between past and future. But where the voices of hope had seen equality born out of war and reconstruction, life insurers saw a dying race unable to survive outside of slavery. Inside and outside of life insurers' offices, statistical fatalism could not compete with visions of rupture and change. Many ordinary Americans put their hope in rupture. But life insurance corporations showed them that rejecting fate, as much as trusting fate, could be an ugly business.

76. Emily A. Ransom, "Life Insurance for Women," a reprint of a speech delivered on 6 November 1895 at the Women's Congress of the Atlanta, GA, Exposition, in Folder "Sales Aids 1895," Box 1008, MONY Papers.

THREE

Writing

On 12 April 1909, an agent of New-York Life in Tennessee submitted an application for life insurance, according to a corporate promotional pamphlet published that year. Thus began the application's journey through the insurer's massive "machinery," from the Nashville branch office to the New York headquarters, through the Index Division with its three million cards of past applicants, to the Medical Department, the Rating Bureau, and (via pneumatic tube) the Policy Issues Division. Once "adopted into the family of the New-York Life and christened '4,133,587,' the new baby was sent to the Photo Bureau" to be copied and then mailed, while clerks (probably from the army of women employed by all the large life insurers) produced new cards for the Index Division, the Division of Policy Briefs, the Comptrollers' Department, the Actuary's Department, and the Agency Department. In a single day, that one life insurance application set in motion thirty-six sequential processes and a flurry of paperwork, the bureaucratic birthing of a risk: a statistical individual named 4,133,587.[1]

The pamphlet's description evokes an assembly line of sorts—the pamphleteer used the term "treadmill"—and that is appropriate. The pamphlet describes a system for manufacturing what had by then become a mass commodity: life insurance. By the first decade of the twentieth century, corporations manufacturing that commodity required extensive operations in large part to accommodate their growing volume of business, but also because the making of life insurance had—in recent years—become premised on

1. New-York Life republished a series of these pamphlets discussing their home office operations in a volume called *A Temple of Humanity* (New York: New-York Life, 1909), 27–38, especially 37.

more individualized risk making. This chapter explores this seeming paradox: that life insurers, in order to make a mass commodity, first had to make more finely differentiated statistical individuals.

As in the first two chapters of this book, this chapter concerns itself with the questions nineteenth-century Americans asked about life insurers' project of making statistical individuals and with the critiques and challenges that accompanied such questioning. In the first chapter, the most acute challenge to risk makers came from within the life insurance industry as one firm (Lambert's) attacked the statistical methods of its competitors. Lambert's challenge faded away as he went off to prison and his company unwound, but the tension he exposed between classing and smoothing remained unresolved. In the second chapter, the people being made into risks rose up in critique and convinced northern legislatures to outlaw fatalizing as far as African American applicants were concerned. Life insurers responded, abandoning African Americans and adding one more caveat to their faith in fatalizing. Yet, again, the fundamental question they raised— should risk makers be able to use the statistical past to predict individual futures—remained unanswered. Life insurers went on fatalizing in day-to-day practice, just as the tension between classing and smoothing continued to drive their risk making.

The challenge presented in this chapter lacks the drama of Lambert's trial or Chappelle's legislative victory. It manifests itself in the day-in and day-out struggles of those at life insurers' headquarters to get the kind of information they wanted from those in corporate peripheries. It concerns office politics expressed through the occasional privately published paper or muted hallway confrontation. But a dearth of drama hardly implies a lack of significance. On the contrary, the systematic challenges presented here had a more lasting—and surprising—effect than did their showier predecessors. As Theodore Porter has argued, office politics and the systematic challenges involved in operating at long distances often play determinative roles in giving numbers more power.[2] In this case, they helped bring about a significant

---

2. Porter makes this argument in *Trust in Numbers*. He sketches out its implications for life insurance in Theodore M. Porter, "Life Insurance, Medical Testing, and the Management of Mortality," in *Biographies of Scientific Objects*, ed. Lorraine Daston (Chicago: University of Chicago Press, 2000), 226–246. This chapter draws on Porter's argument but offers finer-grained analyses of the practices that preceded the changes Porter describes, expands the scope of the study to include the development of the Medical Information Bureau, and offers more insight into the particular political problems that brought about innovations. For a consideration of how similar systematic challenges played out in Europe as well as the United States, see Lengwiler, "Double Standards."

individualization of risk makers' statistical methods and thereby preserved (for at least a little while longer) life insurers' vast, costly, and—for those concerned with growth—often annoying networks devoted to writing, mailing, sorting, analyzing, and filing the details of Americans' lives.

To accompany prose praising New-York Life's "great treadmill," the pamphlet's creators included a series of remarkable photographs illustrating the corporate spaces where risks were made. Five such photographs set the scene for each section of this chapter. The first two photographs capture the Medical Department and the Inspection Department in action. These sites resemble what Bruno Latour has called "centers of calculation," although calling them centers of processing or sorting could be just as or more apt.[3] Each site concerned itself with bringing together various reports about an individual, summarizing them, and assessing them. The third photograph presents the mailroom (the "Letter Division"), the place where all of the reports to be processed first arrived. By the turn of the twentieth century, most applicants never set foot in a life insurers' home office—instead, they arrived as pieces of paper, via the mail. The fourth photograph is of the Index Division, the place where risks resided as paper cards once they had been reduced to their essentials. The final photograph comes from another pamphlet and does not belong, strictly speaking, to the treadmill—but in showing us the gilt Meeting Room of the Finance Committee, it evokes the ultimate power that led life insurers' risk-making systems to write about individuals the ways that they did at the turn of the century.

These photographs, each beautiful in its own way, show multiple dozens of people reading, writing, and filing. They show a distributed bureaucratic system for risk making and they show, very clearly, the primary medium for the treadmill's "brain work": paper.[4] What drives this chapter, however, are the particular experiences—the frustrations and innovations—of a small cast of characters: the MONY medical director Brandreth Symonds; his counterpart at New-York Life, Oscar Rogers; the mercantile agency chief, Charles B. Holmes; and, briefly, the actuary Emory McClintock.[5] These characters serve

---

3. Bruno Latour, *Science in Action: How to Follow Scientists and Engineers through Society* (Cambridge, MA: Harvard University Press, 1987), chapter 6.

4. The pamphlet writer claimed that New-York Life's treadmill was "essentially human, and although moving almost automatically, brain work is ever present and is a dominating force." New-York Life, *Temple of Humanity*, 38. On the history of "paperwork," more generally, see Ben Kafka, "Paperwork: The State of the Discipline," *Book History* 12 (2009): 340–353.

5. Since this is a chapter titled "Writing," I cannot help but mention that Holmes and Symonds garnered a small amount of attention for literary endeavors. Holmes wrote two novels. The first began with a fire burning down a Methodist church and a tornado destroying an

dual purposes: they first stand in as typical of their peers in the final decade of the nineteenth century, allowing us to see what it looked like to reduce people to risks at what is the start of this chapter's story; secondly, their experiences explain, to varying degrees, how and why life insurers devoted new energies to making statistical individuals at the turn of the century.

* * *

The New-York Life's pamphlet writer expected to awe the company's agents and its businessmen clients with pictures of modern office efficiency like that in figure 3.1. Looking at the medical department's offices in 1909—with its female clerks in bright white blouses reading and typing, their male counterparts poring over their own papers, and at least one manager overseeing all the reading and writing from the back corner—a viewer witnessed the corporate organization of mind-work, a version of what Charles Babbage had many years earlier called the "division of mental labor."[6] It may be difficult to understand the power of this photograph today, but the pamphlet's author counted on it to convey to viewers the thoughtful method by which the company evaluated each individual who applied for life insurance, a thoughtfulness that did not depend on individual evaluations but on systematic cooperation. Indeed, the photograph spoke an important truth: by the early twentieth century, New-York Life really did rely on the automated processing of application papers and medical reports by its many clerks. It made Americans into risks—in the majority of cases—by reducing piles of papers about each case to a few details or numbers on an index card, by applying algorithms and rules to each case, by setting those cards into motion.

A few decades earlier, no company came anywhere close to New-York Life's model of automation. But most companies did at least employ some kind of statistical reasoning—albeit less sophisticated—like that

---

Orthodox church. Neither appears to have carried any insurance. But Holmes's God had a purpose: unifying the town's (Protestant) churches. Charles B. Holmes, *Elsieville: A Tale of Yesterday* (New York: Charles B. Holmes, 1903). Holmes himself noted of the second, even less successful than the first, that it "possesses no literary or commercial value and is merely a country yarn," albeit a country yarn that hinged on a "stranded Malay sailorman." See "A Rejected Manuscript: A Country Girl. By Charles B. Holmes, Author of 'Elsieville.' Limited edition," *New York Times*, 14 January 1905. Symonds, meanwhile, appeared in a flattering profile for his work tracing mythologies surrounding Santa Claus. See George MacAdam, "Christmas Now and Centuries Ago. 'Santa Claus, as We Recognize Him Today, Is an American Conception,'" *New York Times Book Review*, 25 November 1923.

6. Charles Babbage, *On the Economy of Machinery and Manufactures* (London: John Murray, 1846), chapter 20.

3.1. New-York Life's Medical Department, in New-York Life, *Temple of Humanity*, 32.

which would eventually help make automation possible. When Brandreth Symonds joined MONY, another of the so-called Big 3 companies, along with New-York Life and Equitable, he encountered an industry with vast networks of informants employed to report on Americans across the nation and possessing powerful new methods for interpreting individuals in light of statistical insights. But translating an individual's paperwork into a risk still required extensive specialized training on top of prior medical exper-tise; it was not yet trusted to an army of clerks. Making risks looked more like a craft enterprise than an industrial one, even as the volume of risks that life insurers churned out was already exploding. Symonds joined MONY with a doctor's training and began to learn his new craft. We can take his induction into the methods of risk making as in many ways typical.

Symonds came to life insurance in fragile health. It isn't clear exactly what happened to him in 1889, but as he explained much later, he was left with "a busted heart." That busted heart drove Symonds, a promising young doctor, to seek an alternative to private practice. "He counted it wise," explained his Hobart College friend and eulogist J. W. Van Ingen, "to accept a position on the medical staff of the Mutual Life Insurance Company of New York, rather than risk the uncertain hours and physical strains inherent in the life of the regular practitioner."[7] Symonds began as an assistant medical director, working under Walter Gillette and G. S. Winston. His reeducation began.

He came to the company already a well-rounded, well-educated man. He was precocious and active in his youth, entering Hobart College at the age of fifteen and graduating in 1881 as the president of the Hermean literary society, a member of the football squad, and a Sigma Phi man.[8] He went on directly to the College of Physicians and Surgeons, the medical division of Columbia College in New York, for three years.[9] There he encountered lectures offered by professors of surgery, obstetrics, gynecology, anatomy, chemistry, pathology, physiology, materia medica, and a range of clinical fields. Much more important to his medical education would have been the mentoring relationship established with his "preceptor," a practicing physician who oversaw each student's practical education. Symonds began his education working under Dr. W. H. Helm, who practiced in his hometown of Ossining. For his second and third years, Symonds became a student of the distinguished surgeon John A. Wyeth.[10] After graduation, Symonds served an internship at Bellevue Hospital.[11]

But what really captured Symonds's attention during his time at the College of Physicians and Surgeons was the laboratory. Chemistry became his specialty. Symonds's dissertation on the "Origin and Destiny of Glucose in Blood" pointed him toward the burgeoning fields of blood and urine analysis, which were, unbeknownst to Symonds, becoming important tools for insurance medical examiners.[12] Symonds was not alone in his fascination with the laboratory. The College's Physiological and Pathological

7. J. W. Van Ingen's eulogy and Symonds Information Sheet, in Brandreth Symonds Alumni House file, Hobart and William Smith Colleges Archives.

8. From 1881 "Echo of the Seneca" college yearbook, Hobart and William Smith Colleges Archives.

9. J. W. Van Ingen's eulogy claims Symonds's degree was held back because of his age, in Brandreth Symonds Alumni House file, Hobart and William Smith Colleges Archives. But Symonds was actually one of many students who began taking three-year courses ahead of the college's decision to make three years compulsory in 1887–1888. See John C. Dalton, *History of the College of Physicians and Surgeons in the City of New York: Medical Department of Columbia College* (New York: College of Physicians and Surgeons, 1888), 195.

10. College of Physicians and Surgeons, *Annual Catalog* (1883–1884), Columbia University Medical Center Archives.

11. J. W. Van Ingen's eulogy and Symonds Information Sheet in Brandreth Symonds Alumni House file, Hobart and William Smith Colleges Archives.

12. College of Physicians and Surgeons, *Annual Catalog* (1883–1884), Columbia University Medical Center Archives; on blood and urine examinations becoming more important around this time in the United States and Britain, see Alborn, *Regulated Lives*, 265–268.

Laboratory of the Alumni Association (est. 1878) had only existed for a few years by the time he arrived, but it was already bursting at the seams with excited young students. A year after Symonds left, the school's director bemoaned a constant crush of experimenters that the facilities could hardly sustain.[13] Symonds's experience reflected the college's privileged position as a leader in sponsoring physiological research in the laboratory, but it reflected more generally the proliferation and exaltation of the laboratory all over: laboratories were becoming *the* privileged spaces for the production of scientific knowledge across disciplines.[14]

Still intending a normal medical career, Symonds became an assistant physician in the out-patient department of Roosevelt Hospital and served as an attending physician at the Northwestern Dispensary, a New York clinic.[15] He also found time to tutor students in general chemistry and even wrote a manual on the subject. A chapter on "urine" in the book suggests that Symonds had expanded his dissertation expertise from blood analysis to encompass excretions too.[16]

The one subject Symonds probably never encountered in any depth, prior to joining MONY, was statistics. While doctors in the mid-nineteenth century would have still been trained in the numerical method of medicine or public health statistics, laboratory-based topics like experimental physiology had displaced statistical studies by the second half of the century.[17] Yet the claim that doctors and medicine stopped being statistical is not quite

13. Dalton, *History of the College of Physicians and Surgeons*, 126.

14. Claude Bernard is often highlighted as the champion of the laboratory in medical and biological research. See, for instance, William Coleman, "Cognitive Basis of the Discipline: Claude Bernard on Physiology," *Isis* 76, no. 1 (1985): 49–70. Bernard's methods of experimental physiology were introduced at the College of Physicians and Surgeons—and perhaps in the United States for the first time—by John C. Dalton in 1854. See Dalton, *A History of Columbia University 1754–1904* (New York: Columbia University Press, 1904), 318, 327–328. On the introduction of the laboratory and its revolutionary significance in Bernard's wake, especially from the 1870s on—in fields as diverse as physiology and physics—see Gerald Geison, *Michael Foster and the Cambridge School of Physiology* (Princeton, NJ: Princeton University Press, 1978), chapter 4. Robert Kohler posits that the laboratory and its twin, the "field," both came to prominence as sites of scientific endeavor in the period from 1840 to 1890. See Kohler, *Labscapes and Landscapes: Exploring the Lab-Field Border in Biology* (Chicago: University of Chicago Press, 2002), 1–11. Industrial research labs originated around the same time, first in Edison's Menlo Park in 1876 and then Germany's BASF and Bayer labs, and in DuPont's turn-of-the-century experimental station. See David Hounshell and John Kenly Smith, *Science and Corporate Strategy: DuPont R&D, 1902–1980* (New York: Cambridge University Press, 1988), 1–10.

15. Brandreth Symonds, *A Manual of Chemistry for the Use of Medical Students* (Philadelphia: P. Blakiston, Son, 1891), title page.

16. Symonds, *Manual of Chemistry*, preface.

17. See Porter, *Rise of Statistical Thinking*, 159–161, 237–238.

right.[18] Instead we should say that the site of medical statistics shifted, out of medical schools and internships, and into other institutions, like asylums, hospitals, and especially life insurers.[19] Insurers taught doctors serving as examiners to keep detailed records (as will be discussed more in a few pages) and they collected private statistical caches kept by doctors.[20] Just as importantly, they demonstrated for the first time that doctors could use statistical findings derived from groups to diagnose or predict disease in individuals.

Those doctors, like Symonds, who became insurance medical directors learned not only how to record and collect statistics, but how to analyze them. On the shelves of his MONY office, Symonds would have found an 1875 report called *Preliminary Mortuary Experience of the Mutual Life*.[21] Written by MONY medical directors E. J. Marsh and G.S. Winston, the report subjected the company's medical records to careful scrutiny, primarily with the purpose of making medical examinations more effective. Marsh and Winston searched out correlations between factors like age and the eventual discovery of disease (and found that examiners put too much stock in the idea that tuberculosis was a young man's disease). They also compared the distribution of disease incidence against the years since policyholders were granted insurance and decided, because many people who passed medical examinations developed alcoholism, cancer, and tuberculosis only a year or two later, that their examinations had some serious flaws.[22] Marsh and Winston were hardly alone in making such studies. They followed similar efforts of at least eleven British medical directors.[23]

Medical directors used mortality investigations to train themselves in new methods for translating an applicant's paper records into a new risk. They seldom evaluated applicants directly; instead, they spent much more time evaluating applicants only on paper. Evaluating what we might call

18. Cassedy approaches this claim in *American Medicine and Statistical Thinking*, 238.

19. On asylums and statistical records, see Theodore M. Porter, "Funny Numbers," *Culture Unbound* 4 (2012): 585–598. Medical statistics had also persisted in some areas. The southern physicians whose work on racial inferiority grounded in statistical data presaged Hoffman's analysis highlight one strand of this ongoing tradition. See John S. Haller, Jr., "Physician versus Negro," in *Outcasts from Evolution: Scientific Attitudes of Racial Inferiority 1859–1900* (Carbondale: Southern Illinois University Press, 1996), 40–68.

20. For example, MONY's Frederick Winston bought doctors' statistical records when he came across them in his southern and western tours. See, for instance, 25 May 1877 meeting, Minutes of the Mortuary Committee (Book B), MONY Papers.

21. Mutual Life Insurance Company of New York, *Preliminary Report of the Mortuary Experience of the Mutual Life Insurance Company of New-York* (New York: Mutual Life of New York, 1875).

22. Mutual Life, *Preliminary Report*, 21–24, 38–45.

23. Alborn, *Regulated Lives*, 284–285.

"paper people" required developing and learning a new kind of medical judgment.[24] Sometimes that judgment conflicted with established medical ideas. To take an example offered by Symonds, the average person and even most physicians paid little attention to occasional fluid discharging from the ears. But as Symonds told potential examiners, "we have found out, however, from experience, that these cases, as a class, are not good insurance risks. Some of them die from the extension of the inflammation to the adjacent regions; but quite a fair percentage develop tuberculosis."[25] Insurers' statistics allowed such analytical leaps—from ear fluid to the most feared disease of nineteenth-century America—pointing out correlations that no one else had the capacity to see. They did not point to causation necessarily, but that was not a problem. Insurers sought cheap signs of a bad risk and did not care about medical causation in and of itself.[26] For them, it paid to think about applicants—provisionally at least—as statistical individuals. Life insurance medical directors like Symonds learned to evaluate risks by looking at individuals (on paper) with eyes trained by looking at (statistical) groups.

* * *

Medical men and later medical departments had enjoyed a long association with life insurers. They had even—as in the case of Lambert's American Popular Life—become leading figures in insurance corporations. Each doctor in a life insurance corporation served, in Brandreth Symonds's words, as "a sentinel at the gate to prevent the ingress of those who would only

24. Sociologists of science and of surveillance have developed some excellent concepts for thinking about the power of human representations in modern bureaucratic systems. Haggarty and Ericson offer us the "data double" as a representation that maps onto a real-life individual, but lives its own life to some extent, even as the life it lives in various bureaucracies can have a powerful effect on its fleshly counterpart, and vice versa. Think here of the relationship between an individual and his credit rating or criminal record. Kevin D. Haggerty and Richard V. Ericson, "The Surveillant Assemblage," *British Journal of Sociology* 51, no. 4 (2000): 605–622. Ellen Balka and Susan Leigh Star proposed the "shadow body" as a way of thinking about the various ephemeral accretions of self that infrastructures create, and which persist, even as individuals change. Think of the online records we leave, the Facebook profiles we ignore, or even the medical records we leave after a hospital stay. Balka and Star, "Mapping the Body across Diverse Information Systems: Shadow Bodies and How They Make Us Human" (paper presented at the 4S annual meeting, Cleveland, OH, 2 November 2011). In making a risk, insurers created both data doubles and shadow bodies, on paper.

25. Brandreth Symonds, *Life Insurance Examinations: A Manual for the Medical Examiner and for All Interested in Life Insurance* (New York: G. P. Putnam's Sons, 1905), 55–56.

26. Rothstein draws a similar conclusion in William G. Rothstein, *Public Health and the Risk Factor: A History of an Uneven Medical Revolution* (Rochester, NY: University of Rochester Press, 2003), 64.

destroy the structure."[27] Doctors tried to pick the particularly healthy, in hopes of increasing profits or of keeping premium rates down (to attract more customers and build larger capital reserves to be invested in the process). They kept out the morally suspect, those designated undeserving. (The most obvious moralizing of life insurers came around alcohol: companies tended to punish any taste for or association with beer or spirits.) And doctors kept out the sick, who might be seeking insurance because they suspected they would soon die.

Still, some applicants concealed their illnesses, hid their bad habits, or lied about their pasts or their financial security or the health of their parents. Such frauds—"material concealment and misrepresentation" in contemporary legal speak—drew less attention than their more spectacular brethren. Pretending that one had a more secure job than one did, or that one *did not* occasionally spit blood (a possible sign of tuberculosis), failed to attract the same kind of attention as insuring one's husband only to poison him or attempting suicide by smoking oneself to death (with cigars!).[28] But insurers worried about those who lied about their finances: might they have an incentive to die young, escaping financial woes, and thus securing a better future for their families, or might they attempt for similar reasons to fake their own deaths? And they worried about those who lied about their medical or family histories: might they be hiding a reason that no life insurer would accept them?

Inspection departments, like that pictured in figure 3.2, surfaced in the late 1880s and 1890s to provide a second line of defense against those who might attempt to deceive the life insurer by means grand or subtle. Post-1873 expansions played a role in their rise: as the volume of applicants increased, the opportunities for fraud did too. But the larger factor was the advent of "incontestability."[29] Incontestable clauses became popular features of life insurance contracts from the 1880s. They guaranteed payment for a claim on any policy that had been in force for a few years, even if the company found lies or errors in the original application, and even if suicide was suspected.[30]

---

27. Brandreth Symonds, "The Medical Jurisprudence of Life Insurance," in *A System of Legal Medicine*, ed. Allan McLane Hamilton and Lawrence Godkin, (New York: E. B. Treat, 1895), 1:493–582 at 510.

28. On spectacular and fantastic attempts to defraud life insurers, see John B. Lewis and Charles C. Bombaugh, *Stratagems and Conspiracies to Defraud Life Insurance Companies: An Authentic Record of Memorable Cases* (Baltimore, MD: James H. McClellan, 1896).

29. Also sometimes called "indisputability."

30. Despite some examples of incontestable policies from the 1860s, it was Henry Hyde's

3.2. New-York Life's Inspection Department, in New-York Life, *Temple of Humanity*, 36.

Incontestability made it more important to ferret out untruths and misstatements in insurance policies before they were written (since an insurer could no longer contest a policy in court later if it were found to contain possible falsehoods), and it gave insurers more reason to worry about whether an applicant might in fact be a confidence man or charlatan. Life insurers shared that worry with credit reporters, who had developed methods targeted toward addressing exactly such fears and whose help life insurers began seeking in the 1880s. Before incontestability, life insurers had only dabbled in the sort of detective work in which credit reporters specialized. In 1876, for instance, Dan Gillette, the brother of MONY's medical director Walter Gillette, founded a new "Department of Revision," whose purpose was to sniff out liars and cheats among MONY policyholders and to "revise" policies written from fraudulent bases. But from the 1880s on, Gillette's

Equitable that championed the incontestability clause two decades later and essentially forced it on the rest of the industry. See Stalson, *Marketing Life Insurance*, 317, 501. Equitable was explicit regarding the new demands placed here on selection: they would make selection tougher in order to make getting a claim easier. See "Indisputability in Life Insurance," *New York Times*, 4 July 1879. My thanks to Jonathan Coss, formerly of AXA, for encouraging me to think about incontestability in this context.

Department of Revision began to concern itself as much with predicting fraud and preventing error as with finding such evils after the fact. Other companies followed suit.

Credit reporters' archives of individual data, as well as their methods of assessing financial risk and judging individual character, suddenly became essential to life insurance corporations. Those papers burdening a large wooden table in figure 3.2 were one result. Life insurers started trading in new kinds of paper people, in the writing of American lives in one more report—yet another simplistic biography, the reduction of a life to the credit reporters' categories: capacity, character, and capital.[31]

Charles B. Holmes, whose early life is now lost to the past, surfaced in 1889 to play a crucial role in bringing the techniques of credit reporting—which were comparatively stronger in their capacity for gathering, centralizing, and circulating information about individuals—to the problem of predicting and preventing fraud. While older mercantile agencies continued to serve large wholesalers, and newer, local consumer credit reporting societies developed sophisticated techniques for tracking a much wider swath of the population in major urban areas, Holmes—who must have had some experience in a more traditional mercantile agency—carved out his own space in the market: specializing in the surveillance needs of accident and life insurance companies.[32] His company, the Holmes Mercantile Agency, had a detective's license, since in the eyes of New York State Holmes was just one more private investigator.[33] For one dollar, Holmes promised insurers a complete report on any individual in the United States, delivered by mail to the insurer, or by telegraph for urgent requests.[34] Eventually, Holmes offered the service to any and everyone.[35]

But what made Holmes particularly valuable was his capacity to translate credit reporting methods to fit life insurers' particular investigatory needs. One of his early innovations brought the cross-company blacklist out of

31. As Scott Sandage has emphasized, these categories adhered mainly "in the man" and gave short shrift to business circumstances or larger social or economic contexts. They also ignored other means of assessing a life's worth, as Sandage makes abundantly clear. See Sandage, *Born Losers*, chapter 5.

32. On early consumer credit operations, their local character, and their increasingly sophisticated methods, including the aggregation of "ledger experience," see Josh Lauer, "The Good Consumer: Credit Reporting and the Invention of Financial Identity in the United States, 1840–1940" (PhD diss., University of Pennsylvania, 2008), 91, 96–98, 101.

33. *Annual Report of the Comptroller of the State of New York* (Albany: J. B. Lyon, 1902), 689.

34. "Holmes Mercantile Agency" (advertisement), *Spectator* 73, no. 8 (25 August 1904): vi.

35. "Holmes' Mercantile Agency" (classified ad), *American Monthly Review of Reviews* 34 (December 1906), classified advertising section: 157.

credit reporting and into insurance. Holmes aggregated notices of both claims and policy rejections from his clients and used the resulting database to flag potential problem applicants. Holmes designed a code system that insurers used to notify him of rejections.[36] He stored their notices—perhaps in a ledger or in one of the new filing systems beginning to be used in mercantile agencies, banks, and soon in insurance—but his precise method or technologies are not known. When, later, a company asked for information about an applicant or claimant, Holmes consulted his files to see if that individual had a history of suspicious claims or had previously been rejected by another company.

Over time, Holmes expanded his services, or repackaged them. His specially trained correspondents could report on applicants for insurance, as always. But by the turn of the twentieth century, Holmes emphasized that he could also report on *Medical Examiners, Deaths, Agents, or General Information* in any locality in the *Country.*"[37] Credit reporting became for insurers a general mechanism for establishing and reinforcing trust and confirming facts. Holmes's investigators offered clients local knowledge of "deaths" and "general information." But they also gave companies an independent eye by which to oversee their field workers: their agents and medical examiners.

Holmes made a success of himself. He signed exclusive deals with some companies, insisting they use his service alone.[38] He bragged that his "patrons" included "most of the insurance companies of the country."[39] At the same time, he insisted on his own independence—a valuable quality for one claiming to sell unbiased information.[40] Over the years, Holmes built a valuable collection of files, organizing techniques, and, of course, correspondents.[41] When he retired in 1907, Holmes sold the agency's materials to a young credit bureau specializing in insurance and founded in 1899 as the National Insurance Information Bureau, but incorporated in 1906 as Hooper Holmes, named for its founders, the librarian turned actuary William DeMattos Hooper and the detective Bayard P. Holmes. It isn't clear if the two Holmes were related.[42]

36. "The Progress of Fourteen Years," *Our Society Journal* 12, no. 70 (July 1889): 1–16 at 15.
37. "Holmes Mercantile Agency" (advertisement), vi.
38. "Life, Fire and Miscellaneous Notes," *Spectator* 73, no. 3 (21 July 1904): 32.
39. "Holmes Mercantile Agency," *Spectator* 74, no. 24 (15 June 1905): 331.
40. "Holmes' Mercantile Agency" (classified ad): 157.
41. "In and about New York," *Spectator* 78, no. 6 (7 February 1907): 69.
42. Terri Mozzone, "Hooper Holmes, Inc.," in *International Directory of Company Histories,* ed. Thomas Derdak (New York: St. James Press, 1998), 22:264–267 at 264–265.

The details of the purchase make clear just how large Holmes Mercantile Agency had become and the grand scale that Hooper Holmes aspired to. Before Hooper Holmes could move Holmes Mercantile Agency's files three blocks south, from 132 Nassau Street to 87 Nassau (bringing the files to within a few blocks of MONY's magnificent headquarters at 34 Nassau)[43], it had first to secure more rooms next to its present offices. The acquisition enlarged Hooper Holmes's insurance clearinghouse to encompass "over a million records of accident and sickness claims, rejections, cancellation and other valuable data, covering the United States and Canada." And that was only the files. The purchase also allowed Hooper Holmes to vastly increase its small inspection department. The files would not gather dust, but instead were "constantly supplemented by reports from local correspondents in all the cities and towns throughout the continent."[44] Holmes and his successors built careers out of their recognition that life insurers in the age of incontestability valued a set of practices—surveillance practices particularly—developed within the larger credit reporting industry.

While life insurers came to depend upon credit reporting for valuable services (and still do), some insurers from very early on began moving credit reporting practices in-house, into what would come to be known as inspection departments. MONY, for instance, acted swiftly. In 1890, only a year after it hired Brandreth Symonds, the company bought a copy of Charles B. Holmes's files and access to his correspondent network. Holmes Mercantile Agency had only recently opened its doors, but MONY thought it could use Holmes's intellectual capital as the foundation for its own National Commercial Agency, which would eventually develop the capacity to look into about 20 percent of all applicants' financial and personal pasts.[45] There, allied to the Department of Revision—which in 1894 alone investigated nearly twenty-five hundred medical examiners too[46]—MONY built credit reporters' systems into their risk-making process. Most life insurers came to rely on inspection reports to make risks.[47] Some joined MONY in building

43. Keller, *Life Insurance Enterprise*, 39.
44. "Information Bureau Adds to its Facilities," *Standard* 60, no. 7 (16 February 1907): 195; see also "The Hooper-Holmes Information Bureau," *Spectator* 78, no. 6 (7 February 1907): 80.
45. Clough, *Century of American Life Insurance*, 173–174.
46. Clough, *Century of American Life Insurance*, 174.
47. Some companies valued inspection reports more than others. Most that did use them bought them from outside commercial agencies. See Harry Toulmin, "Some Details of Office Methods," *Abstract of the Proceedings of the Association of Life Insurance Medical Directors of America* 18 (1907): 98–109 at 103, 105–106.

their own inspection services. By 1909, New-York Life's inspectors could call for reports from about sixteen thousand informants.[48]

* * *

Brandreth Symonds and Charles B. Holmes had one key thing in common: the work each did depended on maintaining a steady influx of reliable information from across the nation. To assess individual risks, they needed a lot of writing that told narrowly focused stories of Americans' lives. The problem they faced was simple and systemic: maintaining that steady influx took serious effort at significant costs, and as we'll see in the next section, not everyone thought such extensive writing networks were worth it. Holmes's agency, like other mercantile agencies, relied on the nation's lawyers to provide details about their neighbors' lives, while Symonds and his fellow medical directors turned to the nation's doctors. Professionals attracted corporate attention because they possessed local knowledge. But their professionalism presented its own problems. Lawyers and doctors could be expensive to employ and difficult to control. Still, Symonds, Holmes, and their peers endeavored to build writing networks that would serve as their eyes and ears around the nation.

Letters arrived in the thousands daily at large life insurers in the nineteenth century, as we see in New-York Life's case in figure 3.3. Many of those came from company agents, who gave insurers the capacity to act at a distance: soliciting new applicants, checking in on existing policyholders, paying claims.[49] They wrote an awful lot of letters. As did medical examiners and credit reporters. At the turn of the twentieth century, life insurers engaged approximately fifty thousand medical examiners.[50] Credit reporting agencies in the 1870s employed about a fifth as many attorney-informants.[51]

Life insurers developed systems for making risks that leaned heavily upon professionals and the personalized reports they wrote. Symonds and Holmes's medical and inspection departments required piles of paper in order to translate each life into a risk. The more disorderly, idiosyncratic, or haphazard those piles, the longer the translation took. So as life insurers moved toward the mass production of risks, they strived also to make their

48. New-York Life, *Temple of Humanity*, 52–53.

49. On the crucial roles of agents in selling life insurance, see Stalson, *Marketing Life Insurance*; and Zelizer, *Morals and Markets*, 119–147. Murphy argues that agents in the 1830s and 1840s acted not just as sales people, but as gatekeepers and observing correspondents. See *Investing in Life*, 47–76.

50. Symonds, *Life Insurance Examinations*, 1.

51. Lauer, "From Rumor to Written Record," 308.

3.3. New-York Life's Letter Division, in New-York Life, *Temple of Humanity*, 28.

correspondents more efficient and consistent. That meant disciplining doctors and lawyers so that together they formed observational communities trained to see and write what they saw in increasingly standardized ways. It meant, essentially, exerting corporate power over ostensibly independent professionals.

"Discipline" does not equate to "punish" in this case. Following Foucault's distinction, it instead means something closer to "teach" or "train."[52] To be disciplined is to be taught how to see, how to understand, and how to behave in a particular way. Every university student undergoes such training upon being inducted into the mysteries and practices of an academic "discipline." But disciplines exist in all realms and not only in the academy. Insurers and credit reporters could get more from their network correspondents if they disciplined them. Similarly, historians of science following work by Lorraine Daston have recently considered the importance

52. Michel Foucault, *Discipline and Punish: The Birth of the Prison*, trans. Alan Sheridan (New York: Random House, 1995).

of "observational communities" to scientific endeavors. Such communities coalesce around shared habits, instruments, questions, and techniques.[53] They too, like disciplines, can exist in venues outside the traditional sites of science: corporate writing networks fit the definition just as well as do scientific communities.

When insurers called upon doctors to write reports, they asked them to change many of their habits, assumptions, tools, and techniques. Doctors accustomed to patients who would talk openly with them—detailing every ache and pain in excruciating detail—had to learn to read the bodies of more guarded patients, who might tell outright lies or who might be lying to themselves. (Brandreth Symonds, for instance, told doctors to imagine a patient convincing himself that his venereal ulcer couldn't be syphilis.)[54] They also had to learn to judge from a baseline of "average health" instead of from perfection. Physicians, Symonds complained, could be too discriminating, too concerned by small faults.[55]

Life insurers made some efforts to retrain doctors, to teach them to think more like risk makers. Symonds, eager to see improved medical curricula and the advent of a more robust observational community, wrote a textbook on life insurance medicine in 1905 for use in new medical school courses.[56] Life insurance medical directors had some leverage to use in arguing for changes to medical training. They paid fees to medical examiners and had done so for decades. One insurer estimate put the total income to the medical profession from life insurers at several million dollars a year, "undoubtedly larger than any one item received by the whole profession put together outside of their regular practice."[57] Dr. John W. Riecke, of Grand Rapids Medical College, argued, "Insurance companies certainly pay out enough money for examinations annually to entitle them to some consideration from medical schools and doctors of medicine generally."[58] Some medical schools did accommodate life insurers, and occasionally doctors even endorsed insurance work as a worthy scientific enterprise in its

53. Daston, "On Scientific Observation," *Isis* 99, no. 1 (2008): 91–110; Lorraine Daston and Elizabeth Lunbeck, eds. *Histories of Scientific Observation* (Chicago: University of Chicago Press, 2011).

54. Brandreth Symonds, "A Plea for Under Graduate Instruction in Making Life Insurance Examinations," *Medical Examiner* 9, no. 7 (July 1899): 207–208 at 208.

55. Symonds, "Plea," 207.

56. Symonds, *Life Insurance Examinations*.

57. "Editorial," *Medical Examiner* 9, no. 7 (July 1899): 197–204 at 197.

58. "Editorial," 199.

own right.[59] But getting doctors to think differently was not truly essential to risk makers' project.

They could make doctors serve as members of an observational community just as well or even better by imposing new objective tests and measures on the doctors in their networks. Medical directors instituted formal medical and family history questionnaires, insisted on height and weight measurements, dictated the use of tools like the stethoscope, and introduced new ways to analyze urine chemically.[60] The stuff that made new objective measures possible—the urinary test kits, stethoscopes, thermometers, and books of medical examination blanks, "useful also for any physical examination, such as for Police and Fire Department, the Militia, Civil Service, etc."—had to be bought from third-party suppliers (in some cases owned by medical directors making money on the side).[61]

Still, even standardized forms and objective tests could fail if doctors did not cooperate. Medical examiners couldn't be trusted to fill out forms completely, correctly, or promptly. "The busy practitioner, strange to say, is just the sort of a physician who neglects the little things connected with an examination," complained a writer in the trade journal, *Medical Examiner*. He continued: "These remarks have been suggested to us by the habit of some examiners of omitting to answer questions or of only half answering them."[62] When insurers received incomplete examination forms, they could and often did point out the error and seek a correction from the examiner, although doctors who often kept poor records of their own did not always have the necessary data to fix the problem.[63] Insurance medical directors responded to these challenges with stricter medical forms (suffering no omissions or revisions) and instructions rendered in minute detail. "Be

---

59. "Medical Examiners Organizing," *Medical Examiner* 5, no. 46 (1895): 138. "It has occurred to me that they [medical examiners] should organize for the purpose of discussing the innumerable scientific questions which daily arise in the business."

60. Davis, "Life Insurance and the Physical Examination," 398. On the turn toward objective measures, see Porter, "Life Insurance, Medical Testing, and the Management of Mortality," 226–246.

61. All were advertised frequently in *Medical Examiner* in the 1890s. See, for instance, *Medical Examiner* 9, no. 7 (July 1899): 196. On the "job printers" who mass-produced the various blank forms that allowed modern bureaucracies to flourish, see Lisa Gitelman, *Paper Knowledge: Toward a Media History of Documents* (Durham, NC: Duke University Press, 2014), 21–52.

62. "Practical Object Lesson," *Medical Examiner* 5, no. 12 (1895): 255.

63. One doctor suggested that insurers supply blanks to medical examiners who could then keep a copy of the exam for themselves. "Medical Examiners' Blanks," *Medical Examiner* 5, no. 46 (1895): 138.

very careful to record everything that [the applicant] tells you," Symonds instructed potential examiners. "No matter how unimportant or trivial a question may seem to be to you," warned Symonds, "it must be answered by at least 'Yes' or 'No.' "[64] Communities of observation required reliable and consistent reporting as much as or more than shared observational skills. Part of creating an observational community was convincing informants to obey instructions and accept limitations to the scope of their individual judgments.

This could be a tough sell. Dispirited by doctors' unwillingness to be disciplined, Manhattan Life medical director George Wells argued that "some clerks are better qualified barring strictly technical medical training, to make a good medical examination than some doctors ever will be."[65] Many doctors resisted being turned into mere clerks, the tools of an insurance medical director or statistician. At the same time, other voices worried that life insurers had been too successful in making doctors' skills unnecessary. Leo Crafts of Hamlin University, occupant of a new chair for Preventative and Legal Medicine and Medical Insurance, lamented that too many doctors "now usually look upon [an insurance exam] as outside their lines of practice, and simply as filling out blanks in a perfunctory way, for so many dollars in fees." Crafts concluded gloomily: "This inevitably means that the examination is superficial and almost valueless. In my observation, the physical examination made by examiners for life insurance amounts to nothing."[66] In so far as medical directors like Symonds continued to support advanced medical training in insurance medicine, they must have been at least somewhat sympathetic to such arguments. To make an observational community, Symonds and his peers sought to limit doctors' professional judgment without eliminating it.

Many doctors, however, did not think such a fine balance could be struck. They resisted the entire idea of corporate disciplining because they felt it interfered with the profession's more central disciplining role. Such concerns rattled through the profession's fight in the 1890s to keep fees for examinations at five dollars instead of the three dollars that some leading companies were beginning to pay. Doctors defended being paid by insurers even as they voiced fears that their expertise was being cheapened by

64. Symonds, *Life Insurance Medical Examinations*, 47.
65. "Editorial," 202.
66. "Editorial," 197.

commercial pressure.[67] The medical director George Wells countered that commercial values could not help but intrude in medicine: "If the doctor does not possess that knowledge he cannot sell it, and insurance companies cannot buy it, and will not, therefore, be of profit to the doctor. If this is commercialism, then be it so. Doctors are not practicing medicine for their health."[68] Other medical directors turned discussions away from money, emphasizing instead the essential service insurance work provided in preserving institutions responsible for taking care of widows and children. "Unless the medical examiner does his duty in barring out undesirable risks and accepting only those who may reasonably be expected to live out the theoretical expectation of life, the company is predestined to loss and ruin," argued Symonds.[69]

Another hindrance to corporate disciplining came from some agents who had their own ideas about how doctors should make their evaluations. Agents sometimes wanted to train doctors to their own ends. They offered medical examiners incentives to issue favorable reports and got examiners fired—by complaining to the home office—when they failed to cooperate. In extreme cases, agents and doctors went out together filling out falsified documents for any warm body they encountered. Agents knew the key words—"vertigo," for instance—that could kill an application if it showed up on a medical history, so they coached applicants on what to say, and most likely coached doctors too.[70] Dr. John D. Landford's 1895 request for a way to submit his report without going through the agent or applicant was hardly atypical and suggests the struggle of a doctor subject to disciplining pressure from opposing interests.[71] The editor of *Medical Examiner*, a life insurance medical director, bemoaned what he considered "an unusual epidemic of evil doing on the part of the [medical] profession."[72] But by "evil," he meant to some degree the refusal of doctors to bend to insurers' dictates.

Credit reporters intent on disciplining lawyers did not have a much easier time. Upon founding his Mercantile Agency in the 1840s, the credit

---

67. For a taste of the furor over lowering medical examination fees, see "Reduced Fees," *Medical Examiner* 5, no. 12 (1895): 253; "Proper Price for Life Insurance Examination," *Medical Examiner* 5, no. 46 (1895): 137; and "Medical Examination Fees," *Medical Examiner* 5, no. 12 (1895): 259–260.

68. "Editorial," 204.

69. Symonds, "Medical Jurisprudence of Life Insurance," 510.

70. G. S. Stebbins, "Practical Suggestions concerning Life Insurance," *Medical Examiner* 5, no. 46 (1895): 131–133.

71. "Correspondence," *Medical Examiner* 5, no. 55 (1895): 221.

72. "Editorial Notes," *Medical Examiner* 5, no. 12 (1895): 251–252.

reporting pioneer Lewis Tappan had claimed that his attorney correspondents required no special observational disciplining. But it seems likely that he mainly wanted to ease the minds of those "reporters" who worried that their deal with the Mercantile Agency—to provide local intelligence in return for debt collection referrals—might amount to unneighborly or unprofessional behavior.[73] In practice, once Tappan and his competitors began a correspondence with an attorney—seldom recruited for any special qualifications—they set about molding him to meet their needs.

Beginning with the first visit from a recruiting agent who taught the recruit the expected form and content of communications to the company, mercantile agencies cultivated special habits of observation and reporting.[74] Tappan exhorted his reporters while in the field to "record all facts that come to your knowledge, of persons changing their business, failing, moving away, new partnerships, &c., &c."[75] In times of economic trouble, the company reminded its correspondents to be especially complete.[76] On their own, local lawyers undoubtedly kept loose tabs on the businesses in their area, but under the patronage and tutelage of the mercantile agencies, they became systematic observers recording long-form descriptions of their neighbors' business dealings and daily lives. Tappan enthused that each correspondent "having his eye upon every trader of importance in his county, and noting it down as it occurs, every circumstance affecting his credit, favorably or unfavorably, becomes better acquainted with his actual condition than any stranger can be."[77] He might have also argued

73. Rowena Olegario, *A Culture of Credit: Embedding Trust and Transparency in American Business* (Cambridge, MA: Harvard University Press, 2006), 57. Sure enough, the ethical validity of Tappan's argument came into almost immediate question and even worked its way through the courts, although never to any particular conclusion. Contemporaries complained especially vigorously about the tendency of credit reporters (and insurers too, for that matter) to systematize and centralize "gossip": talk about drinking habits, church politics, and sexual relations. Thomas F. Meagher made this critique most forcefully in Meagher, *The Commercial Agency "System" of the United States and Canada Exposed* (New York, 1876), 67–73. Gossip about (the supposed failure of) John Beardsley's marriage set off a court battle that imperiled Tappan's fledgling system. See Sandage, *Born Losers*, chapter 6.

74. James D. Norris, *R. G. Dun and Co. 1841–1900: The Development of Credit-Reporting in the Nineteenth Century* (Westport, CT: Greenwood Press, 1978), 22.

75. Quoted in Olegario, *Culture of Credit*, 64.

76. Norris, *R.G. Dun and Co. 1841–1900*, 45–46. What did "complete" mean? According to an 1858 article, "no report is considered full unless it embraces, in regard to each trader, his business, the length of time he has pursued it, his success, or the contrary, his age, character, habits, capacity, means, prospects, property out of business, real estate, judgments, mortgages or other liens upon his property." Quoted in Edward Vose, *Seventy-Five Years of the Mercantile Agency R. G. Dun and Co. 1841–1916* (Brooklyn, NY: R. G. Dun, 1916), 66.

77. Quoted in Lauer, "From Rumor to Written Record," 308.

that such systematic observation set reporters apart from their non-network neighbors.

Yet Tappan's repeated urgings suggest that his network did not naturally take to regular, complete, and systematic reporting. Moreover, when Tappan's company came under scrutiny in a very tiresome, not to mention embarrassing, lawsuit, he blamed a foolish (or corrupt) reporter for supplying the company with bad information.[78] Over time, problems disciplining attorney-observers did not go away. If anything, they became more acute. Toward the end of the century, mercantile agencies employed specialized reporters to oversee their larger networks of correspondents, checking on their work in questionable cases.[79] The watchers had to be watched.

When Charles B. Holmes opened his agency in the 1880s, he faced a new challenge: how could he get lawyers to join his network when he had no work to trade for their information? Tappan's agency offered lawyers jobs as debt collectors in exchange for providing credit information. But Holmes sold information to corporations who had their own lawyers and thus had little or no work to farm out. So instead of promising to refer clients to his correspondents, Holmes paid them. This appears to have been a welcome move—at least in the eyes of some attorneys. "The overburdened ass, the country lawyer, has at last kicked," wrote one of Holmes's correspondents, "and for the future the 'mercantile agency' can only succeed by doing like the rest of the world, 'pay as you go,' as you have always done with me."[80] An added benefit for Holmes in paying cash may have been that it gave him (and his peers) more leverage over their lawyer networks. Such leverage would have mattered, since Holmes apparently expected different skills from his informants than did other mercantile agencies. He advertised "Special trained Inspectors in my employ in all large cities." A competitor promised "Experienced Insurance Inspectors."[81] One suspects, however, that even the carrot of cash did not spare Holmes and his competitors

78. This was the famous Beardsley libel case. See Sandage, *Born Losers*, 179.

79. Lauer, "From Rumor to Written Record," 315. Olegario reports early "traveling agents" who served this purpose, but she notes that they may have been more useful for recruitment than for oversight. Olegario, *Culture of Credit*, 55.

80. "Life Insurance Notes," *Spectator* 63, no. 19 (1899): 211. Holmes may have had the quote printed in return for his consistent purchase of advertising space. We know he must have supplied the quote, so we should take the correspondent's gratitude with some skepticism.

81. "Holmes Mercantile Agency" (advertisement), *Spectator* 63, no. 1 (1899): v; "The Insurance Bureau" (advertisement), *Spectator* 63, no. 1 (1899): v.

from the challenges associated with maintaining networks of professional correspondents.

On the whole, disciplining observational communities took much effort and met with mixed results. Life insurers and credit reporters relied on local professionals, and those professionals resisted being reformed. But they only resisted to a point. In the end, thousands of local doctors and lawyers took corporate money and joined a larger observational community. Even if they did not always fill out forms correctly or answer queries promptly, they succumbed to some amount of discipline. And these were not inconsequential men. Among the lawyers who moved through mercantile agencies were future presidents like Abraham Lincoln, Chester Arthur, Grover Cleveland, and William McKinley. Among those examining for life insurers were the field's most respected figures. One Yale medical school professor bragged of "being the examiner for about ten companies, and having examined about eight thousand applicants."[82] Through changes to professional curricula, corporate and commercial disciplining could even reach those never paid a cent by a life insurer or credit reporter.

The late nineteenth century certainly challenged the traditional professions in a host of ways: it limited their autonomy and may have even provoked "status anxiety."[83] At the same time, some of the new corporate middle class thrilled at the enhanced power they came to wield.[84] But when we look past the overt politics of corporate expansion, when we look to the level of day-to-day practices, we find little room for doubt that corporations, and the men in charge of national corporate networks of professionals, co-opted the traditional professions, while also supporting them financially. Insurers, credit reporters, and other corporations used professionals—whether they resisted or not—to systematically observe and record American society. They enlisted professionals to help make risks.

* * *

---

82. "Editorial," 199.

83. Haskell explores the changing status of traditional professions in Thomas Haskell, *The Emergence of Professional Social Science: The American Social Science Association and the Nineteenth-Century Crisis of Authority* (Baltimore, MD: Johns Hopkins University Press, 2000); on status anxiety, see Richard Hofstadter, *Age of Reform* (New York: Vintage, 1960).

84. Olivier Zunz argues that midlevel managers often willingly gave up the autonomy of the professions or similar positions in exchange for access to corporate power in the mid-nineteenth century. Zunz, *Making America Corporate 1870–1920* (Chicago: University of Chicago Press, 1990), 37–66.

Around the same time that Holmes and Symonds became associated with the life insurance industry, some risk makers started to wonder out loud if co-opting professionals and maintaining national writing networks warranted the time, trouble, and expense. In 1903 MONY reported spending over half a million dollars to pay its medical examiners across the country. Of the portion of each first-year premium set aside for company expenses, medical examinations claimed a quarter of the total.[85] But critics did not just point to medical departments' expenses. They wondered if focusing on *excluding* potential risks made sense at a time when insurance officers obsessed with growth and agents driven by high commissions wanted most to find ways to *include* more Americans. Medical directors had good reason to be nervous about such critiques. American industrial insurers and some British ordinary insurers had already cut back substantially on medical exams and were thinking seriously about eliminating them.[86] Medical directors' tricks for assessing individualized risk, inspectors' extensive files of potential fraudsters, lawyers' reports on ordinary Americans' characters, and doctors' observations of American bodies—the entirety of the industry's writing system—needed to be defended.

The defense began with index cards. Cards first emerged as a useful tool with which leading life insurers could process their records (and keep an eye out for irregularities—New-York Life used cards to sniff out medical examiners who approved a suspiciously large number of applicants who then went on to die in a couple years, or examiners who regularly failed to fill out forms correctly).[87] They could also use cards to pool resources and information, making their correspondent networks more efficient. But over time, cards facilitated an unexpected revolution in risk making. When New-York Life explained its machinery in 1909, it revealed that the "index division" as displayed in figure 3.4 made the entire system possible. Insurance medical directors (banding together to save their departments) teamed up with actuaries (who thought they had exhausted the possibilities for smoothing across large groups) to individualize risk to a greater extent than ever before.[88] Out of challenges to life insurers' systems of writing emerged new

85. New York Legislature, *Testimony*, 2:1737.

86. D. E. Kilgour, "Life Insurance without Medical Examination," *Transactions of the Actuarial Society of America* 22, no. 65 (1921): 120–139, 145–175. See also Alborn, *Regulated Lives*, 246–247.

87. New-York Life, *Temple of Humanity*, 52.

88. Lengwiler claims that actuarial ambitions to edge out medical directors led to some of the important innovations discussed in this final section. While there is some evidence suggesting jostling between the two industry groups, I argue that their cooperative activities were more

3.4. New-York Life's Index Division, in New-York Life, *Temple of Humanity*, 29.

methods for personalizing risk assessments and a new centralized file (a proto-database) for storing thoroughly statisticized individuals in steel case files.

The idea of using cards to reform life insurance writing systems came from the outside. New-York Life, along with Equitable and John Hancock, became pioneer customers of the Library Bureau, a midsize company founded by Melvil Dewey, supplying index card filing systems around the time of Symonds's hiring. Card systems would overthrow the ledger as the dominant office technology at the end of the nineteenth century—just as the computer would overthrow the card system nearly a century later. But first, card makers had to convince corporations and states that new technology had something special to offer. Life insurers, committed to keeping

important. Besides, as chapter 6 will discuss further, actuaries in many cases ended up *more skeptical* than the medical directors of the individualizing of risk calculations. See Lengwiler, "Double Standards."

extensive, individualized records, proved a receptive audience and opened doors to the larger world of finance.[89]

In 1890, the Library Bureau's president had an idea. His company had been working with libraries on what they called "cooperative cataloging." Large libraries, hoping to avoid unnecessary duplications of labor, contracted with the Library Bureau to share copies of bibliographic cards for works found in multiple libraries. Why not do the same with life insurers and their risks? Presented with the offer, Frank Wells, medical director at John Hancock and secretary to the fledgling Association of Life Insurance Medical Directors of America (ALIMDA), agreed.[90] Companies with representatives in ALIMDA—an exclusive group of the nation's most powerful insurers—agreed to share, through the Library Bureau, cards for any and all applicants for insurance who were rejected for medical reasons.

The so-called Rejection Exchange, or Rejection League, amounted to a blacklist. Insurers had toyed with the idea of creating blacklists before, but dropped the idea because of logistical difficulties, resistance to systems that might make it harder to sell insurance, some concerns about individual fairness, and a fear that customers might find out.[91] Holmes' blacklist for rejected claims had only surfaced in 1889. The Rejection Exchange extended that same idea much further. The Library Bureau served as a central clearinghouse to which insurers sent coded notices of rejections and the reasons for rejection, receiving complete files of recent rejections in return. ALIMDA organized the exchange and set its rules and standards. The exchange brought card filing systems into the offices of every major insurer.[92] Insurers' risks became creatures of the card and simultaneously began circulating. The exchange knitted together insurers' correspondent networks for the first time into a wide, if still porous, web. But card systems and a new clearinghouse did not silence criticisms on their own.

Skepticism toward insurance medical men and their networks of doctors made a distinct impression on a young doctor in Symonds's cohort:

---

89. That is indeed the same Dewey of Dewey decimal system fame. Gerri Flanzraich, "The Library Bureau and Office Technology," *Libraries and Culture* 28, no. 4 (Fall 1993): 403–429. See especially 413. For a useful effort to integrate Flanzraich's argument into a larger narrative on the evolution of paper systems for organizing information, see Markus Krajewski, *Paper Machines: About Cards and Catalogs, 1548–1929*, trans. Peter Krapp (Cambridge, MA: MIT Press, 2011), 87–106.

90. Herbert E. Davidson to Frank Wells, 5 May 1890, cited in Flanzraich, "Library Bureau," 406–407, 421, note 19.

91. See, for instance, "Insurance Invalid Corps," *New York Times*, 15 January 1872; and "Caution That Defeats Itself," *New York Tribune*, 27 November 1875.

92. Flanzraich, "Library Bureau," 406–407.

New-York Life's Oscar Rogers. Like Symonds, Rogers came to insurance with many important, nonstatistical skills. He had studied civil engineering at Union College and surveyed for the Delaware and Hudson Canal Company after graduating in 1877. He later did survey work for the Second Avenue elevated railroad and worked for the Parks Department while finishing a medical degree at the College of Physicians and Surgeons.[93] His dissertation had none of Symonds's laboratory focus. Instead he wrote up a case study on, of all things, a tapeworm.[94] After some time in private practice, Rogers took a job as a full-time medical examiner in 1890.[95]

Two years in, Rogers experienced firsthand his and his peers' vulnerability. An angry agent cornered Rogers, a then junior officer in the medical department, and accused him in front of the chief medical director of being too harsh in his judgments of applicants, of rejecting applicants who were in fact better risks than those Rogers recommended accepting.[96] The source of their tension was obvious: Rogers worked for the company directly and wanted to keep it safe from bad risks; his incentives inclined him to err on the side of being conservative. In contrast, the agent's commissions—which could be hundreds of dollars apiece[97]—depended on getting applicants approved. Rogers's frustration and humiliation resonated among his colleagues, who increasingly talked of agents as unscrupulous adversaries.[98]

Rejections became a fascination for medical directors and, more importantly, framed the debate over the utility of insurers' medical networks. By definition, insurers did not know what happened to the people they rejected and could not compare them to those they insured. Resourceful medical directors in the early 1890s, however, found exceptions—they looked for the already insured who applied for more insurance and were rejected,

93. "Rogers, Oscar Harrison," *National Cyclopaedia of American Biography*, 564–565, clipping and Union College Alumni Record, both in Alumni File of Oscar Rogers, Union College Class of 1877, Union College Special Collections.

94. Personal communication with Stephen Novak, Archives and Special Collections for Columbia University Medical Center, 20 September 2012.

95. "Rogers, Oscar Harrison," *National Cyclopaedia of American Biography*, 564–565, clipping in Alumni File of Oscar Rogers, Union College Class of 1877, Union College Special Collections.

96. This story appears in Lawrence F. Abbott, *The Story of NYLIC: A History of the Origin and Development of the New York Life Insurance Company from 1845 to 1929* (New York: New York Life Insurance Company, 1930), 280–282.

97. First year commissions at the turn of the century ran from 60–80 percent of the first annual premium. Stalson, *Marketing Life Insurance*, 528–530.

98. See, for instance, Edgar Holden, "The Object of the Association," *Abstract of the Proceedings of the Association of Life Insurance Medical Directors of America* 6 (1895): 50–57. Porter expands on the tensions between medical directors and agents as they played out in the early twentieth century in "Life Insurance, Medical Testing, and the Management of Mortality."

for instance—and argued from the data that those rejections made sense. At least one actuary—a medical selection skeptic—attempted an overt experiment, accepting a sample of rejected applicants so that they could be tracked by the company's risk-writing machine.[99] Rogers undertook the most aggressive rejection study in response to his 1892 encounter. He and the staff his superiors afforded him for the purpose followed up on (approximately) twenty-five thousand declined policyholders—an enormous task requiring sending out inquiries and, ironically enough, relying on the help of agents, other employees, and correspondents around the nation to determine whether individuals with only a trace presence in New-York Life's card files were still living or when they had died.[100]

All the data Rogers collected by repurposing New-York Life's networks convinced him and his bosses that the medical department did important work. But it also raised new questions. The more Rogers looked at "rejections," the more he became concerned with "impairments," with those characteristics detailed on an application that led to the worst (or best) mortality rates.[101] Rogers had ulterior motives: he wanted to defuse agent-medical examiner relationships by making it possible to reject fewer applicants. His mechanism for this, already well established in American industrial insurance and among British life insurers more widely, was "sub-standard" insurance.[102] Most ordinary life insurers, like New-York Life, rejected around 15 percent of all their applicants, but Rogers calculated that nearly half of those applicants could be approved with some sort of higher premium or other restriction if the excess danger posed to the company by those individuals could only be accurately and precisely priced.[103] To better price

99. See the discussion following Walter S. Nichols, "The Value of Medical Examinations in Industrial Insurance," *Transactions of the Actuarial Society of America* 3, no. 10 (1893): 225–231; and 3, no. 11 (1894): 410–418 at 411–415; Hoffman, *History of the Prudential*, 200–201.

100. Arthur Hunter, "Insurance on Sub-Standard Lives," *Annals of the American Academy of Political and Social Science* 70 (March 1917): 38–53 at 41–42.

101. Oscar H. Rogers, "Medical Selection and Substandard Business," *Abstract of the Proceedings of the Association of Life Insurance Medical Directors of America* 18 (1907): 81–95 at 82.

102. For example, Metropolitan allowed doctors to classify policyholders within one of four classes, a system likely borrowed from British life insurers, and higher premiums were set for those applicants below the first class. "A Meeting of the Medical Examiners for the Industrial Branch of Metropolitan Life Insurance Company," 17 February 1880, 5–7, Folder "Medical Examiners 1880–1972," Box M11, RG/13, MetLife Archives. For an enlarged discussion, see Bouk, "The Science of Difference: Developing Tools for Discrimination in the American Life Insurance Industry, 1830–1930," (PhD diss., Princeton University, 2009), 210–213. On British life insurers' practices, see Alborn, *Regulated Lives*, 271–295.

103. Rogers, "Medical Selection and Substandard Business," 91.

"borderline lives," Rogers blended the two usually distinct analytical traditions housed in life insurers' offices: the actuarial and the medical. He
reformed the way insurers made risks in the process.

Rogers's statistical training came from life insurance, but he achieved a
more catholic education than Symonds. He talked to actuaries—including a
British actuary named Frederick Frankland from a renowned scientific family, hired specifically to help Rogers in his studies[104]—and read their reports
alongside insurers' medical studies. Actuaries' statistical methods—unlike
doctors', which focused on mortality alone—encompassed all policyholders, alive and dead, and attempted to determine the risk that a person of
any given age would die. For that calculation to work, one needed to know
how many people died at a certain age, but also how many continued to
live. Rogers adopted this technique and paired it with the discriminating
instincts of a medical director to ask new questions of his company's "experience," its vast collection of risks now written on paper cards. Rogers
directed "a systematic study of insured lives, taking them from our records
according to build [height and weight], family history, occupation, personal
history, and the like, and studying mortalities which occurred in the various classes."[105] Doctors classified while actuaries averaged—Rogers saw the
power in doing both at once. New-York Life soon saw the value too, and in
1896 sold its first substandard life insurance policy—by 1906, substandard
policies made up more than 10 percent of the company's business.[106]

In the 1890s, Rogers had made insurers' data more adaptable to individual difference. He had made it, in other words, a better tool for writing personal stories about each risk. Emory McClintock, the most widely respected
actuary of his day, took to the floor in Paris on the twenty-eighth of June
1900 at the Third International Actuarial Congress to call actuaries to do
the same. He urged his peers to spend less time aggregating and more time
classifying. "It is more important for the future interests of life insurance,"
he said, "to learn how fishermen compare with farmers, how physicians
compare with clergymen, how brewers compare with manufacturers of
soda-water, and the like, than it is to gather together all these heterogeneous

104. Frankland's father, the chemist Sir Edward Frankland, belonged to the famous "X-
Club," a group of natural scientists including T. H. Huxley. Abbott, *Story of NYLIC*, 280–281.

105. Rogers, "Medical Selection and Substandard Business," 82.

106. Rogers cites 130,400 substandard policies with $228,253,000 in insurance. Rogers,
"Medical Selection and Substandard Business," 92–93. In 1908, the company carried a total
of just under $2 billion in insurance and 978,209 policies. New-York Life Insurance Company,
Annual Report (1908): 36.

materials into one grand average in the form of a new life table."[107] Actuaries had been model sharers of data, with important cooperative studies stretching back to the 1840s in Britain.[108] But most cooperative studies had been intended to create general laws of mortality applied to the average policyholder. McClintock proposed new cooperative efforts that drew on many companies' shared data with the intent of shedding new light on "special classes of policies": those policies on people engaged in hazardous occupations, residing in "special localities," covered by exotic insurance plans, possessing disadvantageous "personal characteristics," or burdened by poor "family records." In the process, actuaries could join medical directors like Rogers in improving medical selection and, presumably, also make possible the accurate pricing of subprime risks.[109]

McClintock's American colleagues listened carefully and commissioned a study to do what McClintock proposed—an action that Oscar B. Ireland, president of the Actuarial Society of America, judged "one of the most important steps which we have ever taken in the history of our Society."[110] The *Specialized Mortality Investigation*, published in 1903, encompassed ninety-eight "classes of risks"—classes that ranged from individuals carrying more than $20,000 in insurance, to sawmill workers, to railway mail clerks, to asthmatics, to those whose siblings died of cancer, to residents of Warren County, Mississippi.[111] This, according to McClintock, was the future of actuarial science. Actuaries became more invested in insurers' correspondence networks, the data they produced, and the theory of risk they relied upon. They laid the groundwork for a proliferation of risk classes and the more precise personalization of risk rating.

Rogers and his fellow medical directors, for their part, embraced the idea that all lives should be understood in terms of individualized risk evaluations. That embrace came in the context of an extension and reimagination of their blacklist. The Rejection Exchange had worked well for years, but never without friction and internal controversy. Medical directors battled

107. Emory McClintock, "On the Objects to Be Attained in Future Investigations of Mortality and Death Loss," *Transactions of the Actuarial Society of America* 6, no. 24 (1900): 373–379 at 374.

108. On early actuarial collaborations, see Alborn, "Calculating Profession."

109. McClintock, "On the Objects," 373–374.

110. Actuarial Society of America, "Abstract from the Minutes of the Annual Meeting," *Transactions of the Actuarial Society of America* 6, no. 24 (1900): 449; Actuarial Society of America, "Abstract from the Minutes of the Annual Meeting," *Transactions of the Actuarial Society of America* 7, no. 25 (1901): 64–69. Ireland quoted from page 68.

111. Actuarial Society of America, *Experience of Thirty-Four Life Companies upon Ninety-Eight Special Classes of Risks* (New York: Actuarial Society of America, 1903).

with one another over how or whether to share more individual files.[112] They also argued over the definition of a "medical" cause for rejection, since many of the characteristics that went into a doctor's decision to reject (domicile, reputation, etc.) were seldom discussed in medical school.[113] And they faced a constant struggle to keep the exchange secret.[114] In 1902, Symonds and Rogers, now leading men, instigated a reformation.

A new Medical Information Bureau or "M.I.B." replaced the Rejection Exchange and responded to its faults. M.I.B. expanded the amount of data in circulation enormously: where companies had previously only shared cards for those individuals whom they chose to reject, they now shared cards for all individuals—even policyholders—burdened by any "impairment." As a category, impairments belonged to insurance and as such avoided the "medical" problem: they included all characteristics generally considered a problem by medical directors, encompassing any class of risks, without worrying too much about medical tradition. If there were any questions, the Special Committee on Medical Information Bureau, which included a handful of doctors with Symonds among them and Rogers as chair, standardized the names and codes for all impairments.

M.I.B. shifted the focus away from the extremes, from the 15 percent who might be rejected, toward the broad middle—that majority of the population possessing some impairment (a history of tuberculosis in the family, overweight, a taste for beer, childhood illness, traces of albumin in their urine) who nonetheless qualified for insurance. It democratized surveillance and simultaneously made secrecy all the more important. Rogers, Symonds, and their colleagues believed that many Americans would find this power threatening if it were more widely understood. So they determined to protect themselves and their data. After voting to create M.I.B., the Association of Life Insurance Medical Directors incorporated, with Symonds and Rogers serving as two of the four directors. The *New York Times* article explaining the incorporation did not mention M.I.B.[115] The Rejection Exchange had valued secrecy, but M.I.B. belonged to another league. Members were strictly forbidden from referring to M.I.B. at all, when talking or corresponding with other companies or with applicants. We would expect such measures. But

---

112. "Record of M. I. B. In the Proceedings of the Association of Life Insurance Medical Directors from May 1892- to December 1916," 1 in Folder "Medical Information Bureau 1916," Box RG/13-Subject Files M11, MetLife Archives.

113. "Record of M. I. B.," 4.

114. "Record of M. I. B.," 1–5.

115. M.I.B. did quite a good job of avoiding the newspapers. "Life Insurance Medical Men Incorporate," *New York Times*, 11 July 1902.

the medical directors went one step further. Companies risked penalties if a medical director talked about M.I.B. even with his own medical examiners, his company's agents, or its managers.[116] The special committee understood the value of the risks that M.I.B. circulated and wanted to keep impairment data exclusive to members. But they also understood just how disastrous it would be if too many people understood precisely what they were up to.

* * *

The ultimate decision makers in a late nineteenth-century life insurance company—those at the heart of corporate power—resided in finance or investment committees that met in rooms like the one pictured in figure 3.5. Often chaired by the president and including the most powerful trustees, investment committees decided how to invest the enormous capital that their companies accrued. That capital correlated directly with insurance in force: the more insurance, the more to invest. Investment committees' power in the wider world correlated directly with the capital they controlled, with the capital they could funnel to friends in investment banks (with whose leaders they had personal relationships)[117] to prop up new railroad bonds, to finance mergers, or to support new trusts (with whose directorates they often interlocked).[118]

Medical directors and actuaries understood this, as did everyone else in the industry. They knew that, in the end, company executives often cared less about profit (indeed, a number of the leading companies were "mutual," and essentially acted as nonprofits) than about volume and growth. Executives wanted more policyholders, larger reserves, and no distractions from achieving those ends.

The stewards of life insurers' correspondent networks wanted, above all, to attract as little finance committee attention as possible. Rogers and his peers hoped in the end to make a risk-making system that would not stimulate complaints from agents or policyholders. One New-York Life pamphlet, focusing on medical department activities, touted the company's new commitment to "the rating of ALL lives . . . a method which could be formulated

116. "Record of M.I.B.," 7–8.

117. New-York Life's finance committee chair, George Perkins, doubled as a J. P. Morgan partner, to take the most egregious example. Keller, *Life Insurance Enterprise*, 177.

118. Keller writes: "With involvement in high finance as much an expression of personal executive desires as of inescapable business requirements, it was difficult for these leaders to draw a line between private and corporate interest. Officers and directors were involved as individuals in many of the ventures to which they committed their companies' funds." See *Life Insurance Enterprise*, 143.

3.5. New-York Life's Meeting Room of the Finance Committee, in New-York Life, *Temple of Humanity*, 16.

and expressed in rules and be quickly applied to any life."[119] Such rhetoric emphasized mechanical fairness. The medical directors hoped that they could not be blamed for simply following the rules. The sharing of risk data across companies through M.I.B. could also decrease the likelihood of embarrassing situations where one company accepted a risk as standard that another rejected.[120]

Rogers, Symonds, and McClintock all hoped to make risks faster and at less cost (which was all the finance committees desired) in a way that preserved life insurers' writing networks. With the aid of card technologies, the M.I.B., and improved statistical insights into reading individualized risk off of a paper person, life insurers could more quickly determine "automatically" (meaning they used trained clerks instead of doctors) which of their

119. New-York Life, *Temple of Humanity*, 49.
120. Porter elaborates on these advantages of relying on rules in "Life Insurance, Medical Testing, and the Management of Mortality."

thousands upon thousands of applicants warranted deeper scrutiny and which could pass through quickly. As an added bonus, these new methods promised to make even more applicants ultimately insurable through substandard insurance.

Efforts to individualize and standardize risk making—and, when possible, automate it—on a national scale renewed the lease on life for the industry's writing networks because it did what executives wanted. It increased the speed and efficiency of medical selection. It limited distractions in the form of griping policyholders or agents. And it opened life insurance to a wider swath of the population. In short, it made it possible for medical directors to go on making more and more of the nation into risks, while investment committees went on centralizing economic power. That more Americans became more widely and intensively interpreted through the statistical lenses of the impairment and risk factor was a secondary effect.

Then, in 1905, the centralization of economic power in life insurers' offices became a major distraction in itself.

Smoothing

On the stand in New York City Hall's Aldermanic Chamber sat Richard McCurdy, president of MONY and a key witness in the 1905 state investigation (commonly called the Armstrong investigation) into life insurance companies' Gilded Age misdeeds. But McCurdy refused to cooperate. "I couldn't say" or "I haven't any recollection of it"—those were his refrains. Even: "I have never formulated an idea in my mind on the subject."[1]

The investigation's "chief inquisitor"—as his enemies labeled him—was Charles Evans Hughes, a New York lawyer whose primary qualification was that he had (miraculously) escaped prior employment by a life insurer.[2] It was his job to look into charges that life insurers had in recent years used policyholder money (an extraordinarily large and growing pot of it) improperly to inflate their own salaries, buy political influence, serve the needs of big investment banks, and otherwise betray the public's trust—charges that he would, more or less, substantiate.

Hughes put insurers on trial for, essentially, corrupting American capitalism. The corruptions Hughes revealed could be obvious, like those that the financier-turned-muckraker Thomas W. Lawson illustrated in his *Everyday Magazine* exposés (later collected as *Frenzied Finance*).[3] Lawson invoked what he called the "System," a structural corruption of finance that led to the pooling of capital in the hands of a tight-knit cabal who gambled

---

1. New York Legislature, *Testimony*, 2:1303–1304. Two printers made editions of the testimony. Succeeding references to the Lyon editions in this chapter will use the shortened form "Vol. X, pp."

2. "Philbin or Rand May Run Insurance Inquiry," *New York Times*, 4 August 1905.

3. Thomas W. Lawson, *Frenzied Finance: The Crime of the Amalgamated*, vol. 1 (New York: Ridgway-Thayer, 1905).

with the masses' money for their own benefit and purchased politicians to protect themselves and their schemes. Lawson's allegations, in no small part, pushed Hughes into the Aldermanic Chamber in 1905.[4] Some very important corruptions, however, took on more complicated and obscure forms than those that Lawson revealed. The "System" had many twists and turns.

One of those twists drew Hughes's extended attention after his encounter with McCurdy. As had happened repeatedly in the nineteenth century, life insurers' efforts to make risks out of individuals raised questions and objections. But this time, "smoothing," or averaging (rather than classing, fatalizing, or writing) became the subject to be interrogated, while actuaries (rather than medical directors, statisticians, or correspondents) had to explain themselves. Emory McClintock, the most celebrated actuary of his day, made the case for smoothing. Insurers smoothed data all the time, he explained. They smoothed mortality experience to better predict future death rates and to retrospectively fix past predictions. They smoothed earnings to predict their corporate finances. They considered smoothing to be the most natural thing in the world. And they saw no reason why their enterprises should not base their dealings with individuals on a smoothed reality.

Hughes, New York legislators, journalists, and policyholders did not consider it nearly so natural to favor risk over circumstance, smoothed reality over lived reality. They questioned, doubted, and condemned to varying degrees. Smoothing belonged to the making of risks—as much or more than classing, fatalizing, or writing. So when Hughes and other outsiders called smoothing into question, they could not help but wonder: did corporate risk making corrupt capitalism?

* * *

The question that MONY president McCurdy refused to answer implied one of the more obvious corruptions of capitalism. Hughes wanted to know: why had McCurdy's salary gone up from $100,000 in 1900 to $150,000 in 1901, while the dividends paid to policyholders fell (and kept falling)?[5]

4. For the best treatment of the Armstrong hearings and their causes, see Keller, *Life Insurance Enterprise*. Even more instrumental than Lawson's muckraking was a messy and public fight to control one massive insurer's (Equitable's) vast assets following an ill-advised ball thrown by the son of Equitable's founder. See Beard, *After the Ball*.

5. Vol. 2, 1303–1304, 1399–1400, 1408. In a wonderful exchange, McCurdy launched into a long discussion of the "missionary" character of a mutual life insurance company, to which the concept of "profit" was entirely foreign. Hughes replied: "Well, you have made a very full explanation and treated it as a missionary enterprise. The question comes back to the salaries

In other words, was McCurdy, who held the purse strings of a gargantuan institution—it reported over $400 million in assets, stood first in the world among "moneyed corporations" and bragged that it possessed more capital than all the banks of Manhattan combined[6]—stealing from policyholders to pay his salary? Hughes believed, with many others, that the accumulation of dividends facilitated executive misdeeds—that it gave men like McCurdy giant slush funds to do with as they pleased.[7] Here in his "inquisition chamber,"[8] Hughes tried to determine if executive salaries were rising at the expense of those dividends, but his bigger question was: were executives misusing policyholder money? These were the stakes behind the question that McCurdy refused to answer.

What opened McCurdy to such charges was a system built by insurers that required the payment of dividends to policyholders (instead of or in some cases alongside dividends to stockholders). By the turn of the century, those dividends had become an absolutely essential tool in insurers' quest to lure policyholders away from competitors. The most popular form of insurance at the time charged policyholders a premium assumed to be unnecessarily high and then paid regular dividends back to the policyholders either in cash or as additions to insurance, making insurance policies serve as savings instruments. The driver of life insurance company growth over the last two decades had been a special form of insurance called a deferred dividend policy,[9] which accumulated those dividends for a period of ten, fifteen, or twenty years. Anyone who died or lapsed his policy before that period forfeited the accumulated dividends to the remaining policyholders. Investors in these policies enjoyed the safety of life insurance; no matter what, their policies promised to support dependents or pay off creditors if the policyholder died. They simultaneously enjoyed a speculative thrill: if

---

of the missionaries?" Vol. 2, 1412. On McCurdy's ideology of insurance as philanthropy and similar ideas held by his peers, see Keller, *Life Insurance Enterprise*, 26–33.

6. Mutual Life of New York, *Annual Report* (1903): 17–19.

7. Clough offers a nice summary of the investigation's aims in *Century of American Life Insurance*, 222–224.

8. "Belief in the People Is Hughes's Reliance," *New York Times*, 27 September 1906.

9. Beginning in the 1880s, deferred dividends became essential to a company's growth. One of the few companies to abstain on principle from selling them, Connecticut Mutual, went from being the second largest company in 1870 to a minor insurer as a result of the decision. Stalson, *Marketing Life Insurance*, 487–490. Roger L. Ransom and Richard Sutch estimate that two-thirds of all life insurance in force in 1905 was on a deferred dividend plan, amounting to something like nine million policies at a time when there were eighteen million households. See Ransom and Sutch, "Tontine Insurance and the Armstrong Investigation: A Case of Stifled Innovation, 1868–1905," *Journal of Economic History* 47, no. 2 (1987): 379–390 at 385–386.

they could live out the twenty-year term and somehow, in the age of panics recurring every decade, avoid going broke and defaulting on their policies—if they could win, and the odds for winning weren't bad—they won
their own savings plus a share of all the losers' savings.[10] Insurers nearly
always won in this deal, since with deferred dividend policies they got to
hold on to (and invest) those savings for a very long time. And during that
time, Richard McCurdy got a raise.

McCurdy insisted that he had no knowledge of how his salary was set—
that job belonged to a mysterious subcommittee of the board of trustees.[11]
He further claimed to know very little about what he called the "abstruse"
question of dividends. When Hughes pressed him for details, McCurdy pro-
tested: "You are trying to prove me a fool."[12]

The domain of the "abstruse" belonged to MONY's actuary, Emory Mc-
Clintock. Hughes set out looking for one kind of capitalist corruption: steal-
ing from the small for the sake of the big. But, through McClintock, he
stumbled upon what looked to him like another kind of corruption, one
born of corporate practice married to a probabilistic, statistical worldview
played out in individual lives.

* * *

In his first two mornings on the stand, during which he fielded friendly questions from the associate counsel for the investigation, McClintock delved

---

10. Ransom and Sutch insist it was a fair bet, with odds set actuarially and only a small
percentage off the top for administration. Ransom and Sutch, "Tontine Insurance and the Armstrong Investigation," 389. It may have become a better bet with time. When Equitable pioneered it, its president expected two out of three policies to lapse before dividends would be
distributed, increasing the final pot. But the disincentive of losing one's dividends joined to the
appeal of a big payoff appears to have helped lower the rate of lapse (and thereby made the final
pot smaller). Buley, *American Life Convention*, 1:94–96.

11. Vol. 2, 1416–1418. McCurdy, after much pressing and with the aid of his own counsel,
had mounted a halfhearted justification of his raise, noting the growth in company assets, insurance in force, and a variety of other measures of company size and strength from the mid-1880s
when he took over the presidency through his raise in 1901. McCurdy noted that the company
"was practically stationary, if not retrograding" when he took it over. Of course, his decision to
move to selling deferred dividend policies must have played a large role in subsequent growth.
Vol. 2, 1399–1407.

12. Vol. 2, 1433. Contemporary newspapers often reported McCurdy's behavior straight,
but some attributed it variously to ignorance, spite, or savvy. A *Wall Street Journal* reporter wrote:
"Actuary McClintock, to whom Mr. McCurdy, when on the stand, always referred when asked
any hard question"; the *Salt Lake Herald* gave him some credit for his responses: "When pressed
for an answer, Mr. McCurdy had repeatedly 'declined to discuss the question.' He did not admit,
however, that he did not know." "Why Dividends Grow Less: Actuary McClintock Put through
His Paces," *Salt Lake Herald*, 25 October 1905.

into technical details. He explained, crucially, that MONY's investments had suffered from decreasing returns for years and that this accounted for much of the drop in dividends.[13] He also touched on the difficulties of actuarial work and the centrality to actuarial practice of "smoothing" or averaging data. In smoothing lay the heart of what for Hughes seemed a new corruption of capitalism.

Smoothing helped life insurers set their initial rates, but it also did crucial work in the calculation of dividends, a highly structured process governed by a rigorous method known as the "contribution plan." The contribution plan, by then a de facto industry standard, required the actuary setting the dividend to calculate for individual classes of policyholders and for all policyholders the amounts saved when (1) fewer people died than expected, when (2) the company spent less than its allocated expense budget in a given year, and when (3) investments returned more than the expected 4 percent.[14]

But actuaries, McClintock said, did not use in their calculations (1) the actual number of deaths or lapses or (2) accounts of actual expenses or even (3) the actual rate of interest on investments for any given calendar year. Instead, for each of those elements the actuary used a factor he judged to express the average over a range of years. As McClintock put it, "I can only state this, that in estimating these percentages for lapses and losses, our experience is not uniform, our experience varies from year to year, and those policies are issued at all times in the year, some on the 1st of January, and some on the last day of December, so we cannot go according to the experience of calendar years."[15]

Such practical difficulties with no easy, precise solution made smoothing attractive. As McClintock explained, it gave actuaries a method for deciding

13. In 1865, MONY counted on a 6.5 percent return on investments. By 1885, that had fallen to 6 percent; by 1890, 5 percent; down to 4.6666 percent around 1902 and 4.3 percent in 1903 and finally down to 4.1 percent in 1904. Vol. 2, 1735. Lester Zartman's independent analysis came up with slightly higher average numbers but expressed the same trend. His numbers may have been higher because they excluded the costs of investment. Nevertheless, he cites an average across companies of 6.5 percent for the early seventies, 6 percent for the late seventies, 5.5 percent for the eighties, 5 percent for the nineties, and 4.5 percent for the early 1900s. Lester Zartman, *The Investments of Life Insurance Companies* (New York: Henry Holt, 1906), 82. Similar trends troubled British life insurers around the same time. See Alborn, "First Fund Managers," 86.

14. On the contribution plan, see Homans, "On the Equitable Distribution of Surplus." For evidence of its ubiquity, see Massachusetts Insurance Department, *Annual Report of the Commissioner of Insurance of the Commonwealth of Massachusetts* (1868), reprinted in Mutual Life Insurance Company of New York, *Annual Report* (1869): 26.

15. Vol. 2, 1805.

how much of the dividend each policyholder had "contributed" to the over-all surplus (calculated for the calendar year) when policyholders bought policies throughout the year.

Calculations presented a problem too. Even in an era and industry benefiting from the rise of fast mechanical calculators and tabulators, it still took months to fix dividends for MONY's more than six hundred thou-sand policies.[16] Actuaries like McClintock lacked the luxury of waiting for exact data. Indeed, since dividend calculations began every November so they could be announced early the next year, the actuary could not possibly acquire complete data about the previous year to rely on in making his calculations.[17]

And yet, practical necessity did not—all by itself, or even primarily—provoke actuaries to practice smoothing. Emory McClintock, like many of his colleagues, valued smoothing as a good in itself—one rooted in deep actuarial traditions.[18] Actuaries emerged in the late eighteenth and early nineteenth century as another of a long series of productive blendings of finance and science.[19] So too, smoothing betrayed bifurcated and tangled roots, having one foot in banks' books and the other in the stars.

The Bank of England pioneered smoothing (as a state-sanctioned art of "financial subterfuge," explains Judy Klein) in 1833.[20] Faced with members of Parliament eager to rein in the bank by forcing it to divulge the exact amounts of its "treasure," bank officials struck a deal that allowed them to present purposely muddled data. Bank officials worried that bank deposi-tors would panic every time they saw the bank's not infrequent down and up fluctuations. But by publishing every month the average bullion hold-ings of the previous three months—what we'd call a moving average—bank

16. On life insurers as a key market for (and force shaping the development of) mechani-cal tabulators, and then electronic computers, see JoAnne Yates, *Structuring the Information Age: Life Insurance and Technology in the Twentieth Century* (Baltimore, MD: Johns Hopkins University Press, 2005). As of 31 December 1904, MONY had 659,344 policies and just over $1.5 billion of insurance in force with assets of $440 million and a $71 million surplus. See *Report of the Joint Committee of the Senate and Assembly of the State of New York: Appointed to Investigate the Affairs of Life Insurance Companies* (New York: M. B. Brown, 1906), 4.

17. Vol. 2, 1840.

18. Rufus Weeks of New-York Life, for instance, used different methods but still relied un-abashedly on smoothed data. Vol. 2, 1126–1127.

19. Jacob Soll points to a number of interesting transfers of note-keeping practices between finance, state administration, medicine, and science in "From Note-Taking to Data Banks."

20. Ironically explains Klein, economists in later years would adopt smoothing methods—as would actuaries—not to deceive or obfuscate but as "a theoretical model of how things really work in a capitalist economy." Judy L. Klein, *Statistical Visions in Time: A History of Time Series Analysis, 1662–1938* (New York: Cambridge University Press, 1997), 73–102 at 102.

officials smoothed their jumpy data and aimed to calm the watching public's jitters, hopefully averting panicked withdrawals in the process.[21]

Around the same time, smoothing practices also appeared in astronomy. John Herschel led his fellow astronomers in the 1830s in adopting graphical methods, relying on a recently invented paper technology—graph paper—to help him determine celestial orbits from data comprising few points of disparate quality.[22] Herschel plotted his data and then drew a line through his data points, fitting his curve closest to the strongest data, while insisting that the true curve would exhibit "large and graceful sinuosity, which must be maintained at all hazards."[23] Herschel expected natural laws to behave regularly and gradually and he analyzed accordingly.

Astronomers' efforts to reveal natural laws appear to have shaped actuarial practice more than bankers' essays at obfuscation, but both sources undoubtedly influenced later smoothing. Figure 4.1 from around 1861 points to the ways that actuaries in Britain and then America adopted astronomers' graphical methods for constructing mortality curves.[24] The figure shows an American actuary's hand-drawing of a new mortality curve (in red and titled Homans, see the cutout for an enlargement) through a maze of data and existing tables.[25] When Samuel Brown, an eminent British actuary, evaluated the resulting mortality curve, he praised it for being "admirably graduated."[26]

By the time of the Armstrong investigation in 1905, actuaries seldom hand-drew curves. Just as astronomers largely shifted to analytical methods for approximating curves from data, actuaries in the second half of the nineteenth century in Britain and then the United States turned to "Makeham's

---

21. "Smoothing was born," argues Judy Klein. But she also admits that smoothing likely had a near simultaneous alternate birth in the observational sciences. On smoothing's origins in banks and finance, see Klein, *Statistical Visions in Time*, 73–102, especially 88, 95.

22. Thomas L. Hankins, "A 'Large and Graceful Sinuosity': John Herschel's Graphical Method," *Isis* 97, no. 4 (December 2006): 605–633, see especially 606–610.

23. Hankins, "A 'Large and Graceful Sinuosity,'" 610.

24. On the complex interconnections between astronomy and actuarial practice in Britain, see William J. Ashworth, "The Calculating Eye: Baily, Herschel, Babbage and the Business of Astronomy," *British Journal for the History of Science* 27, no. 4 (1994): 409–441; and Alborn, "Calculating Profession."

25. This is the table that would become the American experience table—see figure 1.1. The table marked "Mutual Life—C. Gill" had also been drawn by hand years earlier, as the actuary Charles Gill explained in his accompanying notes. See *Gill's Table*, Box 1515, MONY Papers. Undoubtedly, many of the others had been smoothed by the same method.

26. "For the Private Use of the Trustees of Mutual Life Insurance Company of New-York," bound into Minutes of the Insurance Committee (Book 1): 379–380, MONY Papers. For more context, see Bouk, "Science of Difference," 144–148.

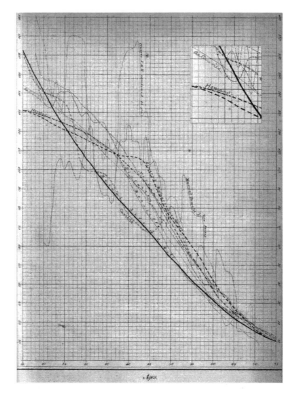

4.1. An actuarial smoothing, inserted into Minutes of the Insurance Committee (Book 1), MONY Papers. The horizontal axis shows increasing age of the policyholder, while the vertical axis shows the number of policyholders out of which one would be expected to die in any given year. Thus, the lower the likelihood of death at any age, the higher the value on the vertical axis. Enlarged cutout added by author to highlight the hand-drawn curve labeled "Homans." Courtesy of AXA.

law of mortality," a mathematical formula that could be fit to various collections of mortality experience.[27] Yet actuaries' reverence for sinuosity persisted even as their methods changed. One of the great advantages of Makeham's law was that it consistently produced smooth gradations.

The art of smoothing had a profound effect on McClintock's understanding of his job and of the "economic chance-world" he inhabited.[28]

27. See Alborn, "Calculating Profession," 458–461.
28. The phrase belongs originally to William Dean Howells's *A Hazard of New Fortunes* (New York: Harper and Brothers, 1889), but Jonathan Levy has recently reintroduced it as one way Americans thought about nineteenth-century capitalism. See Levy, *Freaks of Fortune*, 17.

McClintock explained his data smoothing this way: "It is part of our business to prevent fluctuation, and make things go as smoothly as possible."[29]

* * *

Unfortunately for McClintock, the early afternoon of his second day of testimony did not go smoothly. The investigation's chairman, Senator William W. Armstrong, pushed Charles Evans Hughes to replace his softball-lobbing colleague (who was rewarded for his incompetence with a job at MONY after the investigation).[30] "You must take charge of this," Armstrong told Hughes.[31] Hughes did. He sped up the questioning. Where McClintock's first day and a half involved gentle questions and very long answers, he now nearly stopped talking. Hughes asked a question, usually only a sentence long, maybe two. McClintock answered with one or two words, occasionally a sentence. The pace quickened and the long, lazy morning turned busy, staccato. When McClintock pushed to say more, when he broke rhythm, it meant that Hughes had jarred him.

In his first day and a half, McClintock played a role just then appearing in American life: the rational, salaried manager. Timothy Alborn has previously referred to actuaries in mid-nineteenth-century Britain as "the first fund managers."[32] But by the turn of the century in the United States, actuaries no longer ran their companies. Financiers did that, while actuaries attended to the "abstruse" details. Speaking anachronistically, McClintock was a quant, not a fund manager.[33]

McClintock played the role well. He gave detailed, "technical" answers (a word frequently used in the chamber and in the papers to talk about

29. The full quote reads: "It is part of our business to prevent fluctuation, and make things go as smoothly as possible in order that policyholders shall get their results according to the average cost, and not according to chance annual fluctuations that take from the average." Vol. 2, 1805.

30. Keller, *Life Insurance Enterprise*, 251.

31. Quoted in Merlo J. Pusey, *Charles Evans Hughes* (New York: Macmillan, 1951), 1:145.

32. Alborn, "First Fund Managers." On the status of actuaries, see also Theodore M. Porter, "Precision and Trust: Early Victorian Insurance and the Politics of Calculation," in *The Values of Precision*, ed. M. Norton Wise (Princeton, NJ: Princeton University Press, 1995), 173–197.

33. The term "quant" derives from quantitative trader or analyst and applies to a wide range of individuals, but especially to mathematically sophisticated analysts. By the turn of the twentieth century, some quants had actually become the bosses and managers. With the financial crisis of 2007, the public rather suddenly discovered the quants. For an overview, see Scott Patterson, *The Quants: How a New Breed of Math Whizzes Conquered Wall Street and Nearly Destroyed It* (New York: Crown Business, 2010).

McClintock's testimony),[34] but he also salted his time on the stand with not a few references to his personal quirks. McClintock depicted himself as a kind of iconoclast or even, in his words, "a crank," while betraying a fascination and love of Britain—yet his crankiness and Anglophilia only made his character more convincing. He looked like an independent thinker who inhabited a cosmopolitan network of experts. He had the credentials to back that image up: McClintock brought to the stand thirty-seven years of experience and the respect of fellow actuaries around the world. He had done graduate work in chemistry at Göttingen University in Germany; received honorary degrees from Columbia, Yale, and Wisconsin; and published many mathematical papers. He also helped found the American Mathematical Society in the 1890s.[35] A fellow mathematician deemed him "practically alone among notable American contributors to pure mathematics" outside the universities.[36] Playing the managerial role gave McClintock real advantages, not the least of which was conferring on him the obscurity that comes with technicality—the way that the technical hides in plain sight. This obscurity helped actuaries *and especially their risk-making methods* survive the investigations (marginally) better than financiers or other officers and their "System" with a capital S.[37]

McClintock's testimony contrasted sharply with that of McCurdy, who, in his obstinacy and ignorance had played to a T the role of the heedless financier. We know these roles today because Alfred Chandler immortalized them in his history, *The Visible Hand*, in which he explained how rational, salaried managers staged a managerial revolution that swept financiers out of their corner offices.[38] These were not, however, stable roles. Technically,

---

34. See, for instance, vol. 2, 1698, or "More about Dividends," *Boston Evening Transcript*, 25 October 1905.

35. Thomas S. Fiske, "Emory McClintock," *Bulletin of the American Mathematical Society* 23, no. 8 (May 1917): 353–357; Walter S. Nichols, "Dr. Emory McClintock as a Great Creative Mathematician—The Calculus of Enlargement," *Transactions of the Actuarial Society of America* 17, no. 56 (1916): 290–302; William A. Hutcheson, "In Memoriam: Emory McClintock," *Transactions of the Actuarial Society of America* 17, no. 56 (1916): 373–381. On McClintock and the American Mathematical Society in its larger context, see Andrew Fiss, "Professing Mathematics: Science and Education in Nineteenth-Century America" (PhD diss., Indiana University, 2011), 92–96.

36. Fiske, "Emory McClintock," 355.

37. "Technical" meant simultaneously (1) difficult to understand and (2) belonging to a specialized trade or profession and therefore outside the realm of knowledge that ordinary individuals would be expected to understand. See Theodore M. Porter, "How Science Became Technical," *Isis* 100, no. 2 (2009): 292–309 at 293–294, 298.

38. Alfred D. Chandler, Jr., *The Visible Hand: The Managerial Revolution in American Business* (Cambridge, MA: Belknap Press of Harvard University Press, 1977).

McCurdy and McClintock were both salaried managers and McClintock's $25,000 a year salary—while only a sixth of McCurdy's—still amounted to an extraordinary five to ten times the salary of an American worker doing very well. Hughes, for his part, tried and often succeeded (as the papers judged it) in making McClintock look less than rational. Hughes admitted no fundamental difference between McCurdy and McClintock.

As Hughes sped things up, he troubled McClintock's managerial presentation. He also refocused the examination on the question he had asked Richard McCurdy: why had salaries gone up and dividends gone down? McClintock could not speak to salaries, but he could speak to dividends—and the conversation veered again back to smoothing.

* * *

Under Hughes's insistent questioning, McClintock admitted some less flattering reasons for actuaries' recourse to smoothing, one of which involved various problems with MONY's corporate statistics. Asked by Hughes what the actual cost of maintaining company investments was, he replied: "Well, nobody can tell." Asked why, he said only, "Because there is no account kept of it."[39] MONY, one of the great corporations of its day, a model of corporate organization, did not bother to keep tabs on its actual expenses. State regulators sanctioned and enabled such ignorance by allowing companies to follow general accounting rules for estimating those expenses instead. These were the realities faced by and created by Chandler's rational managers. As Richard White has shown in his history of the transcontinental railroads around this same time, the most advanced American corporations with the most extensive, professional managements still had a surprisingly tenuous grasp on their own inner workings.[40]

McClintock also revealed his appreciation for smoothing's deceptive powers. Smoothing protected actuaries from consumers, he argued, who might not understand or tolerate the ups and downs that the company actually experienced. "Dissatisfaction" was McClintock's word for what would result if an actuary did not smooth.[41] And even so well respected an actuary as McClintock knew that he needed to worry about consumer attitudes.[42] If

---

39. Vol. 2, 1848.

40. Richard White, *Railroaded: The Transcontinentals and the Making of Modern America* (New York: Norton, 2011), especially chapter 6 and the story of the accountant William Mahl in 270–277.

41. Vol. 2, 1842.

42. McClintock knew the story of a previous MONY actuary's ouster all too well. Sheppard Homans, the most eminent actuary of his day, ran afoul of Frederick Winston, MONY's

too many policyholders complained, the president or board of trustees—
who wanted, for the most part, to be left alone to use company money in
the world of big finance—might get rid of him.

Yet smoothing had its limits in appeasing or confusing policyholders.
Smoothing worked, until it didn't. The preponderance of critical letters sent
by MONY policyholders to Hughes came from men insured with a kind of
policy that required only ten or twenty years of payments. After premium
payments ended, dividends dropped sharply. Having been accustomed to
seeing gradual movements in their dividends, policyholders found these
drops alarming. And the alarmed parties often realized that their dividends
had been gradually decreasing for years. One "eminent man" in New York,
according to Hughes, explained that he was a "busy man" who "paid no attention" to his dividends. But even he noticed when "they dwindled down
to an absurdedly [sic] low figure."[43] A review of the dividend histories of
those who complained reveals a basic rhythm of deferred dividend policies:
remarkable stability or slow growth for multiple years punctuated by big
adjustments downward.[44]

Still, in the end, McClintock insisted that actuaries smoothed because
smoothing was a "mathematical and ethical" good. McClintock defended
his practices to Hughes on "moral" grounds.[45] He could have made the
same argument with respect to smoothing mortality experience in order to
generate premium rates or justify discriminations among different groups,
but Hughes cared about dividends, so McClintock made his case for that
context.[46] Setting dividends required smoothing both mortality experience
and corporate earnings. With respect to earnings, McClintock described the
goal of smoothing this way: "The object is to have the results of the calendar
years so adjusted, so made uniform, so freed from fluctuation as to produce
a fairly accurate figure which we make use of in the actual dividends to
the policyholders from year to year, and as we find that that average after
getting freed from fluctuations is going up or down then it has to affect

---

president, between 1869 and 1871. Winston engineered a controversy and forced him out. Mc-
Clintock gave his own perspective on the controversy in vol. 2, 1728. Elizur Wright's longer and
more entertaining version appeared in Wright, *Politics and Mysteries of Life Insurance*, 194–204.

43. Vol. 2, 1443.

44. See dividend histories recorded in vol. 2, 1430–1431, 1437–1438, 1444–1445.

45. "An exact determination according to the results of business," he said, would be "mathematically and morally improper." Vol. 2, 90.

46. Whether an actuary is setting rates, comparing classes, or calculating dividends, he must
first smooth existing mortality data and generate a life table.

the dividend to the policyholders similarly."[47] McClintock, waxing transcendental, sought freedom from fluctuations and viewed that freedom in moral terms.

Hughes, skeptical, asked McClintock if freeing data from fluctuations, which depended on his judgment, did not involve "the element of prophecy." To that, McClintock replied: "It involves working according to our previous experience, which is a pretty fair way of prophesying." Prophesying—fatalizing, really—might have been fishy for Hughes, but the actuaries' practices gave them reason to trust in short-term prophecies premised on consistent, smooth laws governing even the financial world. Hughes continued, asking whether in fact "all the calculations of dividends to which [McClintock] referred" partook in prophecy. McClintock hesitated, suggesting "they involve the element of retrospection more than they do prophecy."[48]

Journalists latched onto Hughes's tendentious terminology, citing "prophecy" and the actuaries' "guess work" in the absence of "exact mathematical calculation."[49] Part of the problem was that Hughes, the committee, and reporters generally made the mistake of thinking that calculation and judgment should necessarily be distinct—an assumption that historian Theodore Porter's work on actuaries has shown to be clearly false.[50] Yet the opinion pages, and Hughes too, rightly worried that actuaries were not in a good position to make fair judgments. One editor wrote: "A guessing system is not calculated to check an incompetent, extravagant or predatory management."[51] McClintock and his peers were too exposed to make judgments against policyholders on one side or the financiers who controlled the company on the other, as McClintock had all but admitted. Critics had a point in claiming that actuaries needed rules to protect them.

* * *

After a two-week adjournment McClintock sat for a final day of testimony hoping to redeem himself. The papers had not treated him all that

47. Vol. 2, 1843.

48. Vol. 2, 1851.

49. "Billion Limit on Life Insurance: Only Greed Prevented Fixing Maximum One Company May Carry," *Minneapolis Journal*, 26 October 1905; "M'Curdy Springs Great Surprise: Mutual Will Investigate Itself," *Los Angeles Herald*, 26 October 1905; Also: "M'Curdy Has a Committee: Mutual's Board of Trustees Starts an Investigation," *Los Angeles Times*, 26 October 1905.

50. Porter, "Precision and Trust."

51. "When President McCurdy," *Wilkes-Barre Times*, 30 October 1905.

well the last time: "The chief trouble with Actuary McClintock's explanations of insurance operations," quipped one editor, "is that they do not explain."[52] Another complained that McClintock had "left matters worse confounded."[53] Journalists looking for a rational manager found the real thing disappointing.[54] On his second try, McClintock set out to explain better. It wouldn't be easy.

He provided a table of actual earnings and dividends (figure 4.2), in support of his basic claim that dividends had decreased because interest rates on average had decreased. Looking at earnings for each year next to total dividends (or earnings for one year next to the following year's dividends, which McClintock said was more appropriate given the time lag between the calculation of dividends and the actual annual accounting), one could see a dim correlation. Earnings oscillated wildly up and down, trending up slightly between 1900 and 1904, while the total dividends moved, slowly, smoothly, and uniformly up. Between annual dividends and earnings, however, any relationship seemed quite remote. Even as earnings increased on average, annual dividends fell.

Assemblyman Cox of the committee objected to the lack of direct (and obvious) correspondence between each year's earnings and dividends. He asked: "Why wouldn't it be your policy to give more to the policyholders if the company's earnings, for instance, warranted it?"[55] McClintock responded that such *was* his and the company's policy: "If there is an obvious progressive increase in the surplus compared with the amount of business on which the surplus should be divided, then of course, as you say, it would be more."[56] They talked past one another. McClintock responded to "an obvious progressive increase" rather than to a single earnings figure. He believed the *trend* to be more real than any given *data point*.

---

52. "The Chief Trouble," *Aberdeen Daily News*, 26 October 1905. The quoted line is the entire piece. The editors certainly chose to reprint it, but we should not rule out the possibility that they bought the witticism from one of the other companies' news services.

53. "When President McCurdy," *Wilkes-Barre Times*, 30 October 1905.

54. Newspapers are always tricky—but we have special reasons to be skeptical here. The Armstrong investigation proved insurers to be master manipulators of the press. They paid news services to promulgate their propaganda, often with the intent of disparaging rivals. Hughes made plain that company newsmen were shaping the coverage even of the Armstrong investigation itself by calling men like the former *New York Sun* editor and MONY news "specialist" Charles J. Smith to testify; see vol. 2, page 1738–1754.

55. Vol. 2, 1871.

56. Vol. 2, 1871–1872.

EXHIBIT NO. 303

Mutual Life Insurance Company of New York.
Realized earnings and total dividends as per gain and loss exhibit.

Realized earnings	Total dividends	Of which annual dividends	Balance
1900			
$6,412,717	2,260,863	807,426	1,453,437
1901			
$5,216,171	2,358,028	763,133	1,594,895
1902			
$7,471,023	2,518,382	748,412	1,769,970
1903			
$5,563,861	2,985,061	735,047	2,250,014
1904			
$6,624,677	2,714,547	538,504	2,176,045

4.2. MONY's statement of earnings, total dividends, and annual dividends, as displayed in exhibit no. 303 from the Armstrong investigation, in New York Legislature, *Testimony Taken before the Joint Committee of the Senate and Assembly of the State of New York to Investigate and Examine into the Business and Affairs of Life Insurance Companies Doing Business in the State of New York* (New York: Brandow Printing Company, 1905), 8:421.

The same misunderstanding became evident in discussing McClintock's way of deciding how much "extra profit" to allocate to deferred dividend policies. "Extra profit" was the bonus granted policyholders for staying alive and paying their premiums long enough to collect their deferred dividends; it was their share of the losers' savings. It seemed to Hughes that this value should be pretty easy to determine, if you knew the number of people who lapsed their policies or died in any given group of deferred dividend policyholders. McClintock disagreed. Calculating extra profit, he insisted, required smoothing as well. He explained that what he called "a particular fluctuation"—for instance, the death of a few policyholders with large policies—might make people in one group of policyholders, those who bought policies in 1890, say, get significantly smaller dividends than those in surrounding years. He explained: "The company considers that whatever year the group is dated in that those who are dated in 1889, and 1890, and

1891, have approximately the same rights to compensation for those risks, because the real risks that they have incurred have not been very different."[57] There was, argued McClintock, "more justice in equalization or avoidance of fluctuation than there would be in taking the exact figures."[58]

In other words, MONY and McClintock cared about "risks" more than "accidents," about smoothed averages more than raw statistics. Accidents were fluctuations. They occurred in the real world, in real time, on a human level—and yet they were less "real" than "risks." McClintock elaborated: "It merely happened that there had been larger profits made on one group than the others, but the risks incurred by the individuals of the group of 1890 cannot differ very much from the risks incurred by individuals of the groups of 1889 and 1891."[59] It merely happened. The "merely" says it all.

Risks, on the other hand, emerged from actuaries' analyses of corporate ledgers and accounts. They took all the mere happenings of the world and abstracted them, quantified them, averaged them, smoothed them into something else, something more "real" to use McClintock's adjective.

* * *

Hughes and McClintock shared the same chamber on that day, but they inhabited different worlds and they believed in different capitalisms. Hughes and the Armstrong Committee proceeded from an old Wall Street assumption that chance mattered and could be determinative in business and capitalism.[60] They believed in the economic chance-world that had governed one of the deferred dividend policies' ancestors, the classic tontine. One of the earliest and most important examples of a tontine supported the building of New York's Tontine Coffee House in 1794. Investors bought shares in the coffee house, each nominating a beneficiary to receive regular interest payments as an annuity—this was the tontine. With each beneficiary's death, the annuity payments to all others increased. Chance and luck reigned supreme in this system. No data were smoothed or graduated, nor any fluctuations damped. The luckiest person, the last person living, got it all. As it happened, that tontine did more than create a coffee house; it

---

57. Vol. 2, 1897.

58. The quote continues: "according to the group system which was contemplated originally in the tontine plan." Vol. 2, 1897.

59. Vol. 2, 1897.

60. The resemblance of capitalism to gambling did not escape contemporaries' notice, drawing praise from some, but also much consternation from those in the Protestant elite, like John D. Rockefeller, Sr., who sought to draw a bright line between games of chance and legitimate business. See Lears, *Something for Nothing*, 188–201.

created a home for the New York Stock Exchange.[61] The literal foundations of American capitalism, as Hughes and his colleagues understood them, lay in the unfettered play of chance.

McClintock's smoothing, which Hughes suspected to be a corruption of traditional capitalism, resonated with the ideals of their contemporary, the J. P. Morgan lieutenant and New-York Life officer George W. Perkins. Perkins strained mightily within the manager/financier dichotomy. He worked his way up the New-York Life organization as a reforming manager, intent on making business more efficient and cooperative while eliminating unnecessary uncertainty or competition. As a financier, Perkins oversaw the railroad reorganizations of the turn of the century with an eye toward eliminating competition and securing safe, stable returns on investments.[62] Even after Perkins became a high roller, he devoted much of his energies to the creation of pioneering welfare capitalism programs at US Steel and International Harvester, to safety improvements, and to profit-sharing plans that would implement a corporate socialism, what he called: "a form of socialism of the highest, best and most ideal sort."[63] McClintock and Perkins each in his own way espoused a new corporate approach to capitalism, one that translated smoothing into a general business philosophy.[64]

Both visions of capitalism—Hughes's old Wall Street vision and Mc-Clintock's new corporate vision—took moral positions. Hughes indicted state insurance department investigators precisely for their moral agnosticism; they only worried about a company's solvency, not about whether or not it was corrupt. Hughes wanted them to nose out nepotism, loose spending, misleading bookkeeping, or the use of company funds for personal gains.[65] He advocated a fair game played by fair rules.

McClintock expressed his own "laissez faire" preferences of the fair game/fair rules type when talking about state regulation, but within the corporation he believed that moral ends required scientific, centrally planned

61. William H. Ukers, *All about Coffee* (New York: Tea and Coffee Trade Journal Company, 1922), 122–123; Charles H. Haswell, *Reminiscences of New York by an Octogenarian (1816 to 1860)* (New York: Harper and Brothers, 1896), 29–30; James Grant Wilson, *The Memorial History of the City of New-York* (New York: New York History Company, 1893), 522n1.

62. North, "Life Insurance and Investment Banking," 213.

63. Quoted in Levy, *Freaks of Fortune*, 293.

64. This trend went hand in hand with the "Great Merger Movement" that consolidated American industry into large, oligopolistic trusts at the turn of the twentieth century. Vertical integration of corporations often followed on these horizontal integrations meant to reduce competition and make profits more stable. See Naomi R. Lamoreaux, *The Great Merger Movement in American Business, 1895–1904* (Cambridge: Cambridge University Press, 1985).

65. New York Legislature, *Report of the Joint Committee*, 249–250.

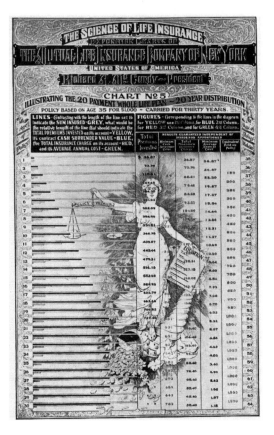

4.3. One of MONY's 1900 Paris Exposition posters, "Chart No. 28: The
Science of Life Insurance," in Mutual Life Insurance Company of New York,
Introduction to the Exhibit of the Mutual Life Insurance Company of New York (Paris:
Mutual Life Insurance Company of New York, 1900), in Folder "Miscellaneous
Publications 1900," Box 630, MONY Papers. Courtesy of AXA.

means and, of course, smoothing.[66] His position extended from the one
portrayed in figure 4.3, a MONY poster from the 1900 Paris Exposition.

In this poster, MONY sold risk making as a fair, rational, and even beautiful good. The policy premiums it charged depended on the "science of
life insurance." This classically garbed woman stands on the cornucopia
of plenty, holding a perfectly balanced scale (the "policy premium") and a
chart labeled "equation." Science, here in the form of smoothly graduated

66. Vol. 2, 1830.

mortality tables, made commercial justice possible, the poster argued. It was an old trope, originally employed to combat fraternal societies in the 1880s that sold policies with premiums that did not vary with age or depend on mortality tables. But by 1905, fraternal societies too employed risk makers' practices in setting their premiums.[67] Very few insured Americans now escaped risk-making practices and thus some form of smoothing.

McClintock and his actuarial colleagues (who all smoothed earnings and dividends in one way or another) extended the reasoning of law-bound mortality to the market. With faith in the science of life insurance, they computed a gradual, regular financial world, an economic chance-world freed from chance and fluctuations.

It isn't clear whether policyholders bought insurers' vision of risk making and capitalism. Certainly those who complained to Hughes seemed more of his mind-set. They may have appreciated their steady dividends and gentle increases over the years, but when the bottom fell out, they expected (and were disappointed not to find) direct correlations between dividends and other business factors.[68] They saw corruption instead of innovation.

Still, whether or not they believed in McClintock's rationale, they experienced its effects. Actuaries' calculations had power. In 1904, McClintock's (admittedly questionable) calculations freed up $440,000 for dividend distributions that otherwise would have stayed in company coffers, and other

67. Well-publicized failures of cooperatives in the 1890s, especially among small, poorly run assessment companies—amplified by angry life insurers stressing their scientific methods and courts who increasingly classed cooperatives alongside other insurers—eventually drove cooperatives to get in the business of fatalizing and smoothing. Having gathered together fraternal societies' experiences, a young actuary named George Dyre Eldridge constructed the National Fraternal Congress table of mortality, an important measure of mortality among middle- and working-class men across the nation, to serve as a price floor for reforming fraternals. As it became more difficult to distinguish cooperatives from mainline insurers, the cooperatives plunged in popularity. On assessment failures, see William George Lehrman II, "Organizational Form and Failure in the Life Insurance Industry" (PhD diss., Princeton University, 1989), 207–213, 265, 289–290. On the assessment societies' indiscretions, see Stalson, *Marketing Life Insurance*, 456. See also S. L. Fleishman, *The Fallacies of the Assessment Plan of Life Insurance* (Philadelphia: Edward Stern, 1897), 5, 11. On New York Insurance Superintendent Payn's investigations see "One More Lesson," *New York Times*, 21 August 1897; "Insurance Law Amendments: To Protect the Policy Holders of Mutual Fire and Assessment Life Companies," *New York Times*, 21 February 1898; "New York Insurance Wrecks: Insolvent Co-operative and Assessment Companies Reported by Superintendent Payn," *New York Times*, 22 February 1898. On courts ruling that cooperatives and their "certificates" looked suspiciously like insurers and their policies, see Levy, *Freaks of Fortune*, 226–228. Most states had adopted the table as a legal price floor for all cooperatives by 1919. Walter Basye, *History and Operation of Fraternal Insurance* (Rochester, NY: Fraternal Monitor, 1919), 95–99; Beito, *From Mutual Aid to the Welfare State*, 142.

68. Vol. 2, 1445, 1447.

actuaries made those sorts of decisions every day: actuaries' market smoothing had practical consequences.

More profoundly, MONY's six hundred thousand policyholders experienced a new kind of market, a rationalized market—one that could not exist on its own, one that had to be made—because of McClintock's smoothing. Years of success predicting mortality by employing practices borrowed from secretive bankers and data-crunching, graph-drawing astronomers led actuaries to believe that the world, even the human-built economic world, was fundamentally regular and smooth. Acting from that belief, McClintock and his actuarial peers, while claiming to be describing existing regularities, actually *made* part of the economy behave regularly and smoothly, in a close analog to the story Donald Mackenzie and Yuval Millo have told of financial quants *making* derivatives markets while purporting to explain them—but the actuaries did it a century earlier.[69] Actuaries' dividend calculations tamed—as best as they could—the nineteenth century's wild, boom and bust economy for America's millions of middle- and upper-class policyholders in the name of making them into risks.

Hughes never really got a satisfactory answer to his question: why did salaries go up while dividends went down? He did, however, decide that the whole business of deferred dividend policies tended toward corruption of all kinds, McClintock's smoothing included. Life insurers, still a political force to be reckoned with, fought off any remedial legislation for a year, but in 1906 Hughes's report and public pressure spurred the New York legislature to pass sweeping reforms that, among other things, banned deferred dividend policies.[70] Meanwhile, Hughes rode a wave of post-investigation popularity to the New York governorship, a narrow loss for the presidency, and a Supreme Court seat.[71]

---

69. Economic sociologists and sociologists of science consider moments where the social sciences (especially economics) create the behaviors they purport to explain using the rubric of "performativity." For a good introduction, see Donald MacKenzie and Yuval Millo, "Constructing a Market, Performing Theory: The Historical Sociology of a Financial Derivatives Exchange," *American Journal of Sociology* 109, no. 1 (July 2003): 107–145; Donald MacKenzie, *An Engine, Not a Camera* (Cambridge, MA: MIT Press, 2008); and Trevor Pinch and Richard Swedberg, eds., *Living in a Material World: Economic Sociology Meets Science and Technology Studies* (Cambridge, MA: MIT Press, 2008).

70. Keller, *Life Insurance Enterprise*, 257; Buley, *American Life Convention*, 1:269–279.

71. For an example of the role the investigation played in justifying Hughes's candidacy, see "Belief in the People Is Hughes's Reliance," *New York Times*, 27 September 1906. Betty Glad, "Hughes, Charles Evans," *American National Biography Online* (2000), http://0-www.anb.org.library.colgate.edu/articles/11/11–00439.html.

In reevaluating (and redeeming) the deferred dividend plan in recent years, economic historians swept away the moral and systemic components of the Armstrong hearings, deeming them unimportant or nonexistent.[72] They might be forgiven for the error. Contemporary journalists did not always realize they were witnessing a moral debate either. In their versions of McClintock's final day, they approximated Assemblyman Cox's question: "But why not give more dividends if the company can afford it?" Then they translated McClintock's testimony into a new, simpler, less technical language. In the *Washington Post*'s version: " 'We fix the rate with due regard to the circumstances of the company,' said the witness. 'If we can afford it, we give more: if not, then we give less.' "[73] To avoid explaining the fundamental difference between McClintock's and Cox's approaches to the world, the difference between caring about accidents and caring about risks, between common sense and probabilistic thinking, between the capitalism of the Tontine Coffee House and the capitalism of risk making, the Post's correspondent excised all talk of "progressive trends" and replaced such talk with a pithy (and wholly invented) second sentence.

The Armstrong hearings did enormous damage to existing life insurers, and especially the "Big 5" at the center of the investigation.[74] Not only did the hearings smear their reputations as upstanding corporate institutions; the legislation that followed strictly limited their capacity to attract new policyholders and invest company funds.[75] A 1907 economic crisis triggered by failures among prominent Wall Street trusts, and tied by some contemporaries to shaken confidence in the entire financial system after the Armstrong revelations, did little to help matters.[76] The hearings and ensuing crises also proved the end for many insurance executives' careers.

72. Ransom and Sutch, "Tontine Insurance and the Armstrong Investigation."

73. "Life Insurance Cost," *Washington Post*, 9 November 1905. The same misquote appears in "Insurance Costs Poor Man Double," *Chicago Daily Tribune*, 9 November 1905.

74. The Big 5 were New-York Life, Equitable, MONY, Metropolitan, and Prudential.

75. Both factors contributed to a boom in new life insurance companies, especially in the South and West. But small companies often became a way for large companies to circumvent limits on new business: once small life insurers acquired policyholders and got into trouble, big life insurers scooped them up. On post-Armstrong remedial legislation and its results, see Buley, *American Life Convention*, 1:247, 269; Stalson, *Marketing Life Insurance*, 552–562.

76. On the 1907 crisis, see Hugh Rockoff, "Banking and Finance, 1789–1914," in *Cambridge Economic History of the United States*, ed. Stanley L. Engerman and Robert E. Gallman (New York: Cambridge University Press, 2000), 2:643–684 at 672. For one prominent Wall Street observer's thoughts on life insurers' embarrassments as one cause of the erosion of trust in the market that brought on the panic, see Henry Clews, *Fifty Years in Wall Street* (New York: Irving Publishing Company, 1908), 797–800.

Even Emory McClintock's. At first it seemed as though his technicality had saved him from enduring the general shame heaped on his non-actuarial colleagues. While most papers ended up largely ignoring or oversimplifying McClintock's complex arguments, the *New York Times* managed to turn him into a managerial hero (although not because of his belief in smoothing). McClintock's reward was a MONY vice presidency in 1906.

His faith in the reality of smoothed experience did not help him in that capacity, however. It did not prepare him well to face the business in its daily messiness, where simplifying assumptions and mathematics tuned to smoothing away troubling fluctuations had distressingly little power. He made some characteristically cranky remarks to the press in 1906, subsequently endured a public pillorying, and then succumbed to a medical breakdown.[77] So ended an illustrious actuarial and mathematical career. One colleague judged him, ultimately, a "martyr to his profession."[78] In coming years, his Armstrong testimony became a textbook, reprinted to teach new actuaries the significance of smoothing as a tool for making risks.[79] As McClintock faded into memory, smoothing lived on.

Corporate methods for predicting individual futures and reading Americans' fates encountered repeated challenges and severe scrutiny in the late nineteenth century. Renegade risk makers had proposed medical judgment as an alternative to statistical prophesying. Northern legislatures had outlawed racial discrimination by calling into question the idea that a statistical past could be relied upon to point toward the future. Skeptical agents and cost-conscious actuaries inside corporate offices had challenged the necessity of nationwide writing networks for personalizing risk assessments. And in 1905, the very concept of smoothing away individual particularities had come under fire. But in each case, life insurers had survived with their systems for risk making largely intact—newly limited in some cases, and no more popular, but still functioning. Risk makers maintained the capacity

77. McClintock reportedly said of an eminent group seeking election to MONY's board that they reminded him of "the college professor's definition of dirt, as 'matter out of place.' " "Emory McClintock Dead," *New York Times*, 12 July 1916.

78. Nichols, "Dr. Emory McClintock," 295. Nichols, a close colleague, blamed the strain of McClintock's work settling controversies in London—and not his own embarrassments—for his breakdown and death. Another colleague thought McClintock's "virtually cloistered life" had not prepared him for post-1905 industry politics. See Miles Menander Dawson, "Mr. McClintock's Services: His Testimony the Father of Many Important Reforms," *New York Times*, 19 July 1916.

79. Miles Menander Dawson, "Mr. McClintock's Services: His Testimony the Father of Many Important Reforms," *New York Times*, 19 July 1916; Hutcheson, "In Memoriam. Emory McClintock," 376.

to predict the futures of a growing population of Americans willing to purchase life insurance.

But a greater expansion of risk-making capacities was still to come as a direct, albeit unforeseen, consequence of the Armstrong investigations and the embarrassment they caused the industry. No one would have guessed that the key to extension of risk making lay in the humiliation of life insurers, and that in corporate instability lay the key to understanding how our days became more widely and thickly numbered.

FIVE

A Modern Conception of Death

Frederick Hoffman spent days upon days riding trains. Already in 1910, a typical year, he had left his Newark, New Jersey office and library to inspect (and critique) a General Electric Company factory;[1] he had made his rounds among professional and progressive conferences including the National Association for the Study and Prevention of Tuberculosis and a meeting of the organizing committee of the International Congress on Hygiene and Demography;[2] and he had given expert testimony on the German compensation system before the Massachusetts Commission on Compensation for Industrial Accidents.[3] The only oddity of the year had been Hoffman's failure to arrange a visit south to tromp through cemeteries typing out birth and death dates off gravestones.[4]

Often enough, Hoffman's insurance colleagues and overlapping progressive circles traveled alongside him on the rails. Such was the case in November 1910, when on a train home after a meeting of the National

1. In West Lynn, Massachusetts—he had visited the Schenectady plant earlier. Frederick L. Hoffman to John F. Dryden, 1 August 1910, in Book 1, Box 1, Hoffman Papers. After his visit he reported: "My time was too limited for a full inspection, but I saw enough to convince me that the claim of the company, that everything possible is done for its employees, is not substantiated by the facts." Hoffman to Dryden, 24 August 1910, in Book 1, Box 1, Hoffman Papers.

2. He attended nearly twenty different meetings in 1910. See *Annual Report of the Statistician's Department for 1910*, 9–11, in Book 3, Box 1, Hoffman Papers.

3. Hoffman to Dryden, 1 August 1910; and Hoffman to Dryden, 19 August 1910, in Book 1, Box 1, Hoffman Papers.

4. In 1909 he visited Alabama; in 1911, Georgia. His grim and Gothic survey of the South lasted more than a decade. The South was on his mind in 1910, however. He was busy analyzing past southern trips and working on a special committee for setting standards for accepting southerners for ordinary insurance policies. *Annual Report of the Statistician's Department for 1910*, 2–3, 7–8, in Book 3, Box 1, Hoffman Papers.

Association for the Prevention of Infant Mortality, Hoffman ran into Metropolitan's new welfare department chief, Lee Frankel, as well as the economist cum health reformer Irving Fisher. Hoffman grilled Frankel—whom he viewed as an interloper, an outsider come to take advantage of a life insurance company during a period of post-Armstrong fragility[5]—about the costs and efficacy of the company's fledgling forays into providing free nursing to industrial policyholders. To his consternation, he found Irving Fisher smitten with Metropolitan's endeavors—not entirely surprisingly, Hoffman admitted, since they had been his idea, in a way. Fisher expressed excitement about a much smaller life insurer as well: Provident Savings Life, whose president, E. E. Rittenhouse, and visionary medical director, Eugene Lyman Fisk, had staked the company's future on the idea of using company medical staff to offer free preventative checkups to all policyholders.[6]

A characteristic tension riddled Hoffman's letter recounting these conversations to his boss, Prudential's president, John F. Dryden. Hoffman felt immense loyalty to Dryden, the man who gave Hoffman his big break and had served as his patron ever since. Moreover, he knew Dryden's mind about recent efforts to remodel life insurance corporations as "social work" institutions directing corporate power and risk-making tools in the service of larger social and health reform movements.[7]

Fisher and Frankel were leading remodelers. Having seen an opportunity after the Armstrong hearings destabilized corporate leadership and put life insurers on the defensive, they seized the moment.[8] Talking about the

5. Hoffman had thought it bad news when the Russell Sage Foundation chose Lee Frankel to go to Europe to make a study of private and public "workingmen's insurance," because Frankel was a reformer and not an insurance man. "It is evident to this class of men, however," wrote Hoffman of Frankel, "that insurance is an unusually interesting subject of such overshadowing social and economic importance that almost any contribution which they may make to the very limited amount of existing general knowledge will attract a relatively large amount of attention." Hoffman to Dryden, 13 May 1908, in Book 1, Box 1, Hoffman Papers. It did not help that Frankel also competed with Hoffman as an expert on German social insurance and that the two came to different conclusions: Frankel liked the German system; Hoffman viewed it skeptically.

6. Hoffman to Dryden, 21 November 1910, in Book 1, Box 1, Hoffman Papers. For some of the few details readily available on Provident Savings Life's work, see also George E. Ide, "Report of the Life Extension Committee of the Association of Life Insurance Presidents," *Proceedings of the Annual Meeting of the Association of Life Insurance Presidents* 3 (1910): 82–85 at 84. "General Discussion—Movement to Prolong Human Life," *Proceedings of the Annual Meeting of the Association of Life Insurance Presidents* 3 (1910): 108–111, 114–118 at 117.

7. Frankel, for instance, once told Hoffman that "he considered it to be the duty of a life insurance company to give primary consideration to social work" as paraphrased in Hoffman to Dryden, 21 November 1910, in Book 1, Box 1, Hoffman Papers.

8. The Armstrong hearings explain how the remodelers got their opportunity. But the remodelers' success must be understood in terms of the larger context of growing labor unrest,

"modern conception of death," as Fisher put it, savvy health reformers argued that insurers wasted their energies by seeking only to measure the risk of mortality. Why only forecast the future, they argued, when risk-makers' methods and tools—built for statistical prediction—could also *change* the future, could *control* the future? Death had once—long ago—been mysterious, Fisher claimed, until insurers helped make its regularities evident. But in its modern conception, death could not only be predicted, it could be controlled and resisted. With some luck in 1909, the modern conception of death gained a foothold in Metropolitan, one of the largest American insurers and the chief competitor to Dryden's Prudential.[9]

Dryden did not like it one bit.

Hoffman affirmed his own disapproval to Dryden. He criticized Frankel for lacking "substantial facts in support of his views" and Metropolitan for "working largely upon pure guesswork, with a chief consideration to advertising and placating public antagonism."[10] To Fisher, he insisted of Prudential: "[We] were not inclined to undertake similar [social or health] work unless upon a substantial basis of trustworthy facts and experience combined with real evidence of the public necessity, which did not seem to me, I said, to call for such undertakings at the present time."[11] To Dryden, Hoffman claimed to be in complete accord with Dryden.

Yet it isn't hard to see Hoffman straining here, as he did often in writing Dryden, to carve out some space for compromise.[12] What if "trustworthy

suspicion of corporate consolidation, and the real threat (as corporate leaders saw it) of state socialism. Megan Wolff explores the broader conjunction of social thought, progressive politics, public health organization, and corporate insecurity—under the umbrella of "political capitalism"—that made it possible for reformers to win places of power in life insurance corporations after the Armstrong shock, which shaped their vision. See Megan Joy Wolff, "The Money Value of Risk: Life Insurance and the Transformation of American Public Health, 1896–1930" (PhD diss., Columbia University, 2011).

9. Metropolitan engaged in a wide variety of other forms of "social work" that do not fit perfectly under the rubric of the "modern conception of death." Their significant investments in developing model housing stands as an excellent example. This chapter does not pretend to be a comprehensive study of Metropolitan welfare work, but instead looks at the reappropriation of risk makers' tools. For an overview of Metropolitan welfare work more generally, see Zunz, *Making America Corporate*, 90–101; on Metropolitan's housing projects, see Samuel Zipp, *Manhattan Projects: The Rise and Fall of Urban Renewal in Cold War New York* (New York: Oxford University Press, 2010).

10. Hoffman to Dryden, 21 November 1910, in Book 1, Box 1, Hoffman Papers.

11. Hoffman to Dryden, 21 November 1910, in Book 1, Box 1, Hoffman Papers.

12. Earlier in the letter, Hoffman related a story in which his colleagues abroad lamented the inability of American state governments to provide reliable vital statistics for much of the United States. Hoffman related to Dryden that he hewed to the party line: Prudential, despite having statistics that could be "the most desired contribution to the congress [on hygiene and demography]

facts" could be made available, or "public necessity" demonstrated? Hoffman offered only conditional refusals. He did not want to look closed-minded. Later events suggest (as we'll see) he probably even hoped to be given a reason to change his mind and the data to change Dryden's mind. Throughout his letters to Dryden, Hoffman talked with varying degrees of concern or enthusiasm about the progressive reforms he saw taking place around him and about the challenges and changes to life insurance that he heard talked about. In each case, with one hand he praised himself and Dryden for their conservative approach, while with the other he warned that their conservatism might get them into trouble. Hoffman—in whom was combined ambition with caution and pride with insecurity—could tell that the modern conception of death would spell new and exciting possibilities for risk makers and their methods. He wanted to be part of the action, to make Prudential part of their movement.

Hoffman spent days upon days riding trains, but the distance between his loyalties and his aspirations could not be closed by rail. He could and did talk with Fisher, Frankel, and others about the modern conception of death. He could and did imagine a larger role for insurers' risk making in progressive reform movements. But he could never learn to live comfortably between Fisher and Dryden, between reform and insurance, between risk as fate and risk as control. His story in this chapter illustrates, in pathetic hues, the tension between the necessary persistence of traditional risk prediction in an industry founded upon fatalizing principles and new dreams—in the wake of the Armstrong hearings—of statistical individuation reimagined and rebranded as a tool for healing.

* * *

Frederick Hoffman's faith in traditional risk making led him to a South Carolina cemetery in 1906 in the wake of the Armstrong hearings' scandals and revelations. Far away from New York and its newspapers, Hoffman walked peacefully and alone, gravestone by gravestone, a typewriter in tow to record the birth and death dates of white southerners as recorded on each monument. In a region of the United States where birth and death registration remained pitifully incomplete and vital statistics evaded even

---

from the United States," did not think it "advisable" to offer an exhibit. Typically, Hoffman did not here ask Dryden to reconsider, but he made it seem to Dryden that he should reconsider. And indeed, Hoffman's strategy seems to have worked in this case. Prudential contributed its statistics to both the 1911 Dresden Exposition of Hygiene and the 1912 International Congress on Hygiene and Demography in Washington. Hoffman to Dryden, 21 November 1910, in Book 1, Box 1, Hoffman Papers; Prudential Insurance Company of America, *Annual Report* (1914), 10.

so persistent a trawler of published data as Hoffman, the Prudential stat-istician thought cemeteries provided some of the best and most reliable records available for white, well-off southerners. Hoffman insisted that such records, supplementing a wealth of other data about southern localities, could allow him to more accurately predict future mortality throughout the South, to accurately discriminate between those places too unhealthy to insure and those primed for expansion.

Hoffman's faith in statistical fatalism had brought him to Prudential in the 1890s, where he completed *Race Traits and Tendencies of the American Negro*. While that work made an argument for rupture—that the Civil War and the destruction of slavery augured a significant change in the fates of African Americans, who Hoffman predicted would soon vanish from North America—it still rested on the presumption that statistics properly inter-preted could reveal the future and that fate was, in normal circumstances, stable and legible. Having imagined away the African American population, Hoffman became company statistician at Prudential, responsible for pro-cessing death claims and analyzing them with methods similar to those employed by Brandreth Symonds at MONY. In 1901 he began building a library to supplement company statistics and facilitate further risk investi-gations.[13] Never particularly talented at mathematics, he not infrequently bemoaned his industry's overreliance on mathematics and probability, in a distant echo of Thomas Scott Lambert's critiques.[14] Hoffman preferred holistic readings of people and places, blending numbers with literature, personal observation with expert advice. And so, when southern agents complained about Prudential's strictures on policies throughout the region, Hoffman convinced the company to book him a train ticket.

Prudential paid for Hoffman to visit a dozen southern states between 1905 and 1916. He wanted to improve rating practices generally—he thought he could best improve Prudential's reputation after the Armstrong hearings by making it more efficient and fair—but the trips won company

13. Frederick L. Hoffman, "Insurance Libraries," in *Essays on Statistics*, vol. 6 (1912–1914) NLM.

14. Mathematical ability, wrote Hoffman in 1910 when looking for a new assistant, could be "more a hindrance than a help." See Hoffman to Dryden, 21 December 1910, in Book 1, Box 1, Hoffman Papers. He fumed, in 1906, over an international congress of actuaries discussion of extra premium charges (i.e., how and when to discriminate in prices): "Perhaps in no other branch of the business is there so much unpardonable ignorance as in this, due to the fact that the writers know only the statistics and mathematics, but *not* the trades and countries them-selves, or the physical and physiological facts necessary for a sound understanding of this branch of insurance medicine and insurance practice." See Hoffman to Dryden, 12 September 1906, in Book 1, Box1, Hoffman Papers.

support because they met the more immediate need of satisfying discontented local managers, agents, and policyholders who had complained to the company about restrictions that made it exceedingly difficult, if not impossible, to sell policies throughout much of the South.[15] Hoffman's visits did not necessarily remove those restrictions, but they justified them, and Hoffman usually went out of his way in a final letter to the local manager to point out some spaces where expansion might safely occur.[16]

Each southern investigation began in Hoffman's library, where he assembled a preliminary report by appending excerpts from extant geographies (some over a hundred years old), medical papers and health reports, census statistics (despite their inaccuracies, which he noted), and any and all books discussing local agriculture and industry for the region. To these he added responses to a survey he made of company medical examiners about the healthfulness of the area. He ended the report with a table showing the amount of insurance already in force in the region—nearly always a very small amount.[17] That was usually the point.

His reports resemble travelogues in more than superficial terms. As Hoffman journeyed south on railroads—speeded by trains but limited also by where the trains went—he observed living conditions from his seat. He groused about poor whites who lived at the low standard of their "colored" neighbors because it made it harder for him to distinguish between black and white neighborhoods as he passed. He stopped in towns to interview prominent merchants, like a prosperous Russian in South Carolina who Hoffman embraced as Prudential's ideal target. And wherever he went, he noted sources of water, sanitary improvements, attitudes toward hygiene, complaints about "chills," the availability of commercial goods, and the prosperity of local agriculture or cotton mills or phosphate mines.[18]

In South Carolina in 1906, Hoffman first recorded local cemetery data. He wrote to the head of Prudential's ordinary insurance department: "I

15. See, for instance, Hoffman to Leslie D. Ward, 7 July 1905, in Prudential Insurance Company of America, *Ordinary Insurance and Health Conditions in Florida, Tennessee, Alabama,* MS B 250 v. 4 NLM.

16. A typical example, from a letter ghostwritten by Hoffman: Leslie D. Ward to J. H. Bruning, 21 March 1906, in Prudential Insurance Company of America, *Ordinary Insurance and Health Conditions in Eastern South Carolina,* MS B 250 v. 7 NLM.

17. See, for instance, Prudential Insurance Company of America, *Preliminary Report on the Health Conditions in Eastern South Carolina 1906,* MS B 250 v. 6 NLM.

18. Prudential Insurance Company of America, *Ordinary Insurance and Health Conditions in Eastern South Carolina,* MS B 250 v. 7 NLM.

had time enough to make a copy of the cemetery record in the town of Summerville, of special interest as the only accurate mortality data for this particular section."[19] The mortality data he gathered—on a paltry thirty-five men and thirty women in Summerville and eleven men and five women in Dorchester—did not suffice to make broad generalizations, but Hoffman appreciated having some objective figures for a region nearly devoid of reliable statistics. Moreover, as he stressed in his report and in subsequent communications with his bosses, he alone possessed this data, which, while not representative of the entire population, perfectly targeted the sort of better-off white southerners (who could afford gravestones) to whom Prudential might want to sell ordinary life insurance policies.[20] In the end, Hoffman visited more than one hundred cemeteries and typed dates, like those in figure 5.1, from around one hundred thousand gravestones.[21] He appears to have been incapable of writing "cemetery records" in his annual reports without preceding the phrase with the word "valuable," and yet he refused to be bound by his own hype.[22] Hoffman made it clear that his final rulings on whether or not to allow insurance to be sold in a city or county rested on the sum total of his evidence. He seldom cited cemetery records alone as determinative.

In the field, Hoffman looked like a naturalist. When, in 1905, he chose to visit Florida in July and August so as to see the state when the conditions were most oppressive and dangerous, he struck the pose of a heroic explorer.[23] He lugged around a typewriter on this journey (and his others) as his primary scientific instrument. Typing allowed him to make carbon copies of all his notes, letters, and findings; he sent one copy of everything to his superiors in Newark and the other he kept and bound for his library records.[24] A

19. Hoffman to Leslie D. Ward, 10 February 1906, page 2 in Prudential Insurance Company of America, *Ordinary Insurance and Health Conditions in Eastern South Carolina*, MS B 250 v. 7 NLM.

20. Appendix to Prudential Insurance Company of America, *Ordinary Insurance and Health Conditions in Eastern South Carolina*, MS B 250 v. 7 NLM; *Annual Report of the Statistician's Department for 1907*, 8–10, in Book 3, Box 1, Hoffman Papers.

21. Hoffman to Merritt H. Egan, New York, 17 June 1940. Hoffman notes a Mr. Harper of the University of Alabama, Tuscaloosa, who also collected cemetery records.

22. See, for example: *Annual Report of the Statistician's Department for 1907*, 8–10 or *Annual Report of the Statistician's Department for 1908*, 6, in Book 3, Box 1, Hoffman Papers.

23. Hoffman to Leslie D. Ward, 7 July 1905, in Prudential Insurance Company of America, *Ordinary Insurance and Health Conditions in Florida, Tennessee, Alabama*, MS B 250 v. 4 NLM.

24. Hoffman also appears to have developed poor handwriting early in his career, which one biographer notes stemmed from increasing difficulty with fine motor control of his hand. F. J. Sypher, ed., *Frederick L. Hoffman: His Life and Works* (Philadelphia: Xlibris, 2002), 74.

Details of Cemetery Records
at
Stockton, Baldwin County, Ala.
(Secured by F.L.H., February 1909)

20

Males			Females		
Date of Birth	Death	Age at Death	Date of Birth	Death	Age at Death
1832	1902	70	1839	1902	63
1888	1908	20	1861	1898	37
1842	1902	60	1869	1905	36
1838	1899	61	1868	1900	32
1867	1903	36	1849	1882	33
1809	1843	34	1831	1889	58
1815	1890	75	1807	1840	33
1869	1900	31	1820	1843	23
1834	1898	64	1807	1881	74
1806	1879	73	1825	1850	25
1827	1905	78	1823	1898	75
1809	1878	69	1876	1901	25
1806	1895	89 Rev.	1837	1886	49
1811	1885	74 Rev.	1811	1882	71
1854	1881	27	1807	1862	55
			1838	1897	59
			1843	1905	62
			1815	1901	86
			1798	1884	86
			1842	1884	42

Total years of life	861			1,024
Total deaths	15			20
Average age at death	57.4			51.2

5.1. Some of Frederick Hoffman's cemetery records, these from his 1909 trip to Alabama, in "Details of Cemetery Records at Stockton, Baldwin County, Ala.," in Prudential Insurance Company of America, *Cemetery Records of Alabama: Appendix to Final Report*, MS B 250 v. 18 NLM. Reprinted with permission from The Prudential Insurance Company of America. Permission also granted by Dr. Francis J. Rigney, Jr. Courtesy of the National Library of Medicine.

year later at James Island, near Charleston, Hoffman recorded mortality data from two cemeteries that housed a group of interrelated families. He thrilled at the find: a long look at a (hopefully) closed and (hopefully) homogenous population.[25] One can imagine similar sentiments from Eugenics Records Office fieldworkers of the next decade, ever on the lookout for long-term

25. Hoffman noted that immigration into and out of the island likely corrupted the data, but he still valued it with that caveat. Hoffman to Leslie D. Ward, 16 February 1906, page 2 in Prudential Insurance Company of America, *Ordinary Insurance and Health Conditions in Eastern South Carolina*, MS B 250 v. 7 NLM.

family data for their studies of heritability.[26] In this exceptional case, the cemetery data made up Hoffman's mind and he deemed the island one of the few healthy spots in eastern South Carolina.

In his statistical collecting outside the South, Hoffman looked less like a naturalist than like a social surveyor in the style of Alice Hamilton, Florence Kelley, and the Hull House investigators or Crystal Eastman and other Russell Sage Foundation researchers—hardly surprising since his earliest statistical work had grown out of the survey tradition at the Bureau of Labor Statistics.[27] When Hoffman took to the shop-room floor, and he did that often, it would have been hard to distinguish him from any of the above contemporaries. Industrial insurers covered workers, but they did not insure those in excessively dangerous jobs. Keeping track of quickly changing industrial jobs, however, and deciding how dangerous they were, proved a challenge. In 1907 alone, Hoffman made rulings on how to rate 289 occupations. That year he made an in-depth survey of the steel industry, collecting data from Pennsylvania sources, including mortality data from a fraternal order that sold mainly to steel workers. He supplemented those statistics with a personal inspection of seven steel plants, including US Steel's Homestead Plant.[28] Nine hundred and eighteen pages of report later, Hoffman had secured enough data that he felt that it would be "possible to materially modify our former practice" as to rating steel workers.[29]

Whatever the context or topic, whether exploring southern backwaters or evaluating GE's production processes, Hoffman sought out some hint of the future. He looked at a local climate or shop floor and judged the mortality risk associated with it. To be sure, he set himself apart from medical

26. See Garland E. Allen, "The Eugenics Record Office at Cold Spring Harbor, 1910–1940: An Essay in Institutional History," *Osiris* 2 (1986): 225–264.

27. On the survey's place in Carroll D. Wright's Bureau of Labor Statistics, see Stapleford, *Cost of Living in America*, chapter 1. Sellers notes that Hamilton, in some sense, followed Hoffman as a Department of Labor shop floor investigator. Christopher C. Sellers, *Hazards of the Job: From Industrial Disease to Environmental Health Science* (Chapel Hill: University of North Carolina Press, 1997), 69–106, especially 70, 88–89; Alice O'Connor, *Social Science for What? Philanthropy and the Social Question in a World Turned Rightside Up* (New York: Russell Sage Foundation, 2007), chapter 1.

28. Steel producers were among those actively reconsidering safety techniques. Managers at US Steel and F. W. Taylor's Midvale Steel, among others, were in the process of implementing new centralized safety programs and employer-paid or contributed accident benefits around the years of Hoffman's visit. Witt, *Accidental Republic*, 116, 122. One of the architects of US Steel's "welfare work" generally was the New-York Life officer and J. P. Morgan partner George Perkins. Levy, *Freaks of Fortune*, 264–307.

29. *Annual Report of the Statistician's Department for 1907*, 11–13, in Book 3, Box 1, Hoffman Papers.

statisticians who put more faith in naked numbers. Toward the end of his career, he explained: "To derive the best results from statistical experience, however, it was clear to me from the outset that it was absolutely necessary to amplify office analyses by filed investigations, chiefly with regard to occupations, race traits, and territory."[30] He further set himself apart from nearly all of his statistical contemporaries in his sheer energy (so many trains) and oddity (all those cemeteries). But his peculiar adaptations of traditional risk-making methods, the centrality of which had only been amplified in his mind by the Armstrong debacle, did not make those methods any less fundamentally traditional. For Hoffman, forecasting smooth futures and mobilizing correspondence networks to equitably class risks—traditional risk making, in short—made insurance possible, and could make insurance more efficient. All other purposes were, for the moment at least, secondary.

* * *

A few prominent voices in and around the life insurance industry began wondering aloud, however, whether such priorities might be mistaken and insurers' sense of the power of making risks too tightly constrained. Brandreth Symonds led by example in this conversation. He became MONY's chief medical director in 1907, his former boss having been a victim of the Armstrong hearings. One of his first major acts proved to be a public relations coup. Appearing before fellow doctors at a meeting of the Medical Society of New Jersey, Symonds revealed one way that risk writing could become a tool for reforming lives and not just for classing them. His subject was the ratio of height to weight and he had a surprise in store for his audience.

Life insurers used height-weight ratios throughout the late nineteenth century in guarding against *underweight* applicants. Insurers in the United States and Britain feared tuberculosis more than nearly any other ailment. Experience and conventional wisdom singled out underweight applicants as TB threats.[31] Then, in 1908, Brandreth Symonds revealed that life insurers

---

30. Hoffman to Forrest F. Dryden, 26 February 1920, in Book 25, Box 8, Hoffman Papers.

31. For a nineteenth-century example, see Allen, *Medical Examinations for Life Insurance* (1869), 68–69. Medical directors were still arguing over the topic in 1906. See George R. Shepherd, "The Relation of Build (*I.E.* Height and Weight) to Longevity," *Abstract of the Proceedings of the Association of Life Insurance Medical Directors of America* 17 (1906): 46–66, especially 57 and 60; on British concerns over TB and the debates over the role of heredity, see Timothy Alborn, "Insurance against Germ Theory: Commerce and Conservatism in Late-Victorian Medicine," *Bulletin of the History of Medicine* 75, no. 3 (2001): 406–445. By the 1890s, some companies

were upending conventional wisdom. "Overweight," he announced, appeared to be a far greater threat to health than underweight, as evidenced by increased rates of mortality among the heavy, associated with increased incidences of heart disease, kidney disease, and diabetes.[32] Early the next year, Symonds announced his findings to the lay public in *McClure's Magazine* and published new height-weight standards for women—under the heading "What Women Should Weigh"—based on tabulations of MONY experience from the United States and Canada.[33] Summing up the findings, the *New York Times* concluded: "It's safer to be thin than fat."[34] Insurers' statistics could not prove that overweight caused greater mortality—they could only correlate it with higher mortality rates. Yet few seem to have been particularly concerned about such niceties. For risk making to become a tool for healing, correlation had to become a basis for action.

To tabulate new weight tables (which insurers would use in making risks, but could also circulate as medical advice), Symonds needed to attract more talent to his staff—particularly after the Armstrong purges—and there he proved momentarily lucky. In 1908, Symonds saw the potential in an underemployed biologist named Louis Dublin, who had earned a PhD from Columbia studying cytology under Edmund B. Wilson and statistics under Franz Boas. For Dublin, the medical department job under Symonds amounted to a revelation of vocation: "My biology, my smattering of medicine, and my knowledge of statistical methods all made sense now, for these interests were the very ones with which the medical department

---

were worried about overweight too. Horace Fletcher, the inventor of the "fletcherizing" eating and dieting method, began his eating experiments after an 1895 encounter with a life insurance company that told him he weighed far too much for his height and could not be insured. See Hillel Schwartz, *Never Satisfied: A Cultural History of Diets, Fantasies and Fat* (New York: Free Press, 1986), 125.

32. Brandreth Symonds, "The Influence of Overweight and Underweight on Vitality," *Journal of the Medical Society of New Jersey* 5, no. 4 (1908): 159–167, reprinted in *International Journal of Epidemiology* 39, no. 4 (2010): 951–957. On weight-height standards and insurance in America, see Emma Seifrit Weigley, "Average? Ideal? Desirable? A Brief Overview of Height-Weight Tables in the United States," *Journal of the American Dietetic Association* 84, no. 4 (1984): 417–423; Amanda M. Czerniawski "From Average to Ideal: The Evolution of the Height and Weight Table in the United States, 1836–1943," *Social Science History* 31, no. 2 (2007): 273–296.

33. Brandreth Symonds, "The Mortality of Overweights and Underweights," *McClure's Magazine* 32, no. 3 (January 1909): 319–327. Symonds names Dr. Faneuil S. Weisse of MONY as the author of the new table.

34. "Weights of Men and Women," *New York Times*, 24 December 1908. S. S. McClure, the magazine's publisher, it might be noted, joined Fisher's health reform league, according to Schwartz, *Never Satisfied*, 127.

of the company was deeply concerned."[35] At MONY, Dublin set to work tabulating height-weight ratios derived from company experience.[36] He encountered there the possibilities of statistically analyzing lives, not merely to rate them, but to improve them. He would soon have an opportunity to explore those possibilities further.

But another outsider had to intervene first. Irving Fisher began to consider the opportunities hidden within insurers' risk-making systems in 1909 as part of his ongoing effort to rejuvenate America. Fisher yearned for the nation to experience the kind of health renewal that had changed his own life.[37] Years earlier, in 1898, Fisher, at the age of thirty-one, ascended to a full professorship at Yale in political economics on the strength of a remarkable body of mathematical and formal work. But personal hardship ran fast after professional success. A mysterious illness led to a dreaded diagnosis: tuberculosis. Fearing the disease that had taken his father all too early, Fisher submitted himself to isolation and treatment in a Saranac, New York, sanatorium and two years of further recuperation in Colorado Springs and Santa Barbara. Fisher beat the disease and professed himself a believer in fresh air and rest treatments.[38] While still recovering, Fisher began to refashion himself as a health advocate and reformer. In succeeding years he communed with health guru John Harvey Kellogg, conducted experiments on "fletcherism"—a diet built on small amounts of protein and very thorough mastication—and popularized his own design for a TB tent to be used in sanatoria. All this he managed while also writing landmark economic works like *The Rate of Interest* and *The Nature of Capital and Income*.[39] His own experience conquering disease (for that was how he saw it) convinced him that

---

35. Louis I. Dublin, *After Eighty Years: The Impact of Life Insurance on the Public Health* (Gainesville: University of Florida Press, 1966), 37–38. Dublin also noted approvingly that Symonds was well versed in Karl Pearson's biometric work.

36. Dublin, *After Eighty Years*, 38.

37. Fisher was hardly alone in his yearnings. Talk of renewal abounded at the century's turn. See Jackson Lears, *Rebirth of a Nation: The Making of Modern America, 1877–1920* (New York: Harper, 2009).

38. He also described himself as a believer in "healthy minded" religion, using a phrase he borrowed from William James, to describe his faith in positive thinking.

39. Fisher is a fascinating character and has finally attracted attention fitting his multifaceted life, with recent book chapters exploring Fisher the economist and economic forecaster, Fisher the statistician, and Fisher the eugenicist health reformer. See Walter A. Friedman, *Fortune Tellers: The Story of America's First Economic Forecasters* (Princeton, NJ: Princeton University Press, 2013), chapter 2; Stapleford, *Cost of Living*, chapters 2 and 3; Nathaniel Comfort, *The Science of Human Perfection: How Genes Became the Heart of American Medicine* (New Haven, CT: Yale University Press, 2012), chapter 2. For the most complete biography of Fisher to date, see Robert Loring Allen, *Irving Fisher: A Biography* (Cambridge, MA: Blackwell, 1993), chapters 3 and 4.

individuals (and society too) possessed the capacity to resist death, if they only chose to resist intelligently.

In the domain of risk making, Fisher saw simultaneously a hindrance and an opportunity. To illustrate the problem with traditional risk makers' approaches to prediction, Fisher drew popular attention to a startling picture, titled *The Bridge of Life*, painted by Maria Sharpe Pearson, as an illustration of her husband Karl Pearson's ruminations on chance and death.[40] Writing in 1909 for *American Health*—the official organ of the twenty-three-thousand-member-strong American Health League, published by Fisher in conjunction with his Committee of One Hundred on National Health, a special committee of the American Association for the Advancement of Science—Fisher explained the Pearsons' image and its risk-derived conception of death. For Pearson, medieval and early modern ideas about death had centered on unknowable, erratic chance, emblemized by the "dance of death." Death came to dance at an unforeseen time and it came to dance with all, without apparent order or logic—rich or poor, good or evil: all succumbed to death.

Then in the eighteenth and nineteenth centuries, actuaries and statisticians discovered regularities accompanying death, regularities in the varying rates and causes of death according to age. As Maria Pearson's painting (reproduced in figure 5.2) indicates: babies succumbed to death because of chances passed down by their ancestors (hence the skulls being flung down); children (subject to high mortality) fell to the equivalent of a machine gunner's mark; young adults had only to escape an arrow; and mature men a blunderbuss, until finally a sharpshooter brought inevitable death to the aged. Here stood mortality risk, as traditionally conceived by life insurers, beautifully illustrated, a life table as etching.

Looking more closely, one glimpsed a new and remarkable understanding of the life table. For in 1895, when Karl Pearson first presented this idea, he argued not only for the regularity of death (a well-established principle, as we have seen) but also for the idea that the life table really consisted of five overlapping curves of mortality. Each curve related to a set of causes of death that most often struck at a given stage of life, some associated with old age (the sharpshooter), some with infancy (the skulls), and so on.[41]

40. On the Pearsons' *Bridge of Life*, see Karl Pearson, *Chances of Death and Other Studies of Evolution* (London: Edward Arnold, 1897); for broader exposition and context, see Klein, *Statistical Visions in Time*, 47–53; and Theodore M. Porter, *Karl Pearson: The Scientific Life in a Statistical Age* (Princeton, NJ: Princeton University Press, 2004).

41. Pearson viewed his life table as most important as a tool for thinking about natural selection and its means of operation in humans. Pearson, *Chances of Death*, 40–41.

5.2. Maria Sharpe Pearson's *The Bridge of Life*, the frontispiece to Pearson, *Chances of Death*.

Fisher took Pearson's innovation a step further, casting Pearson as retrograde in the process. "The conception of Karl Pearson and the earlier actuaries of insurance companies is represented by [*The Bridge of Life*]" explained Fisher, as he began his critique: "A curious reverence seems to have developed in regard to that mortality curve which shows the number of people dying at a given age in a group of ten thousand."[42] Tellingly, Fisher subtitled *The Bridge of Life*: "Repair This Tottering Structure." He thought that actuaries' "reverence" for statistical regularity prevented them from trying to intervene in those regularities, that "the shape of the curve was in some way inevitable and could not be changed." But he countered: "The modern conception of death and morbidity consists of the knowledge that the shape of the curve of mortality is determined by causes, many of which may be controlled by the individual, others by society acting as a whole."[43] Fisher's novelty came

42. This is an unsigned article, but it bears the signs of Fisher's authorship. "Three Conceptions of Death," *American Health* 2, no. 2 (1909): 20–21 at 20.
43. "Three Conceptions of Death," 21.

not really from claiming that death's regularity emerged from constitutive causes—Pearson made that point central to his own conception—but rather from his insistence that analyzing death into its causes somehow defanged it and made it susceptible to control.

Fisher did not reject smoothing or the search for regularities in death statistics. He rejected what he saw as undue fatalism. He did not mind reading fates, but refused to be *bound* by those fates. Fisher demonstrated the potential of risk makers to serve the forces of health reform in a report prepared for Theodore Roosevelt and Gifford Pinchot's national conservation commission. Fisher's 1909 *Report on National Vitality: Its Wastes and Conservation* thought about the American people as a resource to be conserved, just like the nation's streams or forests, and relied on life tables to make its case. But instead of taking life tables as facts, Fisher conceived of them as tools. Drawing up what he called "hypothetical life tables," Fisher demonstrated how many lives could be saved and years added to lives (as well as dollars saved by life insurers) if certain causes of death could be made less potent. He prophesied the easy addition of fifteen years to the average life span, and probably more.[44] The "modern conception of death" made death not only predictable and understandable, but controllable too. It also made it possible to pitch health work as a sound investment.

Fisher made that pitch to the Association of Life Insurance Presidents, a body created in the wake of the Armstrong hearings to improve the industry's reputation and coordinate lobbying. Fisher must have hoped that industry leaders, reeling from scandal, might be open to reimagining their work, to adopting the modern conception of death as their own. Fisher wanted to appropriate risk-making methods for the purpose of health reform. But first and foremost, he wanted life insurers' money and institutional power. To get the ball rolling, claimed Fisher, $200,000 a year would suffice. That investment in the health fund of his Committee of One Hundred on National Health and the American Health League would make possible their endeavor to win popular support for a national health department and nationwide health-education campaigns. With a stronger federal health agency and targeted education, Fisher believed that tuberculosis, typhus, and a handful of other common diseases could be effectively eliminated as causes of death. Less disease and fewer deaths meant more premiums paid to life insurers, Fisher argued, so any lengthening of the life span would

---

44. Irving Fisher, *Bulletin 30 of the Committee of One Hundred on National Health, Being a Report on National Vitality, Its Wastes and Conservation* (Washington, DC: Government Printing Office, 1909), 102–117.

increase revenues while decreasing costs from death claims. Fisher pointed to fire and employers' liability insurers, who already integrated prevention into their standard practice—why should life insurers act differently?[45] The Life Insurance Presidents, as an organization, listened politely to Fisher, published his speech, and then delayed further action.[46]

Metropolitan's vice president, Haley Fiske, decided on his own, however, to radically change his company. Fiske, though nominally the vice president, ran the show after the company's president stood indicted for perjury and forgery as part of the Armstrong inquiry.[47] Here the destabilizing powers of the Armstrong scandals created a unique opportunity for the modern conception of death to take root within a life insurer's skyscraper. Fiske acted out of a deep religious commitment to social justice, but he also knew his company and industry needed better press, that it needed some means to hold off further government intervention.[48]

Fiske saw an opportunity in early 1909 when he heard Lee Frankel reporting on his completed investigations of European social insurance. Frankel started out as a PhD chemist and then managed the United Hebrew Charities before leaving for Europe at the behest of the Russell Sage Foundation. Frankel, inspired by what he had seen and especially by the example of the German social insurance system, sought some agency for bringing social insurance to America's laborers.[49] In 1909, he was already working with Russell Sage to create a social insurance bureau. But Fiske made another offer: why not make over Metropolitan's industrial insurance program in the image of social insurance? Frankel agreed and set about making corporate life insurance into a vehicle for social change.[50]

At first, Frankel struggled to find a good way to use Metropolitan's resources. He invited Louis Dublin—whom he had worked with in Lower

---

45. Irving Fisher, *Economic Aspect of Lengthening Human Life* (New York: Association of Life Insurance Presidents, 5 February 1909), 3, 7–12.

46. See, for instance, "General Discussion–Movement to Prolong Human Life"; and Ide, "Report of the Life Extension Committee," 82–85.

47. Keller, *Life Insurance Enterprise*, 267.

48. Louis Brandeis's successful effort to win legislative backing for savings bank life insurance in Massachusetts frightened many insurers. Keller, *Life Insurance Enterprise*, 255. Hoffman warned Dryden of Brandeis's effort in Hoffman to Dryden, 8 November 1905. On Fiske's personal conviction, see Louis Dublin's description in *After Eighty Years*, 39.

49. Frankel published his findings in Lee K. Frankel and Miles M. Dawson, *Workingmen's Insurance in Europe* (New York: Charities Publication Committee, 1911). See also Rodgers, *Atlantic Crossings*, 248–249.

50. Zunz, *Making America Corporate*, 92–93.

East Side social reform circles[51]—to leave MONY and help him remake Metropolitan. A March 1909 advertisement in Fisher's *American Health* announced Metropolitan's new "Bureau of Cooperation and Information"—a name that wouldn't stick—calling on *"Health Officers, Settlement Workers, Charity Organization Societies, Labor Organizations, Committees for the Prevention of Tuberculosis, Educational Bodies and similar agencies"* to provide literature for distribution by Metropolitan's eleven thousand agents.[52] Very quickly, Frankel and Dublin narrowed their focus to "health" and "welfare work"—a move that left much less room for labor organizations.

In July, Frankel and Dublin launched a "War on Consumption," fought in pamphlets intended to teach industrial policyholders (drawn mainly from America's working classes) how to spot tuberculosis, how to avoid it, and, if worse came to worse, how to treat it. Between 1909 and 1924, Metropolitan distributed around eleven million copies of their first anti-consumption pamphlet, and it was only one of many.[53] Dublin and Frankel worked with medical experts to prepare the pamphlets—apparently building on Fisher's network of health authorities.[54]

Meanwhile, Lillian Wald of the Henry Street Settlement prevailed on Frankel to hire her nursing service to check in on sick policyholders—company agents provided the vital link in notifying the nurses of illness or suffering. The nursing experiment went well in New York and the company expanded it across the nation.[55] On its own terms, Metropolitan considered the whole experiment a success. When, years later, Metropolitan publicity writers tried to explain why "the Attacks have ceased," their answer was simple: the company, through its nursing service, its war on consumption, and its welfare work generally, had become "the friend of the workingman."[56]

But Metropolitan only began thinking seriously about moving beyond these elaborate welfare improvisations and toward repurposing its

---

51. Dublin contributed significant energy to these circles, as did his wife, Augusta, a social worker among eastern European immigrants. Dublin, *After Eighty Years*, 35.

52. Emphasis in the original. Metropolitan Life Insurance Company, "The Metropolitan Life Insurance Co.," *American Health* 2, no. 1 (1909): iv–v.

53. Metropolitan Life Insurance Company, *An Epoch in Life Insurance: A Third of a Century of Achievement* (New York: Metropolitan Life Insurance Company, 1924), 207–211.

54. For an in-depth look at Metropolitan's pamphlet making, see Elizabeth Toon, "Managing the Conduct of the Individual Life: Public Health Education and American Public Health, 1910 to 1940" (PhD diss., University of Pennsylvania, 1998), 212–236.

55. Dublin, *After Eighty Years*, 40–41. On the nursing service, see Diane Hamilton, "The Cost of Caring: The Metropolitan Life Insurance Company's Visiting Nurse Service, 1909–1953," *Bulletin of the History of Medicine* 63, no. 3 (1989): 414–434.

56. Metropolitan Life, *Epoch in Life Insurance* (1924), xxii.

risk-making system when it faced industry criticism that its social work had gone too far. Discussing Fisher's earlier request for life insurance money, a New England Mutual Life medical director had stated the common opinion that "the policyholder's money is not to be expended except for the direct benefit of the policyholder."[57] Such reasoning ultimately killed Fisher's request.[58] Frederick Hoffman, in a November 1909 letter to Dryden, stated the case against Metropolitan's unorthodoxy most forcefully in a related vein: "The drift of Mr. Fiske's remarks showed an absence of a clear grasp of the fundamental nature of insurance and the proper relation of a life insurance company to the state. The whole policy outlined was a policy of drift, directed by men, sentiments and considerations totally different from those which have directed so successfully insurance enterprises in the past. Mr. Fiske seemed to confuse the fundamental nature of contractual obligations with secondary considerations of charity, philanthropy and public policy."[59] Frankel too, as Hoffman made clear after his encounter with him on the train in 1910, too readily ignored traditional risk prediction and industry ideals like equity. Hoffman's critiques, alongside those of other industry insiders, forced Metropolitan to fit its actions more directly within the existing standards of risk making.

Louis Dublin took the lead in making welfare work amenable to risk making, and vice versa. He began with the goal of answering the question implicit in many critiques of Metropolitan's work: "Did it pay?"[60] If Metropolitan's actions could be demonstrated to be essentially free—to pay for themselves in premiums paid over extended lives—then Hoffman's complaints lost much of their bite.[61] To demonstrate effectiveness, Dublin extended the system for making risks to include more than the standard risk

---

57. E. W. Dwight, "Latent Powers of Life Insurance Companies for the Detection and Prevention of Diseases," *Proceedings of the Annual Meeting of the Association of Life Insurance Presidents* 3 (1910): 102–108 at 107.

58. The Association of Life Insurance Presidents settled instead on the cheaper route of printing a series of papers on public health and lobbying for improved public health laws. F. W. Jenkins, "Report of Health Committee of Association of Life Insurance Presidents," *Proceedings of the Annual Meeting of the Association of Life Insurance Presidents* 6 (1912): 78–85, 139. There was some danger in shifting to lobbying. During the Armstrong hearings, much heat had been generated over the misuse of company funds for political ends. As Buley explains: "Contributions to campaign funds were condoned by few; even some of the most conservative papers scored the practice without mercy." *American Life Convention,* 1:227.

59. Hoffman to Dryden, 24 November 1909, in Book 1, Box 1, Hoffman Papers.

60. Dublin, *After Eighty Years,* 41.

61. Indeed, the "Life Extension Committee" of the Association of Life Insurance Presidents, after explaining the narrow trustee obligations of insurance executives, had argued that executives could justify "spending policyholders' money" on health work if they could demonstrate

files that he (like every other medical statistician in the industry) processed every year as head of Metropolitan's Statistical Bureau.[62]

He began by turning nurses into writers of risk correspondence, with a twist: these risk writers tried to change the risk they reported upon. In cooperation with the Henry Street nurses, Dublin worked out an elaborate system of paperwork that aided in daily administration, but more importantly, made possible an ultimate summary of outcomes and effectiveness. The "keystone of the whole record system" was a four-by-six-inch slip on thin paper (so it could be easily duplicated with carbons) called the "New Case and History Slip," reproduced in figure 5.3. Filled out at the bedside (ideally), the slip demanded information as to occupation, birthplace, family history, religion, and the patient's disease details.[63] The new case card began a file that ended with a final history slip (figure 5.4), which summarized the care given and the results seen. Had the patient improved, recovered, stagnated, or died? That answer, when properly tabulated (which Dublin could do, and which he taught the nursing associations to do too), provided one basis for judging the efficacy of the nursing intervention.[64] The number "'recovered' on discharge" mattered most: "There is no better measure of effectiveness of the work done."[65] Dublin, a true believer in the power of statistics, also thought that improving nurses' record systems could make them more effective, easier to manage and allocate, and perhaps even give them data with which they could prove the efficacy of new practices.[66]

---

with data "a reasonable certainty that the expenditure will result in a reduction of the cost of insurance." Ide, "Report of the Life Extension Committee," 83.

62. He became Statistician in 1911. Metropolitan Life, *Epoch in Life Insurance* (1924), 213–214.

63. Dublin deemed occupation and social or economic data to be crucial to understanding disease and health—"almost as essential for the proper handling of a case as the medical items." Louis I. Dublin, "Teaching Nurses in Training the Uses and Value of Sickness Statistics," *American Journal of Nursing* 17, no. 12 (September 1917): 1157–1165 at 1163.

64. Louis I. Dublin, *Records of Public Health Nursing and Their Service in Case Work, Administration and Research* (New York: Metropolitan Life Insurance Company, 1922), 12–16. Originally published in *The Public Health Nurse*. Metropolitan reprinted nearly all of Dublin's addresses for wider distribution. They are available from many research libraries. MetLife maintains a complete collection as well, which its archivist, Daniel B. May, graciously made available to me. In succeeding notes, I cite Dublin's original when it has been possible for me to access it and his reprints when not.

65. Dublin, *Records of Public Health Nursing*, 32.

66. If, he argued, nurses could better treat pneumonia at home than in hospitals, as many believed, then they needed good data to prove it. As he explained: "Before such statements can be made, it is essential that the associations be able to present definite facts giving lower case fatality rates, for cases of equal severity, than are found for other types of care." Dublin, *Records of Public Health Nursing*, 32.

Form 4

..............Visiting Nurse Ass'n. New Case and History Slip.

Name.................... Address........................ Floor................... Ward.................... District...............

Age................ Sex................... Color............ Mar. cond..................... Birthplace.................
Present Occupation:
    Industry.. Kind of Work.............................
Husband or Father:
    Name.................................Birthplace............................Occupation.........................
Wife or Mother:
    Name................................Birthplace...........................Occupation.........................
                                              Basis of { Pay, part pay, free,
Church attended.................Other co-op. agencies.............Payment { no charge, M. L. I. Co.,Industrial....
Policy Number.........................Date, issue.............................Agent's Name and Number...............
Case reported by.......................Physician's Name..................Address.............................
                                                Up and about {
How long was patient ill at time of first visit?.............In Bed { .... Date, first visit...........
Diagnosis...Symptoms observed.........................
Doctor's orders..
Service rendered..
Condition of patient first visit.....................................Nurse.................................

CHILDREN IN FAMILY

Name	Date of Birth	Name	Date of birth

5.3. The New Case and History Slip used by Metropolitan visiting nurses,
as designed by Louis Dublin and the Henry Street Settlement nurses, in
Dublin, *Records of Public Health Nursing*, 13. Courtesy of MetLife.

His office intervened with doctors too, working to standardize their disease diagnoses in claims forms, so that they could be as useful in writing risks at a policyholder's death as they were at the time of application.[67]

Dublin also searched for techniques that revealed changing mortality rates for the company as a whole. He needed to show that Fisher's hypothetical life tables were more than guesses. Comparing company mortality rates against experience prior to the beginning of welfare work provided one possible measure of impact—one that Dublin turned to over and over again throughout his career.[68] Medical departments had once used the same technique as a means of showing the efficacy of their work selecting risks. Now it became a measure of medical intervention. Comparing Metropolitan

67. Louis I. Dublin and Edwin W. Kopf, "The Improvement of Statistics of Cause of Death through Supplementary Inquiries to Physicians," *Publications of the American Statistical Society* 15, no. 114 (1916): 175–191 at 178, 180, 191.

68. See, for example, Louis I. Dublin, *The Effect of Life Conservation on the Mortality of the Metropolitan Life Insurance Company: A Summary of the Experience, Industrial Department, 1914, for Superintendents, Medical Examiners and Visiting Nurses* (New York: Metropolitan Life Insurance Company, 1916).

Form 5

........................Visiting Nurse Association—Final History

Name..Address..District................Ass'n case number........................

Age............Sex................Color................Mar. Cond........................Birthplace..

Present occupation:
Industry..Kind of work..
Husband or father:
Name..Birthplace........................Occupation........................
Wife or mother:
Name..Birthplace........................Occupation........................

Church att'd........................Other co-op. agencies................ Basis of } Pay, pt. pay, free,
Pmt. } no chge., M. L. I. Co.,
Industrial:........................

Reported by..Physician................ Address........................

	Date,	Date,		At last visit was pt.:
At first visit:	At first visit: }	1st visit last visit	No. of }	Able to work?
How long ill?	Up and about? }		visits }	Sick, up and about? }
	In bed? }			Sick, in bed? }

Diagnosis..Complications........................

Nursed
Not nursed } Cond. on } Recovered Transferred } Dispensary
Advised discharge { Improved to { Hospital
Not found } { Unimproved........ } Family
  { Dead Other care........

CHILDREN IN FAMILY				If M. L. I. Co. case give :
Name	Date of Birth	Name	Date of Birth	Policy No........
				Date, issue:............
				M.L.I.Co. case
				number................

5.4. The Final History Slip used by Metropolitan visiting nurses, as designed by Louis Dublin and the Henry Street Settlement nurses, in Dublin, *Records of Public Health Nursing*, 15. Courtesy of MetLife.

mortality details to rates of mortality among the American public or the American working class would have worked better. But American registration systems remained porous. Metropolitan used its agents to proselytize for model registration bills and the Association of Life Insurance Presidents committed itself to similar advocacy in 1912. The advocacy would help expand both the birth and death registration areas, but in the meantime, Dublin had little good data for mortality comparisons.[69] In fact, Dublin

69. By 1920, the "death registration area" spanned the entire nation, while the "birth registration area" covered about half of the population. James H. Cassedy, "The Registration Area and American Vital Statistics," *Bulletin of the History of Medicine* 39 (May–June 1965): 221–231; United States Bureau of the Census, *Birth Statistics for the Birth Registration Area of the United States 1919* (Washington, DC: Government Printing Office, 1921), 7. Cassedy ascribes more importance to the publications of Census officials in standardizing registrars' practices, but pays less attention to the legislative battles necessary to improve registration systems. On the Association of Life Insurance Presidents' debates and eventual decision to back statistical reform, see Jenkins, "Report of Health Committee." On insurance company advocacy and success, see Louis I. Dublin, "The Improvement and Extension of the Registration Area," *Publications of the American Statistical Association* 14, no. 110 (1915): 578–582; Louis I. Dublin, *Vital Statistics in Relation to Life Insurance: Paper Read before Subsection B, Section VIII, "Public Health and Medical Science," of the*

quite often used Metropolitan data to stand in for absent national data. He made a life insurer's community of risks stand in for the national community, but still in the ultimate service of making statistical individuals more accurately.[70]

* * *

In 1913, Irving Fisher united with Fiske, Frankel, and Dublin to launch an even more ambitious expansion and repurposing of the risk-making system. That December, Fisher convinced Metropolitan to sign on as the anchor client to a new organization to be called the Life Extension Institute (LEI). Every Metropolitan policyholder with at least a $3,000 policy would be provided a free preventative medical examination every year by LEI. At the same time, Fisher convinced a wealthy contractor to front the start-up capital; he won William Howard Taft's endorsement for the institute and installed him as its chairman; and he brought in E. E. Rittenhouse and Eugene Lyman Fisk—both previously of Provident Savings Life, that company that had so excited him in 1910 when talking to Hoffman—to set up a medical examination infrastructure.[71]

Fisher intended to do on a grand scale what Provident Savings Life had done for its small body of policyholders. Indeed, the institute hoped to reach out to independent individuals as well as life insurance policyholders. LEI used the same medical examination techniques that life insurers had developed over the years to sort risks. But where examinations had previously classified apparently static bodies, now they became a means to change Americans' bodies and their lives. An "impairment" no longer meant exclusion or a higher premium. Instead, it meant there was a problem to be fixed, and that risks could be made from even those who had never applied for insurance.

LEI seemed to Fisher to provide a crucial path toward his final goal in health reform. He had argued for state eugenics to rid the nation of criminal

---

Second Pan-American Scientific Congress, Washington, December 30, 1915 (New York: Metropolitan Life Insurance Company, 1916), 7. For other works by Dublin on the registration problem, see Louis I. Dublin, *The Registration of Vital Statistics and Good Business: An Address Delivered before the Annual Conference of Health Officers of the State of Indiana, Indianapolis, May 13, 1913* (New York, Metropolitan Life Insurance Company, 1913); Louis I. Dublin, *The Reporting of Disease: The Next Step in Life Conservation* (New York: Association of Life Insurance Presidents, 1914).

70. Dublin called life insurance experience a "nation-wide registration area" in Dublin, *Vital Statistics*, 1, 3, 6.

71. "National Society to Conserve Life: Life Extension Institute Formed to Teach Hygiene and Prevention of Disease; Large Capital behind It; Ex-President Taft, Chairman; Prof. Irving Fisher, E.E. Rittenhouse, and Others Direct It," *New York Times*, 30 December 1913.

and degenerate types. He had pushed for improved public hygiene systems to ensure safe food, clean water, and healthy environments, and a National Department of Health to coordinate such protections. He had called on semipublic institutions, like life insurers, to do their part in improving public hygiene as well.[72] And in all these various efforts to convince society to support health work he had made progress. But the most important changes, he believed, lay in personal hygiene—in the life choices made by individual Americans every day. And before Americans could change their lives, they had to believe there was something worth changing.

That was LEI's primary purpose. Press releases made much of the fact that LEI would be a self-supporting philanthropic institution—that its examination services would pay for its health education and advocacy, and maybe even turn a profit.[73] But it accomplished a more important feat of self-sustenance statistically: it helped precipitate a broadened American health crisis, or, in another sense, helped invent one.[74]

When LEI examiners did their work, their diagnoses were designed to stir their clients to action, to convince them of some latent threat to their health that needed correction, under a doctor's supervision. Gathered together, all those individual warnings suggested a more fundamental problem, which Fisher and Fisk used to their advantage. In 1916, LEI used the press to publicize the results of "the first general physical survey of sample groups of our citizens" revealing "conditions of impairment which are truly astounding." Of all those examined, 59 percent were diagnosed with an

72. See, for a typical example: Irving Fisher, "National Vitality," *American Health* 2, no. 2 (May 1909): 24–28, in Reel 6, Fisher Papers.

73. "National Society to Conserve Life: Life Extension Institute Formed to Teach Hygiene and Prevention of Disease; Large Capital behind It; Ex-President Taft, Chairman; Prof. Irving Fisher, E.E. Rittenhouse, and Others Direct It," *New York Times*, 30 December 1913; "Extension of Life: Taft Enlists in the New Work," *Los Angeles Times*, 30 December 1913.

74. Historians talk about a sickness or industrial accident crisis among workingmen of the early twentieth century, but vary over its cause. On one side, Daniel T. Rodgers argues that no one got much sicker or more often hurt, but that the compensation system broke down, precipitating crisis. On the other, John Fabian Witt lends more weight to the accident crisis itself: more blameless people fell to industrial harm and thus the tort system didn't know what to do with them. See Rodgers, *Atlantic Crossings*, 245–254; Witt, *Accidental Republic*, 1–42. In this case—a case that mattered more for the middle classes than the working classes, but is otherwise closely analogous—Americans did not suddenly became impaired en masse, although it is certainly possible that those bodies did slowly get heavier, or as they aged became more susceptible to chronic diseases (as some evidence suggests). The more important factors in explaining the middle-class health crisis will be found in the social life of ideas. Ideas about disease and impairment changed more than the bodies of Americans did. LEI propounded new ideas and invented the crisis. To locate a crisis in changing ideas, rather than in changing bodies, makes it no less real or important, however.

impairment needing medical attention (including many young people with heart or kidney trouble), while nearly 38 percent were "on the road to impairment because of the use of 'too much alcohol' or 'too much tobacco,' constipation, eye-strain, overweight, diseased mouths, errors of diet, and other ailments."[75] In publicity stunts aimed at attracting individual clients, LEI invited Ford employees, bank clerks, and policemen to avail themselves of free examinations; in each case the result was a shocking level of impairment and defect.[76] Talking about Ford employees—"men who receive more than the average wages and possess more than the average intelligence and mental expertness and presumably more than the average health"—Fisher noted that 99 percent were "more or less impaired." "Of course," he added, "the major part of these impairments were minor, such as defective teeth and slightly high blood pressure, and yet these minor impairments we know are the beginning of serious troubles."[77]

Buoyed by early LEI findings, Fisher joined Fisk in writing a new health book in 1915, called (humbly) *How to Live*.[78] The pair began with Fisher's "scientific" rules of living, which he had been peddling in one form or another since 1906—rules that advocated long exposure to fresh air breathed through the nose, daily baths as "skin gymnastics," "judicious use of enemas," leisurely eating of more "raw foods," less protein, and no alcohol or coffee, daily exercise of "every muscle in the body" and an "optimistic and serene" outlook on life.[79] The final rule bound Fisher's belief in personal agency to his commitment to a social responsibility to protect health.[80] Just like life insurers or governments, every individual had a social responsibil-

75. "The Extension of Human Life: Men and Women of To-day Seem to Die as Soon as They Grow Up," *Washington Post*, 14 May 1916.

76. Laura Davidow Hirshbein, "Masculinity, Work, and the Fountain of Youth: Irving Fisher and the Life Extension Institute, 1914–31," *Canadian Bulletin of Medical History* 16, no. 1 (1999): 89–124 at 97.

77. Irving Fisher, *Life Extension: A Talk at Vassar College* (Poughkeepsie, NY: Vassar College, 1917), 10, in Reel 7, Fisher Papers. Fisher said of LEI statistics: "Through this Institute we have secured for the first time some real statistics as to the extent of physical impairments" (9).

78. Irving Fisher and Eugene Lyman Fisk, *How to Live: Rules for Healthful Living Based on Modern Science* (New York: Funk & Wagnalls, 1920). The first edition appeared in 1915.

79. Irving Fisher, "Rules of Individual Hygiene," *American Health* 2, no. 2 (May 1909): 28–30, in Reel 6, Fisher Papers.

80. In her excellent essay, Helen Veit ponders the apparent paradox of LEI and contemporaries' focus on individual responsibility for health despite so much evidence that decreasing mortality rates had roots in social or economic, as well as personal, contexts. See Helen Zoe Veit, "'Why Do People Die?' Rising Life Expectancy, Aging, and Personal Responsibility," *Journal of Social History* 45, no. 4 (2012): 1026–1048. Looking at Fisher's papers helps us dissolve the paradox: Fisher advocated new personal responsibilities, but those responsibilities existed within a social and institutional nexus that he also endeavored to change.

ity, argued Fisher, to "take part in the movements to secure better public hygiene in city, state and nation."[81] From this core—updated, tightened, distilled (into the form displayed in figure 5.5)—Fisher and Fisk expanded out to fill half of their book, with the help of the "Hygiene Reference Board"—an honor roll of seventy-nine expert men (and three women) from the realms of science, public health, and philanthropy.[82] The second half of the book, devoted to "special subjects," included: a comprehensive list of foods, their costs, and caloric and protein values; a foldout chart showing life insurer data on the impact of weight on mortality; an extended statistical essay on the evils of alcohol; and an introduction to eugenics, apparently penned by Charles Davenport.[83] Fisher and Fisk's book suggests the anxious audiences to whom risk makers had the possibility to speak: to eugenicists nervous about racial degeneration, to men and women concerned about cultural feminization, and to moralists troubled by the weakening and poisoning of modern bodies by drink and environment.[84]

Not everyone believed LEI's pronouncements or deemed their propaganda useful. Faced with mounting criticism, Fisk insisted his critics suffered from "confusion of mind as to what [LEI] was doing." The institute did not propose to extend the life span enormously right away, he explained. And "no advertisement had been issued that could not be read in a medical society or backed up by scientific references, and actual facts of experience."[85] American entry into the Great War helped Fisk's case.[86] Readers of the Provost Marshal's reports (actually, reports about that report)

---

81. Irving Fisher, "Rules of Individual Hygiene," *American Health* 2, no. 2 (May 1909): 28–30 at 30, in Reel 6, Fisher Papers.

82. Contemporaries sometimes charged the board with being entirely a figurehead, a charge that Fisk contested by talking of correspondence folders filled with thousands of communications between himself and the board over the years. See Life Extension Institute, *Addresses at the Banquet of the Life Extension Institute* (New York: Life Extension Institute, 1921), 7. Still, the book clearly hewed closely to Fisher's original 1906–1907 rules.

83. On Davenport's authorship, see Hirshbein, "Masculinity, Work, and the Fountain of Youth," 95.

84. On eugenics, see Daniel J. Kevles, *In the Name of Eugenics: Genetics and the Uses of Human Heredity* (Cambridge, MA: Harvard University Press, 1995); on fears of a crisis of masculinity, see John F. Kasson, *Houdini, Tarzan, and the Perfect Man: The White Male Body and the Challenge of Modernity in America* (New York: Hill & Wang, 2002); on overcivilization and degeneration generally as widely shared concerns in the late nineteenth and early twentieth century on both sides of the Atlantic, see Lears, *Rebirth of a Nation*, 7–8.

85. Life Extension Institute, *Addresses*, 5.

86. Fisher claimed of Fisk, although I have not independently corroborated it: "More than any other man he formulated the examinations used by our army in recruiting for the World War." Irving Fisher, "In Memoriam: Dr. Eugene Lyman Fisk," *How to Live: A Monthly Journal of Health and Hygiene* 14, no. 8 (August 1931): 2–3 at 2, in Reel 7, Fisher Papers. We have a much

The FIFTEEN
Rules of Health

AIR
1 Have Fresh Air where you live and work
2 Wear Light, Loose, Porous Clothes
3 Spend part of your time in the Open Air
4 Have Lots of Fresh Air where you Sleep
5 Breathe Deeply

FOOD
6 Avoid Eating too Much
7 Do Not Eat much Meat and Eggs
8 Eat Various Kinds of Food
9 Eat Slowly

HABITS
10 Have your bowels move at least once
   Each Day
11 Stand, Sit, and Walk Erect
12 Avoid Poisonous Drugs
13 Keep Clean and avoid Catching
   Diseases

ACTIVITY
14 Work hard, but Play and Rest too
15 Be Cheerful and learn not to Worry

5.5. The rules of health as they appeared in a Metropolitan pamphlet glossed
from Fisher and Fisk's *How to Live*, in Irving Fisher, *How to Live Long*
(New York: Metropolitan Life Insurance Company, 1916).

and analysts of draft board exams in 1918 and 1919 wrung their hands over
the volume of American men rejected because of physical defects.[87] And as

---

better developed literature on the way that psychologists took advantage of the draft to employ
new intelligence tests. See Carson, *Measure of Merit*, chapter 6.

87. D. A. Sargent of Harvard's physical training program claimed: "From 25 per cent. to
75 per cent. of our young men were excepted from military service on account of physical dis-
ability." D. A. Sargent, "Men Fit for Soldiers: Draft's Showing Up of Physical Defects of Young
Americans and the Remedy," *New York Times*, 9 March 1918. Later (probably still inflated figures)
fixed the percentage closer to 17 percent. See "Draft Will Save Lives: Social Hygiene Based On
It—Sixth of Registrants Defective," *New York Times*, 9 March 1919 or "Statistics to Save More
Than War Took: Examinations of Men during Draft Expected to Be of Great Benefit," *Los Angeles
Times*, 22 March 1919. These stories point to a New York Draft Board's analysis of its records as
the source for this figure. It might very well have been Fisk who orchestrated and publicized that

William Howard Taft noted in a 1918 foreword, even more men with less severe physical impairments still served.[88]

LEI's propaganda worked in a very practical sense: it attracted sixteen thousand individual subscribers by 1920, more than from its insurer clients.[89] That same year, LEI asked Louis Dublin's help in setting up an internal statistical bureau to analyze its data.[90] While twenty-seven thousand lives, largely limited to middle-class white men, looked small compared to the sweep of the nearly twenty-four million selective service exams or even Metropolitan's yearly processing of 1.5 million new risks, LEI data offered a look at year-by-year changes in individuals' health that could show the efficacy of medical interventions.[91] Yet even before that data came in, the

---

analysis. Historians quote a variety of numbers. Rothstein says 40 percent. William G. Rothstein, *American Medical Schools and the Practice of Medicine: A History* (New York: Oxford University Press, 1987), 123. Reiser gives the following figures: "Of the some 3,764,000 males between 10 and 42 examined, about 550,000 were rejected as entirely unfit for service. Of even the approximately 2,700,000 eventually called into service, 47% had physical impairments." Stanley Joel Reiser, "The Emergence of the Concept of Screening for Disease," *Milbank Memorial Fund Quarterly: Health and Society* 56, no. 4 (1978): 403–425 at 405. He cites what looks like an LEI plant: J. A. Tobey, "The Health Examination Movement," *Nation's Health* 5, no. 9 (1923): 610–611. The Provost Marshal General's official report looks much less discouraging. It points to 23.9 million registrants, of which about 17.6 million were classified, and of which 2.7 million were inducted into the service and 3.6 million were estimated as the total qualified to be inducted and not deferred. It claims that 7.8 million men might have been inducted "without invading any of the deferred classes." Only .899 million men were excluded because of some physical defect, whether remediable or not. See *Final Report of the Provost Marshal General to the Secretary of War on the Operations of the Selective Service System to July 15, 1919* (Washington, DC: Government Printing Office, 1920), 14.

88. Fisher and Fisk, *How to Live*, ix–x.

89. Hirshbein, "Masculinity, Work, and the Fountain of Youth," 96. Hirshbein argues that LEI focused its efforts on middle-class white men. Veit notes that despite this fact, the focus on personal responsibility for longevity probably fell the heaviest on women and African Americans. See Veit, " 'Why Do People Die?' "

90. Life Extension Institute, *Addresses*, 7. Dublin had the institute's records put on Hollerith cards for tabulation.

91. Provost Marshal General, *Final Report*, 9. Provost Marshal Crowder claimed of the selective service rolls produced with the help of 1,319 medical advisory boards: "Never in the history of this or any other nation had a more valuable and comprehensive accumulation of data been assembled upon the physical, economic, industrial, and racial condition of a people." Metropolitan Life Insurance Company, *Annual Report* (1918): 37. Some prominent early results of LEI's statistical apparatus include Edgar Sydenstricker and Rollo H. Britten, "The Physical Impairments of Adult Life: General Results of a Statistical Study of Medical Examinations by the Life Extension Institute of 100,924 White Male Life Insurance Policy Holders since 1921," *American Journal of Hygiene* 11, no. 1 (1930): 73–94; and "The Physical Impairments of Adult Life: Prevalence at Different Ages, Based on Medical Examinations by the Life Extension Institute of 100,924 White Male Life Insurance Policy Holders since 1921," *American Journal of Hygiene* 11, no. 1 (1930): 95–135.

American Medical Association endorsed preventative examinations while health reformers pushed to examine millions of Americans every year.[92] Thus began one of the more significant medicalizations of American society— motivated by broader concerns about individual and public health—but made possible by risk makers' retooling of their old prediction techniques in the spirit of Fisher's modern conception of death.[93] The annual medical checkup had been born.[94]

* * *

When Frederick Hoffman ran into Irving Fisher and Lee Frankel on a train in November 1910, he encountered risk making in transit. As the train rattled on, the institutional and intellectual ground beneath his feet trembled. The Armstrong hearings had—simply by being so unsettling and embarrassing—opened up spaces for new ideas to take root in key life insurance industry sites: in Brandreth Symonds's medical department, in Haley Fiske's Metropolitan. The people—Irving Fisher, Lee Frankel, Louis Dublin, Eugene Lyman Fisk—who took advantage of those momentary ruptures, brought with them a new sense of increased power over death.

Fisher came to that sense out of his own experience of beating tuberculosis. Metropolitan's Frankel in 1913 attributed his faith in human agency to developments in medical science: "To-day, divested as we are of our old superstitions, of our faiths in our old traditions, we stand in the beating rays of the new light in scientific discovery—otherwise and prosaically called the 'germ theory.' "[95] (Frankel admitted, however, to not yet knowing exactly how to live in a post-germ-theory world).[96] William Howard Taft, for

92. Charap cites the 1922 AMA endorsement and 1923 campaign for ten million examinations, while criticizing advocates for uncritically accepting the efficacy of periodic health exams. See Mitchell H. Charap, "The Periodic Health Examination: Genesis of a Myth," *Annals of Internal Medicine* 95, no. 6 (1981): 733–735.

93. Medicalization refers to the expanding scope of doctors' influence and "medical" concepts in modern life, and especially to the role of medical experts as agents of social control. Renee C. Fox presents a useful discussion of the concept in "The Medicalization and Demedicalization of American Society," *Daedalus* 106, no. 1 (1977): 9–22. For an overview of the historiographical debates over medicalization, see Beth Linker, *War's Waste: Rehabilitation in World War I America* (Chicago: University of Chicago Press, 2011), 192–193, note 36.

94. Davis, "Life Insurance and the Physical Examination"; Reiser, "Emergence of the Concept of Screening for Disease."

95. Lee Frankel, *Insurance Companies and Public Health Activities* (New York: Metropolitan Life Insurance Company, 1913), 2.

96. Irving Fisher, it turned out, had a book for him to read—albeit not one that had much to do with germs. On practical responses to the germ theory as it played out in daily lives, see

his part, reminisced over the success of "warfare" against smallpox, yellow fever, and plague that he had witnessed in the Philippines, and looked toward a day when Americans would similarly conquer "the heavy loss from lowered physical efficiency and chronic, preventable disease, a loss exceeding in magnitude that sustained from the more widely feared communicable diseases."[97] W. C. Gorgas's taming of disease in the Panama Canal Zone struck imaginations of the time—including Fisk's—even more forcibly.[98] For all those reasons, Fisher found company in asserting, as in an early LEI pitch: "We now know, however, that the death rate is not the fixed and fatalistic thing it was once supposed to be."[99]

Frederick Hoffman, who after all spent a great deal of time in cemeteries, came around only slowly. Then, in November 1911, Prudential experienced its own disruption: John Dryden died, to be replaced as president by his son Forrest. Around that time, Hoffman began inching toward the modern conception of death and toward imagining the refashioning of risk making and of life insurers' missions. Though initially a critic of Metropolitan's efforts to build tuberculosis sanatoria, by the end of 1911 he had reversed himself.[100] Similarly, he confided to a conference of public health nurses in 1914: "I respectfully submit, however, with a reasonable knowledge of the

Nancy Tomes, *The Gospel of Germs: Men, Women, and the Microbe in American Life* (Cambridge, MA: Harvard University Press, 1998).

97. Fisher and Fisk, *How to Live*, v–vi.

98. Fisk extrapolated from the success in Panama all the way to a belief in unbounded longevity: "Suppose the Canal Zone in its former state had been restricted and cut off from the rest of the world without knowledge of living conditions or death rates beyond its borders," he argued. "There would then have been just as much warrant for scientists to claim that the appalling death rate of that region and the limited life cycle were conditions more or less fixed by nature, and that control of such a situation was beyond scientific power, as there is at present time to claim that conditions affecting the life cycle are fixed and beyond scientific control." Eugene Lyman Fisk, "How Long Will You Live?" *World Wide* (reprint, 24 August 1929): 1337, in Folder "Longevity-(Reprints, Notes, Clippings) 1922–42," Box 7, Dublin Papers. Gorgas and the Canal Zone often received praise bordering on worship. See Paul S. Sutter, "Nature's Agents or Agents of Empire? Entomological Workers and Environmental Change during the Construction of the Panama Canal," *Isis* 98, no. 4 (2007): 724–754 at 725.

99. Irving Fisher, *The Life Extension Institute* (reprint of address delivered to the Annual Banquet of the Insurance Institute of Hartford, Connecticut, on 9 March 1914), 3, in Reel 7, Fisher Papers.

100. Frankel points to Hoffman's early critiques in Lee K. Frankel, *The Influence of Private Life Insurance Companies on Tuberculosis: Reprint of a Paper Prepared for the Eleventh International Tuberculosis Congress Held in Berlin, Germany, October 22–26, 1913* (New York: Metropolitan Life Insurance Company, 1913), 1; Hoffman's new views appear in a 1911 report: *Annual Report of the Statistician's Department for 1911*, 5–6, in Book 3, Box 1, Hoffman Papers.

facts, that the small cost of an official visiting nursing service has paid for itself over and over again, if only in the consciousness, in the satisfaction to the large corporations that they were rendering the right kind of community service, and such as an advanced civilization demanded of them."[101] But Hoffman still worried that he would not be able to convince his new patron without more data. When Hoffman chided those nurses—"In proportion as you outline merely descriptive, picturesque, pretty cases, they are in every report, with a lot of nice photographs of Johnny supporting his mother at nine years old by selling papers, you fail to reach the man who is going to put in a thousand dollars and is trained by his business methods to require facts before he puts money into anything"[102]—he spoke out of his own fears that the man with the money could not be convinced.

Hoffman started offering his analytical services in exchange for data in 1911. He moved out of the cemetery and into places of care and healing. He made a deal with the Loomis Sanatorium. They gave him their record books and he returned his analyses—while keeping a copy of the data, of course. He made similar arrangements with Johns Hopkins Hospital and with the Henry Street nurses.[103] He could justify the deals to Prudential easily: he was securing morbidity and mortality data not generally available that could be used in rating applicants. Yet Hoffman clearly also hoped that his analyses of sanatorium, hospital, and nursing data might be useful in convincing Forrest Dryden that Prudential should get into the health maintenance business.[104] In that endeavor, he failed. By 1917, the furthest Prudential had gone (aside from letting Hoffman publish statistical studies from time to time) was to offer its policyholders the opportunity to submit urine samples for free chemical analysis.[105] Hoffman wanted nurses, treatment facilities, and medical exams, but Prudential only agreed to testing pee and printing pamphlets.

101. Frederick L. Hoffman, "Practical Statistics of Public Health Nursing and Community Sickness Experience," *American Journal of Nursing* 14, no. 1 (1914): 948–960 at 950.

102. Hoffman, "Practical Statistics of Public Health Nursing," 955.

103. *Annual Report of the Statistician's Department for 1911*, 5–6, in Book 3, Box 1, Hoffman Papers; Frederick L. Hoffman, *The Statistical Experience Data of the Johns Hopkins Hospital, Baltimore, MD., 1892–1911* (Baltimore, MD: Johns Hopkins University Press, 1913).

104. Hoffman also used the threat of compulsory state insurance to argue for "a provision for medical attendance" among industrial policyholders. See Beatrix Hoffman, *The Wages of Sickness: The Politics of Health Insurance in Progressive America* (Chapel Hill: University of North Carolina Press, 2001), 111.

105. Eugene Lyman Fisk, "Life Insurance and Life Conservation," *Scientific Monthly* 4, no. 4 (1917): 330–342 at 333.

Hoffman hardly had time to be disappointed before Prudential's core interests suddenly seemed (to him at least) to be under assault. Irving Fisher, working with the American Association for Labor Legislation, had begun championing bills to introduce compulsory state-run health insurance programs around the nation. For Fisher, compulsory health insurance signaled civilization—"At present," he lamented, "the United States has the unenviable distinction of being the only great industrial nation without compulsory health insurance"—and simultaneously improved civilization: it supported industry, fought poverty, lowered mortality, improved health and hygiene, and even made a nation more fit to fight.[106] But compulsory health insurance bills threatened industrial life insurers by including a small burial benefit, in direct competition with companies like Prudential and the industrial (burial) insurance they sold to many wage-earning families—or so Forrest Dryden, Hoffman, and their colleagues judged it. Hoffman chose to support his company's interests—since his company refused to expand those interests, as Metropolitan had done—even if that meant breaking ranks with his professional peers.

He began by breaking the bond between insurance and health, by resisting the idea that Fisher's modern conception of death had any place in life insurance. The "propaganda for compulsory health insurance," argued Hoffman in an influential 1917 *Scientific Monthly* article, rested "upon the obvious fallacy that 'prevention is primarily the purpose of insurance.' Prevention has nothing directly to do with insurance."[107] Health work mattered to America, he agreed, and progressive reforms like extended collective bargaining or improved working conditions should be pursued vigorously. But health insurance had nothing to do with health work or welfare or social economy. He blamed a tiny cabal, including Fisher, for instigating a plan to impose compulsory insurance on Americans, in direct conflict with the "tendencies" and traditions of the American nation.[108] That cabal loved Europe irrationally (the continent was then ensconced in a horrible war)

---

106. See, for instance, Fisher's comments in Irving Fisher, "Public Health as a Social Movement," *Reprints of Reports and Addresses of the National Conference of Social Work* 95 (1917): 4–5, in Reel 7, Fisher Papers.

107. Frederick L. Hoffman, "Some Fallacies of Compulsory Health Insurance," *Scientific Monthly* 4, no. 4 (1917): 306–316 at 311.

108. Hoffman, "Some Fallacies of Compulsory Health Insurance," 307. For the author of *Race Traits and Tendencies of the American Negro*, the word "tendencies" carries added weight. Hoffman's idea of nation might well be translated to "race." And like many of his contemporaries, Hoffman thought that nations/races had deeply ingrained racial characteristics that politics would have a hard time overcoming.

and trusted Germany (the aggressor) blindly. Compulsory health insurance, he insisted, amounted to an "audacious attempt" to mislead Americans. It won its most fervent support from the "Socialist Labor Party."[109]

Hoffman's side won the fight: compulsory state health insurance bills failed everywhere. The economist and health insurance advocate John Commons blamed Hoffman's massive propaganda efforts for a significant share of the defeat.[110] Yet Hoffman had failed to divorce insurance from health and welfare (which, given his ambivalences, might have comforted him). Metropolitan's internal figures suggested that the visiting nurse service paid dividends quickly. When public figures on preventative medical examinations became available in 1922, those showed a nearly 200 percent return on investment.[111] Moreover, the good press the company received transmuted health work into advertising. Metropolitan *expanded* its welfare activities and touted its commitment to social work.[112]

Haley Fiske imagined in 1917, expansively, "the time when, instead of one in every five, four in every five of the population shall be insured in Industrial mutual insurance companies; and in the development of these companies along Welfare lines one may look to the time when the people shall take care of themselves through life insurance in a service covering health in life, care in sickness, indemnity in death, sanitation in community life, the financing of home-owning, of public utilities and civic conveniences—a mutual service of co-operation among such a large proportion of the population that it may be called The New Socialism!"[113] In 1920, *How to Live* entered its seventeenth printing, now a global text having been translated into French, German, Italian, Dutch, Chinese, and Spanish, and found on health course syllabi at Yale, the University of California, and

109. Hoffman, "Some Fallacies of Compulsory Health Insurance," 315.

110. Rodgers, *Atlantic Crossings*, 264.

111. Dublin, *Effect of Life Conservation*; Louis I. Dublin, "Address by Louis I. Dublin, Ph.D.," *Addresses Delivered by August S. Knight, M.D. and Dr. Louis I. Dublin at the Annual Banquet of the Officers and Hygiene Reference Board of the Life Extension Institute* (New York: Life Extension Institute, 1923), 10.

112. As of 1924, it offered LEI medical examinations to all ordinary, intermediate, and special class policyholders, instead of just to those with larger policies. Metropolitan Life, *Epoch in Life Insurance* (1924), xviii.

113. Metropolitan Life, *Epoch in Life Insurance* (1917), xxiv. Buley argues that the "social aims of life insurance" on the rise in 1909 "manifested in the life insurance companies' effort to forestall the 'welfare state,' the central theme of the 'progressive' movement of the period, with a social program of their own." *American Life Convention*, 1:353. Such an interpretation misses a key point: Metropolitan did not seek to forestall the welfare state so much as to become it.

other universities.[114] Thousands of Americans volunteered for preventative medical examinations while nurses made well over a million more visits.[115] Risk makers' reach was extending.

In 1921, Metropolitan entered the health insurance field, bringing risk-rated policies and thus a "scientific basis" to that industry, and further integrating health with insurance.[116] (Prudential began offering group health plans five years later.)[117] Hoffman undertook an extensive trip to South America that same year, where he measured the bodies of nonwhites and questioned whites about their health—a comfortable return to traditional risk prediction as he tried to set guidelines for insuring those responsible for American commercial expansion in the global South.[118] He was, perhaps fortunately, too far away to see Prudential dismantle his statistical library. His patron and protector, John Dryden, long gone, and Dryden's son Forrest embarrassed by scandal, Prudential came under new leadership who saw no need for Hoffman's employ. Nor did they see value in Hoffman's library and its—for the most part—fatalizing data.[119] In 1921, Prudential shunted the entire collection—over one hundred thousand "papers and publications," including all those cemetery records—off on the Army Medical Library (now the National Library of Medicine in the National Institutes of Health,

---

114. Fisher and Fisk, *How to Live* (1920), iv, xi.

115. Hirshbein, "Masculinity, Work, and the Fountain of Youth," 96. I don't have the precise figures for nurse visits in 1920. But in 1918, there were 1.4 million nursing visits. Metropolitan Life, *Annual Report* (1918): 38. In 1916, nurses treated 217,422 cases and made over a million visits. See Lee K. Frankel and Louis I. Dublin, "Visiting Nursing and Life Insurance," *Publications of the American Statistical Association* 16, no. 122 (June 1918): 58–112 at 58.

116. Metropolitan Life, *Epoch in Life Insurance* (1924), 168–172. Deborah A. Stone mourns Metropolitan and its peers' actuarial ("scientific") rating for eventually driving out community rating systems that did not discriminate among individuals or exclude preexisting conditions. See Stone, "The Struggle for the Soul of Health Insurance," *Journal of Health Politics, Policy and Law* 18, no. 2 (1993): 287–317.

117. Beatrix Hoffman (no relation to Frederick) argues: "As with employers, insurance company reaction to the threat of compulsory health insurance was clearly one of the factors that spurred the growth of welfare capitalism in the twenties." In *Wages of Sickness*, 112.

118. See Book 26a: Letters of South American Trip, Nos. 1–50, Book 26B: Letters of South American Trip Nos. 101–150, and Book 26c Letters of the South American Trip Nos. 151–210 and Conclusions. Especially Hoffman to Forrest F. Dryden, 17 January 1922, "South American Scientific Investigations. Conclusions and Recommendations—1," in Book 26c, Box 8, Hoffman Papers.

119. Hoffman to Forrest F. Dryden, 26 February 1920, in Book 25, Box 8, Hoffman Papers. Hoffman resigned his title of Statistician and Third Vice President on May 22, 1922, settling for Consulting Statistician. See Sypher, *Frederick L. Hoffman*, 139.

where they all remain today).[120] When Hoffman started a new library—this time in a consulting role, without help from his former department at Prudential—he directed attention to healing and especially to cancer.[121] The future of risk making lay in life as much as in death, and even Hoffman acknowledged that fact.

120. Prudential and Hoffman reserved the right to borrow these materials without any restrictions. Hoffman to McDonald, 27 October 1933, page 2 in Folder "1930s + 1940s Misc. letters mostly to FLH, some by FLH," Box 26, Hoffman Papers.

121. Ellice McDonald, the director of cancer research at the University of Pennsylvania's graduate school of medicine, eventually inherited that library, when Prudential would no longer house it. McDonald suggested that he install the collection as the "Frederick Hoffman Memorial Library." Hoffman demurred in this case, preferring that the collection be called a "statistical laboratory," not a library. Ellice McDonald to Hoffman, 6 May 1933; Hoffman to McDonald, 17 July 1933, in Folder "1930s + 1940s Misc. letters mostly to FLH, some by FLH," Box 26, Hoffman Papers. Hoffman had published a massive book on cancer in 1915 in which he explained the "menace of cancer" as a threat that increased in significance even as other causes of death became less potent: Frederick L. Hoffman, *The Mortality from Cancer throughout the World* (Newark, NJ: Prudential Press, 1915).

SIX

Valuing Lives, in Four Movements

Charles Ives lived a double life—the life of an odd modernist superhero. By day, he headed the largest MONY life insurance agency in New York, at the side of his longtime partner Julian Myrick. Ives had little taste for work in the field and so focused instead on recruiting and training young agents from places like his alma mater—Yale—to whom he promised a stable income (but "no big fortunes"), a sense of purpose (selling "an essential commodity"), and opportunities for "individuality, ingenuity or initiative." An Emersonian, Ives introduced his job ads with an aphorism from "Self-Reliance": "I appeal from your customs: I must be myself."[1]

By night (and early in the mornings too), Ives appealed from the customs of his time—he strayed from the norms for a corporate manager—and composed music for himself. It had to be for himself, since few of his contemporaries would ever hear it, and since few musicians of the moment could figure out how to play it. Ives hewed modernist compositions, roiled by dissonances and suffused by lovely harmonies, from traditional Americana juxtaposed with the noise of urban life. By 1920, when he joins this story, Ives had a substantial body of work under his belt, including pieces like *Central Park in the Dark* and *Housatonic at Stockbridge*, bound for eventual fame.[2]

1. Ives and Myrick Agency, "I appeal from your customs: I must be myself," in *Yale Alumni Weekly*, 10 January 1930, page 469 clipped in page 51 of Business Scrapbook of Charles Ives 1919–1935, MONY Papers.

2. (Might I suggest the reader include both pieces in her "soundtrack" for this chapter?) In 1920, Ives had also been writing *Essays before a Sonata*, a prose piece that—in trying to explain and defend Emerson and the transcendentalists—explained and defended Ives's *Concord Sonata*. Henry Cowell and Sidney Cowell, *Charles Ives and His Music* (New York: Oxford University Press, 1969), 12, 36, 64–74, 81–86.

In this chapter, I will "perform" an Ives (prose) piece from 1920, revised and reprinted that year under the title *The Amount to Carry—Measuring the Prospect*. Ives wrote "four principal movements" into *The Amount to Carry*, a pamphlet distributed to his agents and to others through industry newspapers.[3] He pitched the piece as its own sort of modernist composition, emphasizing the way it broke from older traditions of life insurance sales to present a "scientific" approach more fitting the twentieth century and the businessmen to whom his agents addressed most of their attentions. In its superficial emphases it differed a bit from older sales pitches—less fear of death and more fear of old-age penury—and it differed because of those expected to employ it: "nice, up-in-the-intellectual-clouds college boys who frown on salesmanship" until they eventually fail or get bored, as critics of Ives's methods sometimes charged.[4] Otherwise, Ives's pitch looked pretty traditional.

For the most part, Ives—who had learned to forecast, fatalize, and smooth from the master, Emory McClintock, in whose actuarial department Ives first landed straight out of Yale[5]—stuck with traditional ideas and methods of risk making. Indeed, he instructed his agents to ground their sales pitches in what he considered to be the certainty of life tables, the surety of statistical methods, and the precision of prediction. The modern conception of death had taken root in the disturbed corporate soil of a life insurance industry unsettled by the Armstrong hearings and their aftermath. But there were other ways to respond to Armstrong instabilities. Ives represents a crowd of life insurance folk who did not repurpose risk-making methods in the service of changing the future, of exerting control over the laws of life. Instead, Ives grounded his work and his agents' efforts in the dogmas of the "science of life insurance" (see figure 4.3) and renewed the argument—the same argument that had struggled, as we've seen, throughout the late nineteenth century—that statistical methods could explain individual Americans.

3. Charles E. Ives, "The Amount to Carry—Measuring the Prospect," *Eastern Underwriter* (17 September 1920): 35–38 at 37. First written and published in 1912, the piece went through multiple revisions and reprintings. Cowell, *Charles Ives and His Music*, 53. A pamphlet version made by Ives and Myrick for distribution among agents exists in the Business Scrapbook of Charles Ives 1919–1935, MONY Papers.

4. These were a Connecticut Mutual agent's words in William Cahn, *A Matter of Life and Death: The Connecticut Mutual Story* (New York: Random House, 1970), 249–250.

5. Cowell, *Charles Ives and His Music*, 38.

In homage to Ives, I will offer alongside Ives's composition two varia-tions on his theme, all answering the question: what is the proper way to value a life? Where Ives offered up a traditionalist's answer, explaining how to predict an individual's economic and vital futures together, each of these accompanying variations shows how life valuations—in dollars or years of longevity—changed as a result of Fisher's modern conception of death. One variation returns us to Oscar Rogers, the brilliant medical director, and in-troduces some of his more skeptical colleagues as they tussled over a sys-tem for rating every potential insurance risk with a single number—only to have their rating methods borrowed, stripped of their original context, and remade into health tools aimed at the general public. In the other, Louis Dublin appears along with *his* more skeptical employee, the brilliant (and like Ives, underappreciated) mathematical biologist, Alfred J. Lotka, who agreed with his boss on the best mathematical method for valuing a life but disputed the true value of a man to society—only to have his life valuations embraced in public discourse as a new way to dignify life.

Ives's theme sets the tempo and structures the entire piece. His is a story of continuities, our baseline for making comparisons. In counterpoint, the two variations allow us to explore patterns of change. They make it possible to ask: what happened when risk making methods moved, when they left life insurers' offices behind to venture into new institutions speaking to dif-ferent audiences? They make it possible to investigate the question: what emerged when risk makers tuned their methods toward extending lives and controlling fates instead of merely predicting life spans and rating insurance applicants?

The answers direct us toward some noteworthy harmonies. In her work on Americans' responses to social scientific surveying (sociological investi-gation, opinion polling, and sex research) in interwar America, the historian Sarah Igo has described a growing openness among the quantified to their own quantification, indeed even an eagerness among some Americans in mass society to see themselves reflected in or mediated by statistical averages.[6] Around that same time, as this chapter will demonstrate, Americans also began to warm to seeing themselves, at least partially, through risk makers' lenses. They became more open to thinking about their lives and the lives of their neighbors in terms of dollars and expected years—and not just when contemplating life insurance. In some ways, risk makers went further than

---

6. Igo, *Averaged American.*

social surveyors, whose data aided mainly in the construction of a sense of mass identity. Risk makers' statistical findings could not only aggregate (by, for instance, collectively valuing the nation), but also individuate (as when they allowed an overweight American to calculate her life expectancy).

The propagation of the modern conception of death's gospel (of control and change trumping prediction) did more than anything else to open new spaces for the statistical to engage (and win over) the individual. Risk makers trumpeting the modern conception shared the by-products of their work more widely, by-products like weight tables and life valuations. And even more importantly, Americans showed a new interest in thinking with such by-products—and submitting themselves to further risk making—now that statistical individuation offered hope for a better future.

It helped that Americans like those described by Igo were becoming more comfortable with social averages. A salesman like Ives tried to take advantage of such trends and the popularity of his method suggests he met with some success—although it will also become clear in this chapter that he and his agents still had quite a bit of convincing to do day to day, and that seldom would statistical argument win out on its own. In the context of reappropriating risk making to extend lives, broadening sympathies for quantifying made risk-making by-products all the more attractive. As numbers moved from one context to another, from the risk makers' hands to awaiting consumers, those numbers thinned, losing their histories and complexities, taking on the appearance of simple truth, of objective facts— the kinds of truths and facts that social scientists were training Americans to understand and appreciate.[7]

Alongside these harmonies, dissonances sounded too. They stemmed mainly from the risk makers, who sometimes ended up *less* comfortable with the making of statistical individuals as they looked on helplessly as others ran with their numbers, as others dreamed up new uses for their old methods. The modern conception of death helped risk making take root in doctors' offices, on penny scales, in popular books, and in the daily press, but the quantifiers squirmed with the realization that their thinned, contextless numbers could mislead and in being misleading become even more popular. With statistical individuation growing more attractive among ordinary Americans, risk makers became their own critics. In some cases,

---

7. How and why numbers become thin, and why thinness became a cardinal virtue in the twentieth century, are questions that deserve further attention. Porter sets out a research agenda for the thick description of thinness in Porter, "Thin Description."

risk makers retreated from the modern conception of death project, turning back to prediction and risk rating; they tried, unsuccessfully, to put the statistical genie back in its corporate bottle. At the same time, old dissonances clanged in traditional risk making too, as Charles Ives made clear. In the end, he also counseled retreat on occasion—away from reason and science, giving in on the mission of making statistical individuals in service of making the sale. Of course, there had to be dissonances. After all, this is, fundamentally, a piece by Charles Ives.

Movement 1

"You either belong to the 8 per cent or the 92 per cent." So began one of Ives's twenty possible openings for his first movement. He did not call it an overture, but it was, one meant to catch the attention of a public often resistant to the life insurance agents' sales pitch. The overture continued: "If to the latter, your income stops at death, and you won't have a cent if you reach age 65. That seems rather an abrupt way to talk to a successful business man. But I'm talking facts based on authoritative data, not personalities. The more securely you're fixed today, the more reason you will be interested in a scientific formula that will offset the chances against you. And there are chances against everybody. Even a millionaire is a fool to take them."[8] In his various openings, Ives promised "facts," "authoritative data," "a scientific formula," "natural laws," and the "law of averages" as guides to the dangers of dying or of living too long. He offered smoothed data and knowable fates (for those in a certain class—the 92 percent), via his instantiation of an old "calculation" that answered an even older question: how much insurance to carry. He expected the traditional ways of thinking about death (and to some extent financial hardship in old age)—as predictable, calculable, and inevitable—to sell insurance, even to millionaires. And in the process, he—like generations of agents before him—sold "prospects" on the insurers' dream of individualized statistical futures.

* * *

In October 1919, Oscar Rogers, the New-York Life medical director, gave a talk with his actuarial colleague, Arthur Hunter, setting forth a different sort of calculation. The two men explained their "numerical method," a

8. Ives, "Amount to Carry," 37.

"scientific" answer to another old life insurance question: *"What is the Value of this life from the standpoint of longevity?"*[9] Since 1904, every application for life insurance to New-York Life had been valued with an eye toward precisely measuring the risk of mortality it posed according to the numerical method. For every applicant, medical department staff—often clerks instead of doctors—generated a single number, a "rating" or score indicating that individual's relative risk of death. (The perfectly average applicant scored 100. A higher number was bad: it meant a shorter expectation of life. Those with a rating around or above 125 thus fell in the substandard category. Lower scores, on the other hand, correlated to longer lives.) An actuary, having each applicant's rating in hand, then decided what sort of policy to offer and on what terms. Substandard applicants would be charged more or be simply declined.[10] Categorizing applicants as acceptable or substandard was hardly new, but the use of precise numerical ratings only appeared with New-York Life in the early twentieth century. By 1919, a handful of companies employed a similar system, while many others used parts of it.[11] By midcentury, nearly all companies had moved to some kind of numerical method.[12]

Numerically rating risks, like all past systems for classing risks, involved writing a life down on paper. But the numerical method resembled an automated process more than most risk-making systems of the past, and so required that lives be even more thoroughly abstracted and quantified than had been necessary before. The system for preparing lives to be processed began with an application, wherein agents and applicants fit a whole person into a series of standardized questions and measurements. Medical clerks translated applications into a shorthand description of relevant details, amounting to something like: "A jeweler, aged 35. Height 5 ft. 10 in. Weight 198 lbs. (20 per cent. over-weight). Family history slightly better than average. Contracted syphilis 14 years ago. He had mild secondaries for

9. Oscar H. Rogers and Arthur Hunter, "The Numerical Method of Determining the Value of Risks for Insurance," *Transactions of the Actuarial Society of America* 20, no. 62 (1919): 273–332 at 274. For the old question, see New-York Life, *Temple of Humanity*, 49. Emphasis in the original.

10. The precise score limits varied from company to company. Rogers and Hunter, "Numerical Method," 298–299. On the division of labor between medical and actuarial staff, see the testimony of Rufus Weeks in New York Legislature, *Testimony*, 2: 1111–1112; and Rogers, "Medical Selection and Substandard Business," 84–85. Charging more could mean a higher premium. It could also mean a "lien" against the policy claim that gradually decreased in value over time.

11. Rogers and Hunter, "Numerical Method," 274.

12. Pearce Shepherd, "Principles and Problems of Selection and Underwriting," in *Life Insurance Trends at Mid-Century*, ed. David McCahan (Philadelphia: University of Pennsylvania Press, 1950), 53.

six months and was under treatment for eighteen months after the disappearance of all symptoms. He was never intoxicated, but used alcohol rather freely up to five years ago, when he married, and he has been practically an abstainer ever since." From that description, clerks guided by specialized tables constructed an arithmetic problem starting with a "basic rating" and adding or subtracting depending on each of the above "factors" (which began to replace the older terms "impairments" and "classes"). In Rogers and Hunter's example, the jeweler began with a basic rating of 115 by virtue of his overweight, gained (bad) 5 points for height, 40 points for "personal history" and 40 points for "habits," while losing (good) 5 points for his good family history, and faced no consequence for his occupation. His final rating of 195 made him all but uninsurable.[13] From the insurer's standpoint of longevity, he had a high rating and a very low value.

Behind numerical ratings stood an extraordinary research program and a growing collection of numerical tables. Traditional risk prediction, it must be clear, continued to develop and evolve even after the modern conception of death opened new avenues of exploration to risk makers. As we saw in chapter 3, Rogers had pioneered the collaboration of actuaries and medical directors to change selection procedures at New-York Life. Emory McClintock had brought the actuarial profession into the project and made possible the 1903 Specialized Mortality Investigation, which first showed the impact of a wide range of classifications on the mortality of men insured by American life insurers.[14] But that investigation did not meet the needs of numerical rating, which depended on isolating the impact of each factor on mortality to make its addition/subtraction scheme work.[15] Arthur Hunter, a young actuary trained in Rogers's company, took control in 1909 of a new study, dubbed the Medico-Actuarial Mortality Investigation, ensuring it would better suit the needs of numerical rating.[16] The study, made possible by Hollerith sorting and tabulating machines, eventually considered almost three hundred possible factors affecting mortality—including 168 hazardous occupations, 99 medical impairments, 15 locales, and 9 classes dealing

13. Rogers and Hunter, "Numerical Method," 297.

14. Actuarial Society of America, *Experience of Thirty-Four Life Companies.*

15. As will become more clear, alternative quantification schemes to addition/subtraction were imagined at this time and they did not require isolating factors. But Rogers insisted that a good study would look at mortality rates among "homogenous groups." The Specialized Mortality Investigation had decided against such an approach, in favor of having larger sample sizes. See Rogers, "Medical Selection and Substandard Business," 83.

16. Association of Life Insurance Medical Directors of America (ALIMDA) and Actuarial Society of America (ASA), *Medico-Actuarial Mortality Investigation* (New York, 1912), 1:4–5.

with race and sex.[17] But to begin with, the study focused on the topic that mattered most to Hunter, Rogers, and numerical rating: "build."[18]

Build, meaning the ratio of height to weight, provided the basis for numerical rating. It was an objective measure and one easily obtained (at least since 1900 when most doctors, in no small part because of life insurers' demands, could be counted on to own an office scale),[19] so clerks looking at build could quickly weed out a large portion of a company's bad applicants for underweight (if tied to tubercular family history) or for overweight.[20] Though Rogers expanded his numerical method beyond build, he kept it as the foundation of all other calculations. The first volume published by the Medico-Actuarial Mortality Investigation established the average weight and the distribution of weights around the average for any given height and age.[21] With that known, the next phase of the investigation drew on 744,672 policies on insured men and 393,032 on women to correlate rates of mortality at every age to degrees of over or under the average weight.[22] Those correlations became the basis for what Rogers called the basic rating, a figure determined by the mortality associated with any given sex, age, height, and weight.

As Rogers, Hunter, and New-York Life continued to develop the numerical method, build grew more complicated. Studies considered whether

17. ALIMDA and ASA, *Medico-Actuarial Mortality Investigation*, 1:8. See also Yates, *Structuring the Information Age*, 48–49.

18. ALIMDA and ASA, *Medico-Actuarial Mortality Investigation*, 1:6. Build, including its origins in life insurance and its progress out into the wider world, has enjoyed the attention of some excellent scholars. This chapter makes use of their work, but looks in ways they did not into the tensions that surrounded numerical rating and the popularization of build standards. See Porter, "Life Insurance, Medical Testing, and the Management of Mortality"; Weigley, "Average? Ideal? Desirable?"; Czerniawski, "From Average to Ideal"; and Deborah Levine, "Managing American Bodies: Diet, Nutrition, and Obesity in America 1840–1920" (PhD diss., Harvard University, 2008).

19. Levine, "Managing American Bodies," 109–110. On the other hand, scale reading doesn't seem to have been that precisely done, if scales were in fact widely used. The Medico-Actuarial Mortality Investigation found that about two-fifths of all weights reported to companies were likely rounded to the nearest five or ten, or were simply estimations. See ALIMDA and ASA, *Medico-Actuarial Mortality Investigation*, 1:16.

20. Rogers, by 1906, saw all overweight over 25 percent to be basically unacceptable at standard rates. See Shepherd, "Relation of Build (*I.E.* Height and Weight) to Longevity," 61–66 for Rogers's comments. His first published work on build dates to 1901: Oscar H. Rogers, "Build as a Factor Influencing Longevity," *Abstract of the Proceedings of the Association of Life Insurance Medical Directors of America* 12 (1901): 280–288.

21. Details of the build study are found in ALIMDA and ASA, *Medico-Actuarial Mortality Investigation*, 1:108–109.

22. ALIMDA and ASA, *Medico-Actuarial Mortality Investigation* (New York, 1913), 2:8, 36.

height might mitigate overweight's negative impact on mortality (it actually did the opposite) or whether credit could be given to those who were indeed just more thickly or muscularly built, as judged by a comparison between chest and abdominal girth (it could).[23] And build even insinuated itself into supposedly homogenous classes, as in studies of mortality based on family history of tuberculosis at various degrees of over- or underweight.[24] Simplifying men (to risks) and valuing them (as numbers) required a new bodily arithmetic: "A bank officer, aged 50. Height 6 ft. Weight 256 lbs. (40 per cent. over-weight). Shows an abdominal girth 3 inches less than chest expanded. Has a good family history." Burdened from the start with a basic rating of 165 for his sex, age, height, and weight, this man gained 10 more points (a bad thing) for being too tall, while losing 10 points (a good thing) for his triangular torso, another 10 points for a family untainted by tuberculosis, and a final 10 points for holding down a job in finance. Rated 145, he couldn't get insurance either.[25]

* * *

Ives's overture began with men's odds of dying in the future (a bodily valuation), but at its end turned to their paternal values (men's values to their families) by way of wives' failures in the present. Ives instructed the agent to begin: "If I ask you a rather personal question, I hope you won't take offence." Having intrigued his prospect, the agent continues: "I can be of help to you in a vital matter. Is your wife spending too little or too much for clothes?" Or, framing the wife as CEO of the household: "Is your wife making her side of the business pay?"[26] When it came to valuing lives, for the salesman of ordinary life insurance, money came inextricably tied to gender and sex.

These "personal questions" hinged on the assumption that a man's first responsibility in life or death was to support his wife and children. Ever since the days of T. S. Lambert, life insurers had been selling ordinary life insurance as a kind of class protection: it kept widows out of the workforce, thereby protecting the family's claim to middle-class (or higher) status.[27]

---

23. Rogers and Hunter, "Numerical Method," 278–289.
24. Rogers and Hunter, "Numerical Method," 293.
25. Rogers and Hunter, "Numerical Method," 297.
26. Ives, "Amount to Carry," 37. For a contemporary, but quite different, view of women as potential sources of domestic efficiency, made possible by science, see Walter Lippmann, *Drift and Mastery* (New York: Mitchell Kennerley, 1914), 213–239.
27. Lambert's better respected contemporary, Elizur Wright, valued protecting wives from labor as one element of a wider free-labor ideology that exalted male independence in the

That "sentimental" message had been sold so long and so well, argued Ives, that "it is now a matter of common judgment that one of the primal duties of man is to make a living for himself and his dependents during his active life-period,—whether he happens to spend all or part of this period in this world or the next."[28] But Ives thought that modern businessmen, while aware of their duties, tended to lowball the magnitude of their familial obligations.

To bring home the inadequacy of most life insurance policy values, Ives turned to data on the cost of living. He gave his agents charts showing the costs of common necessities.[29] And he explained the results of recent studies that fixed a "widow's personal living cost"—at a level that made sure her "standard of living is not materially lowered"—at 30–35 percent of family income. Similar studies yielded the lower but still substantial cost of maintaining children until adulthood (fixed at twenty years, the average of eighteen and twenty-two, to account for those going to college).[30] Once the prospect saw those figures and looked at the cost of household goods, he saw the problem—or so Ives schemed: his $1,000 policy would not be nearly enough. The sale began with shame: "The $1,000 policy-minded man must always be confronted with the fact that he has provided the generous sum of $4.16 per week for his whole family to live on for five years."[31]

Risk makers traditionally fostered the ideal that each man should have enough life insurance to fully replace himself: enough to pay his survivors his current income minus the cost of maintaining himself (that cost being set in Ives's calculation at 36 percent). Translating that desire into life insurance took a bit of arithmetic and an understanding of how to determine the value today of dollars to be earned in the future. The agent, having

---

public sphere and female power in the private sphere. He made special pleas for life insurance as a tool that allowed young marriage, since it allowed men to forgo building up savings against their own chance of death. See Wright, January 1859 report in *Massachusetts Reports on Life Insurance: 1859–1865*, 2–3. One might even see life insurance as a powerful force in the creation of the middle-class ideal of separate spheres, especially after life insurers managed to secure legislation that, in the words of historian Sharon Murphy, assured that "wives, by definition, were economic dependents of their husbands." See Murphy, *Investing in Life*, 145.

28. Ives, "Amount to Carry," 36.

29. That sheet, Ives noted, had worked very well. But inflation in 1919 had made all such numbers increasingly speculative. See Ives, "Amount to Carry," 36. For more on difficulties establishing the cost of living in the United States after World War I, see Stapleford, *Cost of Living in America*, 59–95.

30. Ives, "Amount to Carry," 36.

31. Ives, "Amount to Carry," 36.

completed the calculation, could tell a man the present value of all his future earnings minus his future expenses. This, the agent would argue, constituted each man's ultimate responsibility to his family, his paternal value.

That value could be shockingly high. That's why Ives suggested that agents sow some seeds of doubt about a wife's efficiency. Those cost-of-living figures could be useful for more than convincing a man that he owed his family more. They could also suggest a way to pay the higher premiums that came with his newly discovered insurance responsibility. "The underlying aim," wrote Ives, "was . . . to encourage the prospect (and his wife) to take more interest in domestic science; to help the family income produce more for the family than to produce an effect on the neighbors."[32] In the name of protecting a woman's future place in her home (should her husband grow too old to work or die too young), Ives encouraged men to poke their noses into her domain now. Ives's agents, in their pitches, propounded a modern conception of marriage and efficient living founded on the precise determination of each man's dollar value.

* * *

In 1925, Alfred J. Lotka approached Louis Dublin about beginning a new project for the calculation of paternal values. Lotka had come to Metropolitan a year earlier to serve as the supervisor of mathematical research in the Statistical Department. Lotka and Dublin had already published one paper together—"On the True Rate of Natural Increase"—bound to become foundational to modern demography.[33] Hungry for new challenges, Lotka pointed to an old question: calculating the "value of an individual."[34] Dublin liked the idea and set Lotka and his assistant Bessie Bunzel loose gathering materials and making calculations.[35]

Lotka approached valuing individuals in the traditional spirit of risk making, as had Ives. But Dublin approved the project for different reasons. He wanted to use paternal valuations to *control* the future, instead of as a tool for hedging against it. Since 1909, Dublin had been working to determine:

32. Ives, "Amount to Carry," 36.

33. Louis I. Dublin and Alfred J. Lotka, "On the True Rate of Natural Increase," *Journal of the American Statistical Association* 20, no. 151 (1925): 305–339.

34. Lotka to Dublin, "Subject: *The Value of an Individual*," 18 May 1925, in Folder 4, Box 31, Alfred J. Lotka Papers, Public Policy Papers, Department of Rare Books and Special Collections, Princeton University Library.

35. Lotka, "Subject: *Memorandum of Interview with Dr. Dublin on Study 'The Value of a Man,'*" 29 May 1925, in Folder 4, Box 31, Lotka Papers.

"does it pay?" But the "it" had changed from welfare work in Metropolitan's particular case to public health interventions generally. Dublin's data convinced him that Metropolitan's welfare work—its mobilization of insurer resources to support social improvement and the modern conception of death—resulted in longer lives, lower death claims, and thus more premiums earned. In fact, he regularly bragged in the 1920s of a 200 percent return on welfare investments.[36] Life extension wasn't simply a nice thing to do, in other words—it made business sense. Satisfied with Metropolitan's success in implementing the modern conception of death, he turned his attention to constructing arguments for broader social investments in public health infrastructure, under the slogan "Health Work Pays."

He encountered new problems. "The community's loss by premature death is another problematical value," he explained in 1925, "because we do not as yet know the money value of a human life at various ages."[37] As a stopgap measure, Dublin used $100 per year per person, claiming that to be the amount that the national wealth grew each year. Yet, he was not satisfied.

Exactly one year later, in November 1926, Dublin published an article in *Harper's Monthly Magazine* on "The Economics of World Health." In the intervening year, Dublin and his colleagues had realized that they could use the same methods—with some tweaks—that Ives and his colleagues in insurance sales had been using to value businessmen's lives.[38] The "value of a man as a wage earner" they explained, could be found by calculating the "present worth at age 18 of his future earnings" (more than $41,000, for a man making $2,500 a year) minus "the present worth of his future expenses" (less than $13,000). That gave a total value for a wage-earning man of about $29,000. That was nearly three times the cost to a wage-earning family of raising a child (calculated to be around $10,000). The value of a

36. Louis I. Dublin, *Health Work Pays* (New York: Metropolitan Life Insurance Company, 1925), 1–11 at 10, reprinted from *Survey Graphic* (November 1925).

37. Dublin, *Health Work Pays*, 2.

38. Along with Ives, others including Solomon Huebner and Edward Woods developed techniques for valuing or "capitalizing" human lives. See Solomon S. Huebner, *Life Insurance: A Textbook* (New York: D. Appleton, 1919), 14; Stalson, *Marketing Life Insurance*, 580–582; and Zelizer, *Morals and Markets*, 61–65. These methods also had precursors outside life insurance. Dublin and Lotka pointed to predecessors such as William Petty, William Farr, and Ernst Engel. But even here the ties to life insurance remain complex, since Farr's methods evolved through his work with and around British actuaries. Interestingly enough, they did not explicitly cite the common practices of insurance salesmen. Louis I. Dublin and Alfred J. Lotka, *The Money Value of a Man* (New York: Ronald Press, 1930), 7–21.

wage earner changed with age, peaking at age twenty-five and turning negative at seventy, but even a baby was worth $9,333. Having kids and making sure they grew old, argued Dublin, amounted to one of the best investments any community could make.[39]

Adopting a life insurance salesman's model for valuing lives had an obvious drawback, one that Dublin noted. It gave him no way to value the life of a woman (or anyone else) who did not draw a wage. For argument's sake, Dublin guessed conservatively (or condescendingly) that "the economic value of women in general is only one-half that of men," a figure meant to include savings to the family from women's work at home alongside women's wage work.[40] With that guess on the books, and while lamenting repeatedly his inability to measure a woman's worth so exactly as a man's, Dublin rolled out the big numbers.

A "thousand billion dollars"—$1 trillion: Dublin announced that value for a nation of American wage workers. Add in women and the total swelled to $1.5 trillion, five times the "ordinary material wealth"—the "real property, live stock, machinery, agricultural and mining products" of the nation.[41] Savings from "the application of modern preventative medicine and public-health measures" could exceed $6 billion, not bad at a price of $2 per person or a total of around a quarter of a billion dollars.[42]

Dublin's paternal valuations argued strongly for more paternalist state interventions. His message reached, for starters, *Harper's* two hundred thousand readers.[43] The article struck a nerve, and requests for reprints came quickly, bringing it before the eyes of students in classrooms at the Wharton School, Williams College, or University of Colorado, Boulder, of public health officers from Oakland to Oklahoma to Ontario, of employees of large corporations, and of aldermen, businessmen, and teachers across the country.[44] Susan Blakey Goodykoontz of Boulder received a copy from her pastor

39. Louis I. Dublin, *Economics of World Health* (New York: Metropolitan Life Insurance Company, 1926), 1–8 at 2, reprinted from *Harper's Monthly Magazine* (November 1926).

40. Dublin, *Economics of World Health*, 3.

41. Dublin, *Economics of World Health*, 3.

42. Dublin, *Economics of World Health*, 4.

43. That is *Harper's* figure for circulation. In the same letter that Wells (the editor) gave it, he also asked Dublin to only mention Metropolitan once, lest the article look like an advertisement. Dublin obliged. Thomas B. Wells to Louis I. Dublin, 30 August 1926, in Folder "Economics of World Health," Box 10, Dublin Papers.

44. H .E. Howe to Louis I. Dublin, 1 September 1926; Louis I. Dublin to Thomas B. Wells, 16 September 1926; William Carroll Hill to Louis I. Dublin, 15 October 1926; D. T. Bowden to Louis I. Dublin, 20 November 1926; W. P. Shepard to Louis I. Dublin, 20 November 1926; C. D.

and passed it around among the other "mothers" who were using its arguments to push Congress to renew support through the Sheppard-Towner Act for health centers specializing in treating pregnant women and their young children. Like other aspects of American maternalist social policies that had thrived in the early twentieth century, Sheppard-Towner looked doomed (and was doomed in 1927) in the face of American Medical Association opposition. Goodykoontz urged Dublin to intervene, and he probably would have liked to help.[45] But, he explained, "as a member of this large business organization, it would never do for me to express my personal preferences on a hotly contested political question."[46] Paternal valuations might have supported maternalist policies directly, had corporate policy not forbade it. Metropolitan preferred to use Dublin's piece and its popularity to "skim some of the business cream for ourselves," as a company vice president put it.[47] Dublin adapted paternal valuations with deep roots in salesmanship, and his adaptations still sold life insurance.

Pulling together Ives's theme and its variations from this movement, we see that risk makers' methods for valuing lives in the 1920s could result in a risk score or in a dollar amount. They could be used to sell insurance or to sell public health, or to do both at once. The science of life insurance progressed in its traditional form and in its reform incarnation, all while making more individuals statistical. At least, that was the tune that the risk makers played publicly. In private (as we'll see in the second movement), they sometimes struck notes of caution.

Barrett to Louis I. Dublin, 22 November 1926; Susan Blakey Goodykoontz to Louis I. Dublin, 30 December 1926; Herbert W. Hess to Louis I. Dublin, 14 January 1927; Louis I. Dublin to J. B. Gibson (Safety and Health Director, Western Electric Company), 5 January 1927; the US Chamber of Commerce wanted fourteen hundred copies: James L. Madden to Louis I. Dublin, 31 January 1927; Gordon Bates to Louis I. Dublin, 2 April 1927; F. Adams to Louis I. Dublin, 28 April 1928, in Folder "Economics of World Health," Box 10, Dublin Papers.

45. Dublin lauded Sheppard-Towner after his retirement not only for combating maternal and childhood mortality, but for showing the way that federal funds could spearhead state public health interventions. See Dublin, *After Eighty Years*, 112–114. On Sheppard-Towner, see Paul Starr, *The Social Transformation of American Medicine: The Rise of a Sovereign Profession and the Making of a Vast Industry* (New York: Basic Books, 1982), 260–261. On maternalist social politics, see Seth Koven and Sonya Michel, eds., *Mothers of a New World: Maternalist Politics and the Origins of Welfare States* (New York: Routledge, 1993); and Seth Koven and Sonya Michel, "Womenly Duties: Maternalist Politics and the Origins of Welfare States in France, Germany, Great Britain, and the United States, 1880–1920," *American Historical Review* 95, no. 4 (1990): 1076–1108.

46. Louis I. Dublin to Mrs. C. B. Goodykoontz, 4 January 1927, in Folder "Economics of World Health," Box 10, Dublin Papers.

47. Robert L. Cox to Louis I. Dublin, 1 November 1926, in Folder "Economics of World Health," Box 10, Dublin Papers.

Movement 2

"As soon as the prospect knows 'it's life insurance,' " warned Ives, ". . . he then instinctively starts to defend himself."[48] I'm busy, he'll say. Or: I don't want life insurance, or I already have enough life insurance. Or: I do not talk to agents. Ives insisted that previous generations of agents had succeeded in convincing American men of their moral responsibility, in death as much as life, to care for wife and children. But his second movement admitted another thing that agents had done: made Americans wary of life insurance agents. Hence the popular early twentieth-century signs prohibiting "Peddlers, Solicitors and Insurance Agents."[49]

Ives instructed his agents to "throw" the "central phrase of the objection" right "back into the talk."[50] You don't want to talk life insurance? Well, the agent would say, I'm not talking life insurance in the way you've always heard it. You already carry enough? The agent replies: "In justice to yourself you ought to test the accuracy of that."[51] Ives's agents, part of a growing movement of agents, pitched themselves as a new breed, not selling insurance so much as scientifically proving its necessity (or not) in each individual case.[52] The answer to old objections came from "scientific" solutions. They proclaimed to broad audiences the trustworthiness of risk makers' numbers—this was the theme of the second movement. But here our two variations introduce a counterpoint from the risk makers' private communications. From the jarring conjunction of an outsiders' (Ives's) simplified vision of risk making with the much messier debates on the inside, we encounter serious internal dissonances, some unlikely to resolve.

While Ives preached trust, risk makers could not agree about the best way to value lives. When rating bodies they worried about treating a person as a simple aggregate of independent factors. When pricing people they tripped over the differing values of men and women to their families or the nation.

---

48. Ives, "Amount to Carry," 37.
49. Cahn, *Matter of Life and Death*, 134.
50. Ives, "Amount to Carry," 37.
51. Ives, "Amount to Carry," 37.
52. There were in fact a variety of approaches to "scientific selling" in the late 1910s and 1920s, beginning in the life insurance industry and reaching out to industry more broadly. Ives's peers at Equitable and Mutual Benefit pioneered agent training courses even earlier than he did, although each had different emphases. Edward Woods at Equitable, for instance, concentrated more on selling big policies while others focused on the "psychology of selling." In 1927, the American College of Life Underwriters opened and began credentialing agents or, as they now styled themselves, "underwriters." See Stalson, *Marketing Life Insurance*, 576–596.

Risk makers had reason to doubt their own numbers, as the variations in this movement will now demonstrate.

* * *

The problem of assigning a precise value to an individual's relative risk of mortality proved particularly difficult. Numerical rating promised faster and cheaper sorting of applicants, since clerks could do much of the rating without a doctor's help, and it also promised a rule-based rationalization of what had previously been an "arbitrary" process.[53] But speed and system came at the price of flexibility and complexity. And it could lead to patent absurdities.

Consider these (hypothetical) risks: Suppose an application from a thirty-five-year-old, employed as a steel grinder in McCracken County, Kentucky, weighing near the average with a low pulse rate, who spat blood in the past and whose parents both died at fifty years old from tuberculosis. While such a man would undoubtedly be deemed unfavorable on his face, he would draw a very favorable rating, explained an incredulous actuary. Another actuary went even further. What would be the rating for "an electrician, age 35, 5 ft. 8 in., 155 lbs., both parents reached age 75, pulse rate below 60, total abstainer, residing in Arapahoe County, Colorado," he asked? The answer (with a lower rating indicating a longer life) was negative four. He would be, scoffed the actuary, "subject to no mortality at all!"[54]

If risks—produced by risk makers' treadmills—resisted simple quantification and manipulation by arithmetic, that went double for the lives upon which they were based. The problem was not simply the crudity of adding and subtracting, or even just the complexity of risk making and prediction. Fundamentally, the problem lay in the complexity of individuals. "Are not many of the elements which enter into the estimation of a life risk rather of the nature of chemical compounds, which may totally differ in their effects from the original constituents?" asked an actuary in 1917.[55] Studies of build and its relation to tuberculosis in the Medico-Actuarial Investigation gave force to such a critique. Merely applying simple arithmetic to the case of two twenty-five-year-old men, one overweight and one underweight but

53. Arthur Hunter, "Selection of Risks from the Actuarial Standpoint," *Transactions of the Actuarial Society of America* 12, no. 45 (1911): 1–17, 281–298 at 295.

54. John K. Gore and Thomas B. Macaulay in discussion of Hunter, "Selection of Risks from the Actuarial Standpoint," 289, 291.

55. W. S. Nichols in discussion of Oscar H. Rogers and Arthur Hunter, "The Need in Medical Selection of Standards by Which to Measure Border-Line Risks," *Transactions of the Actuarial Society of America* 17, no. 56 (1916): 281–289; and 18, no. 57 (1917): 164–169 at 168.

each with a consumptive sibling, yielded similar ratings, for instance. Yet the study showed that the underweight man was actually nearly twice as likely to die.[56] Other interrelationships abounded. Rogers and Hunter conceded, for instance, that a heart murmur became much more serious if accompanied by recent rheumatism. They similarly noted that raising the rating of a bartender both for being a bartender and for using alcohol freely made little sense, since most bartenders used alcohol freely.[57] And what about the interrelationships that remained undiscovered?

But for all their intellectual force, such critiques of numerical scoring did not succeed in derailing it and sometimes even aided in risk scoring's ascent to dominance. Each new problem spurred another investigation and, soon enough, more rating tables. By 1919, for instance, build (height/weight) tables took into account age, height, girth, and history of tuberculosis—although not all at once. The relationship of murmurs to articular rheumatism elicited some enthusiasm from another actuary listening to Rogers and Hunter, this one a supporter of rating, who imagined a new set of tables for rheumatic risks with murmuring hearts.[58] Critics continued to point to the problems of arithmetic and correlation, noting the way that small changes to an applicant's history could create large and unreasonable swings.[59] Some thought the system probabilistically naive.[60] But even the most committed opponents of numerical rating had trouble opposing further research.[61] As an unintended consequence, medico-actuarial investigation data (on build especially) abounded and soon became a prime resource for mathematical statisticians seeking to understand and model multiple correlation and joint causation.[62] More importantly, the (ultimately impossible) dream of ascribing a single, precise, and accurate risk value to an individual—of making, in short, a true statistical individual—survived.

56. John K. Gore's example in discussion of Rogers and Hunter, "Need in Medical Selection of Standards by Which to Measure Border-Line Risks," 165–166.

57. Rogers and Hunter, "Numerical Method," 277.

58. Franklin Mead in discussion of Rogers and Hunter, "Numerical Method," 325.

59. E. E. Rhodes in discussion of Rogers and Hunter, "Numerical Method," 317–319.

60. H. N. Sheppard, apparently a young actuary, noted how odd it was to assume the lack of interrelationships in terms of probability theory. See Rogers and Hunter, "Numerical Method," 328.

61. Prudential's John K. Gore, one of the numerical method's fiercer critics, agreed that there would be "general approval" for further Medico-Actuarial analysis, in discussion of Rogers and Hunter, "Need in Medical Selection of Standards by Which to Measure Border-Line Risks," 164.

62. See, for instance, the industrial statistician/economist Andrew T. Court's "Measuring Joint Causation," *Journal of the American Statistical Association* 25, no. 171 (1930): 245–254.

In practice, numerical rating acquitted itself well—New-York Life could process 65 percent of all applicants "automatically" with the system in place, and nearly everyone agreed to the system's special utility in pricing substandard risks.[63] But it still required a medical staff that could exercise judgment in difficult cases, recognize incongruous or absurd results, and take into account the places where medical or actuarial knowledge remained thin. Hunter talked of "a scientific method *with limitations* which can be largely removed by the use of judgment."[64] One actuary, typical of the critics, conceded the point with a crucial caveat: "It [numerical rating] can be a splendid servant, but it would be an exceedingly bad master—in other words, it is a valuable tool, but it requires a skilful hand to use it."[65] Another, more darkly, warned: "In the hands of one who did not clearly comprehend its limitations, the most disastrous results might follow its use."[66] Assigning numbers to lives could work in an insurer's office staffed by capable medical and actuarial staff who knew how to interpret their results (including when to doubt them) and knew not to let those rated know their rating. But outside that context, the numbers' utility lessened and their truth-value suffered. Inside insurers' offices risk ratings could be useful. Elsewhere— whether deployed by other institutions or by those who were being rated— they could be dangerous.

* * *

Alfred Lotka did not intend his boss, Louis Dublin, to use the money-value-of-a-man calculations the way he had. When Lotka saw Dublin's plan to construct a figure valuing the entire American population, he tried to intervene: his numbers had been created to value fathers for their families, not citizens for the state. As his entreaties went largely unheeded he watched, helplessly, as his calculations escaped his grasp, as the numbers he labored over got loose. Numbers, out of their context and away from their creators (who knew their limitations), looked dangerous indeed.

The first public airings of Lotka's project appeared in Metropolitan's *Statistical Bulletin*, a periodical that circulated company data and health

---

63. By "automatically," they meant using only clerks with no medical expertise. Rogers and Hunter, "Numerical Method," 331. MONY avoided the numerical method, but still found ways to use clerks for much of their risk writing too. See discussion of Rogers and Hunter, "Numerical Method," 310.

64. Discussion of Hunter, "Selection of Risks from the Actuarial Standpoint," 295.

65. Thomas B. Macaulay in discussion of Hunter, "Selection of Risks from the Actuarial Standpoint," 294.

66. E. E. Rhodes in discussion of Rogers and Hunter, "Numerical Method," 319.

findings among policyholders and health activists. In a November 1925 article—appearing exactly a year before Dublin's "Economics of World Health" piece—on "The Value of Human Life," Lotka's voice stands out clearly: "Value is relative, and if the word is to have any definite meaning, we must specify not only the person valued, but also for whom in particular the individual has the value in question."[67] The most important relative monetary value of a man was to his wife and children, the article claimed. His value to business associates or his corporate employer came next, and only in the "outermost circle" was "man an asset to the community, to the state, and to the world at large."[68] A man, therefore, had not one but many different values.

Eleven years earlier, while serving as an editor for *Scientific American Supplement*, Lotka had committed himself to the relativity of value. "Value is an essentially *relative* concept, and as such *must* be defined *relatively* to some *one particular type*," he explained. "To speak of the value of a pound of butter, or a bale of hay, or a dozen worms, is meaningless: to complete the statement we must speak of the value of a pound of butter to man, for example, of a bale of hay to a horse, or of a dozen worms to a starling."[69] But Lotka emphasized relativity so that he could discard it, so that he, by naming precisely the value of what and to whom, could ignore all other possible frames of value.

Evolution provided Lotka with the only frame of value he ever really cared about, and one that he believed would allow him to create an "objective" standard for valuing goods and lives. From early on in his eclectic career, Lotka, who worked as a corporate chemist, a government physicist, and a patent examiner before settling at Metropolitan in 1924, devoted himself to understanding life and especially human individuals in evolutionary and mathematical terms.[70] Before a field called mathematical biology existed, Lotka cobbled together a career as a mathematical biologist and became one

---

67. Articles in *Statistical Bulletin* bear no bylines. This one was probably drafted by Lotka and revised by Dublin. Metropolitan Life Insurance Company, "The Value of Human Life," *Statistical Bulletin* 6, no. 11 (1925): 4–5 at 4.

68. Metropolitan Life, "Value of Human Life," 5.

69. Alfred J. Lotka, "An Objective Standard of Value Derived from the Principle of Evolution," *Journal of the Washington Academy of Sciences* 4, no. 14 (1914): 409–418; 4, no. 15 (1914): 447–457; and 4, no. 17 (1914) 499–500 at 455.

70. Lotka never followed a traditional academic career. He studied at the University of Birmingham, at Leipzig, and at Cornell. He eventually submitted a series of notes and articles written and published in his free time for a DSc from Birmingham in 1912. See Sharon E. Kingsland's treatment—the best there is—of Lotka in *Modeling Nature: Episodes in the History of Population Ecology* (Chicago: University of Chicago Press, 1995), specifically 28–29.

of the most influential practitioners of that field (as well as of theoretical ecology and of demography) of all time.[71]

Lotka proposed the value-of-a-man project in 1925 with his prior work on calculating objective values in mind. He adopted a life insurer's concept of value as his guide: a man was worth future earnings minus his cost of upkeep. But even this he viewed as only a special case of his evolutionary theory, for the "biological fundament of life insurance," he argued, lay in the peculiar natural instinct of humans, created by evolution, to care for their kin.[72] Lotka, through insurance and evolution, fused the biological with the economic (and studiously ignored culture).

Metropolitan offered Lotka, along with a good salary and job security, necessary raw materials—data, first and foremost, plus a research and calculating staff—for putting some flesh on the abstract and theoretical bones of his value work. Bessie Bunzel, a graduate of Barnard and Columbia with experience in the settlement house movement, joined Lotka in the project.[73] Though labeled a research assistant, Bunzel appears to have enjoyed a good deal of autonomy. She took special interest in the problem of "the cost of bringing up a child," eventually called the cost of "building the human machine" in later *Statistical Bulletin* articles. Bunzel surveyed New York hospitals, got data from orphanages about their expenditures, looked at surveys of family budgets, and collected school cost figures.[74] In the meantime, Lotka acquired industrial wage statistics from Western Electric statisticians and from public New York data, with an eye toward determining how wages changed over the lifetime of a worker.[75] To that cost and wage data, Lotka

71. In 1925, Lotka published *Elements of Physical Biology* (Baltimore, MD: Williams & Wilkins, 1925). Dover republished the book as *Elements of Mathematical Biology* in 1956.

72. Lotka, in an undated typescript, "Physical Biology—The Biological Value of the Individual," in Folder 8, Box 16, Lotka Papers.

73. Bunzel's 1912 Barnard yearbook photo bore the cryptic subtitle: "Devout, yet cheerful; Active, yet resigned." She belonged to the College Settlements Association at Barnard, which often worked with the Rivington Street Settlement providing social services to New Yorkers. Personal communication with Martha Tenney, Digital Archivist, Barnard College, 28 February 2013. Bunzel earned an MA from Columbia in 1914. See "Columbia Degrees Conferred on 2,000," *New York Times*, 4 June 1914.

74. B. Bunzel to Dublin, 1 June 1925, in Folder 4, Box 31, Lotka Papers.

75. Lotka requested the data, wrote the letters, and did the analysis (which involved a hand-drawn curve smoothing messy data)—he even read the eventual return letters—but all correspondence went out under Dublin's name. E. E. Lincoln shared Western Electric data with Dublin's office as a professional courtesy, one suggesting a larger internal trade in statistical data among the statistical offices then popping up in large corporations. But when publishing time came around, he asked Dublin to be discrete in discussing the data he had shared. See Edmond E. Lincoln to Louis I. Dublin, 1 July 1926, in Folder 4, Box 31, Lotka Papers.

added Metropolitan's extensive mortality data, all indexed by age and limited to wage workers.

Lotka put all the pieces together in late June 1926 and circulated his findings privately among a handful of statisticians and economists. His calculations valued a working man (making, at his peak, about $2,500 a year) at just over $30,000 when he turned twenty-one. Other than a few technical concerns, such as whether $2,500 was too high a wage to assume for most workers, Lotka's calculations stood up to criticism well.[76] But one of the final lines of the paper (perhaps inserted by Dublin) read: "These values indicate the immense assets which a country like the United States, with its high standard of living and production, has in its productive citizenship."[77] Lotka's correspondents refused to make this leap.

The value of a man to his family, as Lotka calculated it, did not equal the value of a man to his nation, they insisted. As one economist put it, "If the man and his family were all to meet sudden death, the nation as a whole would suffer very little economic loss, outside of possible funeral expenses." He continued: "Is it not true that the chief way in which the able-bodied man at 18 is of value to the nation is that he is a prospective taxpayer?"[78] While Dublin intended valuing lives in dollars to help build social institutions, professional economists and rigorous statistics showed how much a dollar value narrowed lives and emphasized how little a life was worth to society.

Dublin took little, if any, notice of this disharmony running through his project, asking Lotka to prepare values for ages fifteen, ten, five, and zero so that he could "estimate the value of human capital in the U.S."[79] Lotka obliged his superior, but not without pressing his own case. "Our computations are entirely justified so long as we restrict our statements to . . . the value of a man *to his family*." It made no sense, however, to add those values together to "figure the total value of the population to the state." Families had a direct interest in a man, while the state's interest was more

76. Chaddock to Dublin, 4 July 1926, in Folder 4, Box 31, Lotka Papers. In 1929, the average Ford family—paid more than the average—only made $1,694, so Dublin/Lotka's figures do look high. David Brody, "The Rise and Decline of Welfare Capitalism," in *Workers in Industrial America: Essays on the Twentieth Century Struggle* (New York: Oxford University Press, 1993), 63.

77. Metropolitan Life Insurance Company, "The Value of a Man as a Wage-Earner," *Statistical Bulletin* 7, no. 6 (1926): 1–4 at 4.

78. W. I. King to Louis I. Dublin, 28 June 1926, in Folder 4, Box 31, Lotka Papers. King was Economist at National Bureau of Economic Research, Inc. See also: Chaddock to Dublin, 4 July 1926, and Edmond E. Lincoln to Louis I. Dublin, 1 July 1926 in Folder 4, Box 31, Lotka Papers.

79. Dublin to Lotka, 11 August 1926, in Folder 4, Box 31, Lotka Papers.

complicated. Lotka emphasized the problem of competition too: men's values could not be simply additive because "whereas a man stands in the relation of a breadwinner, that is a *friendly* relation, to his family, he stands to the rest of the community in the relation of a competitor, that is to say, in a more or less *hostile* relation." In case Dublin did not get the point, he added: "A hundred million people in the United States are competing for food." Lotka's evolutionary worldview, and its Malthusian tint, drove his warnings to Dublin about making any attempt to add up individual values.[80] A month later, Dublin did exactly that for *Harper's Monthly* and its hundreds of thousands of readers.

Aside from its addition problem—an echo across the modern conception of death of numerical rating's addition problem as faced by risk-making traditionalists—Dublin's practical appropriation of Lotka's numbers created a gender problem. Metropolitan planned to release an advertisement and pamphlet at the end of 1926 highlighting the values of individuals in the family. For grown men, that could be done easily enough from relatively strong data, but women and children posed a problem: Lotka never calculated precise values for either group. Dublin settled on valuing women at 50 percent of men at the same age, an assumption that Lotka worried would "arouse the criticism, not to say antagonism, of woman readers, who are liable to use the trenchant weapon of ridicule."[81] Why not just value women equally, he argued, and avoid conflict? Dublin eventually met him halfway, at 75 percent.[82]

That compromise did not satisfy critics, such as the *New York Times* writer who objected that valuing women by wages had not fairly "capitalized . . . the feeling that goes with the service," which "ought to be worth at least the extra 25 per cent."[83] Nor did it satisfy Lotka, for whom women and children

80. Lotka, who zealously guarded his scientific reputation, engaging in frequent and public priority disputes, insisted on being put "on record" in saying that "simple arithmetic" would not work in making a national valuation. Lotka to Dublin, "Subject: *Value of a Man*," 13 October 1926, in Folder 4, Box 31, Lotka Papers. Emphasis in the original. On Malthusian thinking in twentieth-century America, see Derek S. Hoff, *The State and the Stork: The Population Debate and Policy Making in US History* (Chicago: University of Chicago Press, 2012); and Thomas Robertson, *Malthusian Moment: Global Population Growth and the Birth of American Environmentalism* (New Brunswick, NJ: Rutgers University Press, 2012).

81. Lotka to Dublin, 27 October 1926, and Lotka to Dublin, 10 December 1926, in Folder 4, Box 31, Lotka Papers.

82. Dublin and Lotka explained the valuation publicly by noting that homemaking could be quite expensive when purchased, nearing the wages that men received, but that working women generally earned half as much as their male peers. Dublin to Lotka, 18 December 1926, in Folder 4, Box 31, Lotka Papers.

83. "The Worth of a Man," *New York Times*, 30 December 1926.

continued to pose significant valuation problems. How, for instance, did one value a housewife's services? Taking into account the price of a housekeeper only worked to a point, argued Lotka, since a wife received room and board and owed some of her "wages" toward her part of paying to raise her children.[84] A child's value rested on even shakier ground. Since a person had economic value mostly for her children, and a child had no children, how could a child have economic value, asked Lotka?[85] Young women, being women and children simultaneously, seemed to Lotka the deepest mystery of all.[86] To make matters more perplexing, properly valuing a woman's housekeeping work decreased the value of a child: the mother's assets became the child's liabilities.[87] In fact, warned Lotka, properly valuing women's services would likely make the value of a child nothing, or even negative.[88]

Behind Lotka's quarrels with his boss lurked fundamental differences in the way each thought about society. Lotka saw a social system built upon an individualist, biological foundation, one that necessarily entailed competition for scarce resources. In this ecological and evolutionary view, marriage stood out as an exception where nature, custom, and law worked together to create uniquely cooperative relations. As Lotka put it: "A man may have but one wife, for whom he has a positive value; but he is a very exceptional person if he has not at least one thousand competitors, for whom he has a negative value."[89] Dublin, on the other hand, assumed cooperative social relations. When he appropriated Lotka's numbers, he took objects meant to represent a special case (a patriarchal, paternalist family) and asserted their generality. He hoped, by those numbers, to make the special case general,

84. Lotka to Dublin, "In re: *Value of Women*," 17 November 1926, in Folder 4, Box 31, Lotka Papers.

85. Lotka to Dublin, "Re: *Pamphlet: The Value of Human Life at All Ages*," 4 January 1927, in Folder 4, Box 31, Lotka Papers.

86. Lotka, "Method of Figuring Money Value of Preventable Deaths," following "For Use in Completing Book on Value of a Man," 21 June 1928, in Folder 4, Box 31, Lotka Papers.

87. Lotka to Dublin, 10 December 1926, in Folder 4, Box 31, Lotka Papers.

88. Lotka to Dublin, 20 June 1928, in Folder 4, Box 31, Lotka Papers. On changing ideas about valuing children more generally, see Zelizer, *Pricing the Priceless Child*.

89. "Note on the Summation of the Values of Individuals" following Lotka to Dublin, "Re: *Pamphlet: The Value of Human Life at All Ages*," 4 January 1927, in Folder 4, Box 31, Lotka Papers. Lotka later elaborated on the way the marriage relation differed from other economic relations: "This value arises out of the fact that his wife and family have certain rights established by law and custom in the earnings of the husband and father. In this sense, the wife and children have a certain ownership in the wage-earner. This ownership, however, is of a peculiar character and differs in rather important respects from ownership in other commodities." See "For Use in Completing Book on Value of a Man," 21 June 1928, in Folder 4, Box 31, Lotka Papers.

to bring a socially cooperative world more fully into being. Dublin stripped the context from corporate numbers in order to make them more flexible, so that they could work in a new setting and be applied toward very different ends than they had been developed for. Lotka could not stand the sight of seeing his numbers stripped bare in public. Dublin, in other situations, would come to know how he felt.

So would the risk scorers. For while Charles Ives had thought that "science" could answer prospects' questions once and for all, those sitting inside risk makers' offices had a firmer grasp on their science's limits. They understood the internal contradictions and unsettled complexities that vibrated within their calculations and systems: the problems that came with simply adding risk mortality factors to decide an individual's risk or adding individuals' dollar values to figure a community's wealth. They knew how much was lost when numbers grew thinner, when they escaped the grasp of the quantifier. Those on the outside, in contrast, not only did not miss numbers' original thickness; they tended to mistake thinness for truth or even beauty.

Movement 3

Toward the end of his second movement, Ives hinted at his next theme. "All right; I see you are busy," he instructed agents to say:

> just put this down and figure it out tonight. It's as interesting, for a change, as a card game. Just two minutes. You say your wife and you are about the same age—thirty-five? Put down 64 per cent of your normal income. In another column, add your income from business which will go on for 31 years, to your income from property, investments, and from anything you own which will go on for 31 years. Subtract this sum from the 64 per cent; multiply by $18.59. I'll explain why next time. The result is half the answer. I'll come around and get it. I'll then have something more interesting to show you.[90]

Ives's third movement required a pad and a pencil, and some facility with arithmetic. The agent became an instructor in simple probability, in how to think about quantifying one's life and one's future. He came as a herald of the power of numbers, one who expected his businessman prospect to be comfortable with figures and implored the prospect to trust them

90. Ives, "Amount to Carry," 37.

even further. Most of the math goes on behind the scenes, for agent and prospect: the agent finds the life expectancy for a thirty-five-year-old from a table in his rate book—it's sixty-six, meaning the prospect has, on average, thirty-one years ahead. Another table tells him that a dollar a year for thirty-one years, including the present year, compounding at 4 percent interest is worth $18.59 today. Ives tells the agent to suppose a man's family lives on 64 percent of the family's income, while the man lives off the rest. The agent, handing the pencil and paper to his prospect, asks the prospect to put these numbers together. After adding, subtracting, and multiplying, the prospect now has the amount of money that his family would expect from him if he lived as long as an average man. If he wants to replace himself fully should he die too young, that's the amount of insurance he'll need to carry.

What would a prospect learn from this little "game"? Of course, there's the number itself, but Ives put a pencil and paper in a prospect's hands to teach him more, to get him actively involved in treating himself as a statistical individual. Ives supposed he'd learn to think about, though certainly not calculate, life expectancies—at thirty-five, both man and wife had, on average, another three decades to plan for. Ives also wanted him to learn to think about the future and its many uncertainties: What if the prospect died that very day? Would his investments be enough to support his survivors? And what if he didn't die and instead lived a long life—longer even than the average three decades? Would his savings and investments be enough to sustain him *and* his family when work stopped being an option? Ives's calculations were supposed to get prospects thinking concretely about their answers to such questions.

The results could be startling—indeed, Ives expected them to be. Imagining a salaried man making $75 a week (about $4,000 a year), with a wife and three children, Ives illustrated the pad and pencil moment. The man admitted almost no income from property or investments—a few bonds and stocks, and rent from a family farm summing to about $400 a year—and carried a $5,000 life insurance policy as well as a $1,000 burial policy from his fraternal society. After separate calculations for the man's contribution to his wife until she died and his children until they grew up, Ives offered a total. The prospect needed five times more insurance than he had![91] Only that much, Ives taught, could replace him if he died, or support him if he lived.

---

91. Ives, "Amount to Carry," 37.

But some men resisted the premises of this calculation. As generations of life insurance prospects had done before them, they resisted using the average to explain or predict their lives. Some objected that they spent much more on themselves than other men—so much traveling, so many business dinners. Others objected that their incomes or investments fluctuated in value wildly. They couldn't pin them down. To all these objections, Ives replied, mantra-like: "the Law of Average."[92]

Ives believed risk makers had forced many men to see themselves through statistics. He reported that most men (by which he meant businessmen) accepted the validity of a life expectancy. But they still needed prodding to admit that other parts of their lives might be susceptible to smoothing and fatalizing too. Ives's actuarial training and his "devotion to Emersonian doctrine" made common cause in his work, argued his first biographers. He believed, like Emerson, in "the truths revealed by statistical averages, the expression of the Universal Mind, operating in the experience of many individuals."[93] His method did not just set a number; it tried to teach prospects to see themselves as living lives (biological lives, but also social and economic lives) subject to probability and risk statistics. They often resisted, perhaps intuiting the abstract problem inherent in applying statistical knowledge to individual cases, but more likely because they did not trust an average to speak to the richness of their particular lives—lives they knew so intimately and knew to be peculiarly, thickly theirs.

It did not help that the evangelists of probability came in the guise of men predicting deaths and selling a seldom-welcome product. In the variations on Ives's theme—that Americans could and should be taught to see themselves statistically—we find prospects more willing to become statistics if doing so will fend off death.

* * *

The Life Extension Institute set about trying to teach Americans to understand themselves through statistics too, through mortality forecasts tied to readings from weighing scales, tape measures, and risk makers' tables. But

---

92. This appears in the pamphlet version only: Ives, *Amount to Carry* (New York: Myrick and Ives, 1920), 13, in Business Scrapbook of Charles Ives 1919–1935, MONY Papers. On the role of life insurance agents (in Britain, although a similar story clearly fits the United States) as vectors for an ideal of the average man that encompassed a broad portion of the population, see Liz McFall, "A 'good, average man' : Calculation and the Limits of Statistics in Enrolling Insurance Customers," *Sociological Review* 59, no. 4 (2011): 661–684.

93. Cowell, *Charles Ives and His Music*, 39.

they removed these risk-making by-products from the sales pitch and used them instead to convince ordinary Americans to seize new fates. This was what risk scoring looked like through the lens of the modern conception of death. It proved very popular.

Readers of *How to Live* encountered figure 6.1: a foldout chart of medico-actuarial height-weight tables (a risk-making by-product), giving both average weights at given ages and heights and the percentage of increased mortality associated with degrees of variation from the average. (They also encountered very extensive calorie tables covering a wide range of foods.) Even more readers probably saw advertisements in major newspapers (made to look like news stories) bearing the title "The Menace of Overweight." The article/ad featured three tables, each derived from medico-actuarial data, alongside alarming reports of "200,000,000 pounds of excess fat on the bodies of citizens of this country." As LEI explained further: "Life insurance experience has shown that the lowest death rate is not found in adult life among people of average weight but among those otherwise healthy but who are somewhat underweight, according to the tables." When it came to "prolonging the health span and work span, weight regulation is one of the most important factors."[94] To hit his optimal weight, a forty-seven-year-old man standing five foot eight and weighing 160 pounds (the average) needed to shed an eighth of his weight.

"Weight regulation," signaled a newly popular concept, one propagated by life insurers (especially through their influence on doctors) and popularizers like LEI, that leaned heavily on the new technology of the scale. Scales, in turn, depended on weight regulation. Doctors purchased scales in the late nineteenth century at insurers' urging.[95] More and more Americans followed suit in the 1920s with the advent of weight regulation and reducing fads. "A weighing machine may be found in most bathrooms," reported the *New York Times* in 1926, and lines formed in front of penny scales (complete with insurers' weight tables attached) placed in railway stations, theaters, apothecaries' shops, and other public spaces.[96] Manufacturers picked up on the trend and sold Americans on the necessity of owning scales to maintain

94. One instance of this ad appeared as: Life Extension Institute, "The Menace of Overweight," *New York Times*, 16 October 1921.

95. On the importance of insurers for bringing attention to "weight," see Levine, "Managing American Bodies," 102–109; Schwartz, *Never Satisfied*, 153–157.

96. "The Span of Life," *New York Times*, 15 October 1926; Schwartz, *Never Satisfied*, 165–166, 168–171; Dublin, *After Eighty Years*, 192.

INFLUENCE OF WEIGHT ON VITALITY

PERCENTAGE OF NORMAL INSURANCE MORTALITY IN VARIOUS WEIGHT GROUPS (MEN)

Based Upon the Report of the Medico-Actuarial Investigation, 1912.—Period 1885-1908. Underweight, 531,793 Entrants. Overweight, 188,354 Entrants. Average, 24,526 Entrants.

AGES, 25-39. HEIGHTS, 5 FT. 7 IN.—5 FT. 10 IN.

Average weight, age 27, 5 ft., 8 in.—150 lbs.

Pounds under Av. Wt. | Pounds over Av. Wt.

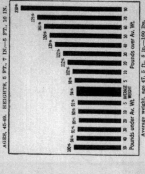

AGES, 45-49. HEIGHTS, 5 FT. 7 IN.—5 FT. 10 IN.

Average weight, age 47, 5 ft., 8 in.—160 lbs.

Pounds under Av. Wt. | Pounds over Av. Wt.

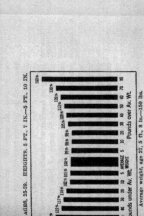

AGES, 57-59. HEIGHTS, 5 FT. 7 IN.—5 FT. 10 IN.

Average weight, age 58, 5 ft., 8 in.—163 lbs.

Pounds under Av. Wt. | Pounds over Av. Wt.

The percentages shown, above indicate the death rate at the designated age and weight as compared to the general death rate among insured risks of the same age. For example, 150% means that in that particular weight group the death rate was 50% higher than among all risks of the same age; 94% means that the death rate in that weight group was 6% lower than among all risks of that age. Note that in the middle-aged and elderly groups there is approximately one point higher death rate for every pound of overweight; that is, 40 lbs. overweight shows 40% extra mortality, 50 lbs. overweight, etc. Also note that middle-aged and elderly lightweights show a very favorable death rate, even lower than among those of average weight, indicating that the man of average weight is either over-fed or under-exercised, or both. It has been possible for the Life Insurance Companies to select a very favorable class of lightweights but impossible, in spite of their care in selection, to secure a favorable type of heavyweights. Apparently all heavyweights, regardless of type, are at a disadvantage as compared to the good lightweights. Lightweight, therefore, after full maturity, is an advantage unless it is due to some form of disease or malnutrition. Unless there be a disease or a long rise, the risk of over-nutrition should be taken, but *after thirty, watch your weight and keep it at the average weight for age thirty*.

TABLE OF HEIGHTS AND WEIGHTS BASED UPON THE REPORT OF THE MEDICO-ACTUARIAL INVESTIGATION, 1912, COVERING AN ANALYSIS OF 221,819 MEN AND 136,504 WOMEN

TABLE OF AVERAGE HEIGHTS AND WEIGHTS—MEN

Age	5 ft. 0 in.	5 ft. 1 in.	5 ft. 2 in.	5 ft. 3 in.	5 ft. 4 in.	5 ft. 5 in.	5 ft. 6 in.	5 ft. 7 in.	5 ft. 8 in.	5 ft. 9 in.	5 ft. 10 in.	5 ft. 11 in.	6 ft. 0 in.	6 ft. 1 in.	6 ft. 2 in.	6 ft. 3 in.
15	107	109	112	115	118	122	126	130	134	138	142	147	152	157	162	167
20	117	119	122	125	128	132	136	140	144	148	153	157	162	167	172	177
25	123	124	126	129	133	136	141	145	149	153	157	161	166	171	176	181
30	126	128	130	133	136	140	144	148	152	156	161	166	170	173	178	184
35	128	130	132	135	138	142	146	150	155	160	165	170	176	182	189	196
40	131	133	135	138	141	145	149	153	158	163	168	174	180	186	193	201
45	133	135	137	140	143	147	151	155	160	165	170	176	182	188	195	202
50	134	136	138	141	144	148	152	156	161	166	171	177	183	190	197	204
55	135	137	139	142	145	149	153	158	163	168	173	178	184	191	198	205

continued

6 ft. 4 in.
172
181
189
195
201
206
209
211
212

6 ft. 5 in.
177
186
194
201
207
213
215
217
219

TABLE OF AVERAGE HEIGHTS AND WEIGHTS—WOMEN

Age	4 ft. 10 in.	4 ft. 11 in.	5 ft. 0 in.	5 ft. 1 in.	5 ft. 2 in.	5 ft. 3 in.	5 ft. 4 in.	5 ft. 5 in.	5 ft. 6 in.	5 ft. 7 in.	5 ft. 8 in.	5 ft. 9 in.	5 ft. 10 in.	5 ft. 11 in.	6 ft. 0 in.
15	101	103	105	107	109	115	118	122	126	130	134	138	142	147	152
20	106	108	110	112	114	116	119	123	128	132	136	140	143	147	151
25	109	111	113	115	117	119	121	124	131	134	138	143	146	150	154
30	112	114	116	118	119	122	125	127	130	134	138	142	146	154	157
35	115	117	119	121	123	125	127	129	132	135	138	142	150	154	160
40	119	121	123	125	127	129	132	135	138	141	145	149	153	157	161
45	122	124	126	128	130	132	135	138	141	145	149	153	157	161	164
50	125	127	129	131	133	135	138	141	144	148	152	156	161	165	169
55	127	129	131	133	135	138	141	144	148	153	158	163	167	171	174

continued

6 ft. 0 in.
155
155
158
161
163
163
167
171
177

(HEIGHT AND WEIGHT TAKEN WITH SHOES ON AND COAT AND VEST OR WAIST OFF)

6.1. Insurers' build tables, tying mortality rates to over- or underweight, presented for public consumption by Irving Fisher and Eugene Lyman Fisk in *How to Live* (1920), foldout.

their health.[97] Those who did not own their own scales could and did "feed pennies into weighing stands," and they would also, in a related develop-ment, "eat by rule of calories."[98] Keeping trim did not depend on owning a scale of course. A mirror could do the trick fine. But in the 1920s, numbers from a scale became integral parts of individuals' conceptions of fitness, as risk makers' tools and the by-products of their work infiltrated Americans' bathrooms and public spaces.

A figure from a scale had certain advantages over figures in a mirror. It of-fered greater precision—with the possibility of making small, gradual trends plain—and it promised objectivity. In the context of build tables, it tied weight to health and even to the risk of mortality. But scales did not neces-sarily compete against mirrors. In *How to Live*, Fisk and Fisher assumed that "proper ideals of health and symmetry" coexisted with life insurance build standards.[99] "The Menace of Overweight," anticipated arguments from some men that their extra ten to fifteen pounds had nothing to do with ill health, but followed from their "heavy-framed type." Such a man, it instructed, should "view himself in the looking-glass 'in the buff' and observe whether there are any sagging outlines on his body, whether his muscles can be out-lined readily under his skin or whether they are overlaid by soft cushions of fat."[100] Mirrors could in the case of middle-aged men's paunches confirm the scale's brutal truth and the risk makers' mortal warning.

America's young women, however, posed a problem for the peace be-tween scale (as a guide to longevity) and mirror. Ideals of beauty could run counter to insurers' tables. A 1929 Metropolitan ad, for instance, featured a stylish "flapper" gazing longingly at a stick-thin department store manne-quin, while the ensuing text warned against succumbing to the use of "stim-ulants, sedatives, or drugs" to reduce. For those women, suggested the ad, the mirror said one thing—you're not thin enough yet. Mirrors, though old technologies, spoke in modern tongues of fashion, while insurance tables here said the opposite in an equally modern vernacular. Medico-actuarial build tables generally argued that Americans suffered higher mortality be-cause of overweight, but they changed their tune for those under thirty. "There are certainly more cases of tuberculosis among young 'underweights' than there are among those of normal weight," explained the ad, glossing

97. Levine, "Managing American Bodies," 110–111.

98. "Our Interest in Health," *New York Times*, 9 January 1928.

99. Fisher and Fisk, *How to Live* (1920), 257.

100. Life Extension Institute, "The Menace of Overweight," *New York Times*, 16 October 1921.

over (on glossy paper) troubling questions of causality and correlation.[101] Such women undoubtedly used scales in their reducing efforts and perhaps resorted to the drugs that Metropolitan criticized, but they broke the bond of the scale to the mortality table, turning it to their own aesthetic ends.

Worries about young women's reducing habits sounded quite a few clashing notes.[102] At an "Adult Weight Conference" of dieticians, physicians, surgeons, and statisticians, some worried about an American "mania for slenderness" and blamed medico-actuarial opinion: "One should not worry if the scales show less or more than certain statistical tables recommend. An intelligent person can almost always determine, by his own experience, what for him seems to be the best, most comfortable and most healthful weight."[103] They defended the scale, arguing for faithful weight regulation and measurement, while rejecting standardized tables and mortality probabilities.

One cautious voice, surprisingly enough, came from a key player in repurposing risk-making tools for healing—Louis Dublin. He worried about the popularization of risk numbers without thorough understanding. "We must not assume that any table based upon a series of averages approximates health," explained Dublin to his fellow conference goers. "Don't assume, as we are all apt to do in our craze for standardization, that human beings fall into a natural category."[104] This from one of the leading human categorizers in the world, a man whose first insurance job had been to calculate standard height-weight tables. There is no paradox here. Dublin's firsthand knowledge of the table-making process made him much more wary than many who only dealt with the finished product.[105]

101. Metropolitan Life Insurance Company, "Stimulants, Sedatives or Food," *Harper's Monthly Magazine* 158 (May 1929). Causation has continued to be a stubborn problem for insurance-derived data and for the concept of the risk factor. Life insurers, looking for a cheap way to exclude bad risks, did not necessarily care about causation. For them, correlation was good enough. But correlation provides a poor guide for those trying to improve lives by reshaping behavior. For more on the problem of risk factors, see Rothstein, *Public Health and the Invention of the Risk Factor*, especially 64–66.

102. Helen Veit notes similar debates and controversies surrounding the use of cosmetics and other common practices of "modern" women. Veit, " 'Why Do People Die?' " 1034–1036.

103. "Weight Reducers Warned of Dangers: Experts Say That Use of Drugs to Produce Slenderness May Lead to Dire Consequences—Conference Discusses Phases of Difficult Problem," *New York Times,* 28 February 1926.

104. "Weight Reducers Warned of Dangers," *New York Times,* 28 February 1926.

105. This example suggests a more general position: numbers' authors trust them less than their consumers. On other more certain relationships of quantification to trust, see Porter, *Trust in Numbers*.

Dublin knew where standard table numbers came from and he knew their limitations. Standard tables, he explained on another occasion, derived from nonrepresentative groups—from the insured or from native-born whites. Even within those groups, he noted, wide variation from the average still fit within the "safety zone," and should not be a cause of concern. Tables still held enormous value—especially when linked to mortality probabilities—but they needed to be used carefully and with judgment. Chiding nutritionists working in large cities for trying to force children from a variety of backgrounds to approximate the standard table averages, Dublin worried that his audience had "forgotten the limitations of which the authors [of the tables] themselves were well aware." Certainly, nutritionists were not alone in trusting tables too much. Dublin and those who compiled build tables knew that "underweight itself is not a final criteria of malnutrition" in children, for instance, but those who came to rely on the tables in their work too easily read them as literal gospel.[106] Just as the actuaries had predicted when discussing numerical rating, insurers' tables lost many layers of meaning (and much of the uncertainty associated with them) when published—whether by LEI, by other insurance departments, or in nutritionists' and doctors' textbooks.[107]

* * *

There is no evidence that Dublin ever saw the similarities between the kind of numerical thinning that he warned against in the late 1920s and the kind that he had done himself not much earlier. When Dublin stepped back from the national values he had built from Lotka's decontextualized numbers, his move had nothing to do with his own internal dissonances. Rather, the *resonances* he heard frightened him.

---

106. In this case, Dublin took issue with improper uses of Child Health Organization of America tables. Metropolitan Life Insurance Company, "Height and Weight Standards in Child Nutrition Work," *Statistical Bulletin* 2, no. 3 (March 1921): 5–6. Only a doctor performing a complete preventative medical examination could reliably diagnose malnutrition, argued Dublin and New York social welfare director John C. Gebhart in *Do Height and Weight Tables Identify Undernourished Children?* (New York: New York Association for Improving the Condition of the Poor, 1923).

107. Aside from *How to Live*, medico-actuarial tables also appeared prominently in a handful of standard textbooks. See Weigley, "Average? Ideal? Desirable?" 419. Metropolitan also thinned Dublin's numbers at least once, claiming that "Overweight," a booklet it distributed, "tells what you should weigh considering your age and height." Metropolitan Life, "Stimulants, Sedatives or Food," *Harper's Monthly Magazine* 158 (May 1929).

Dublin's contemporaries reveled in his big-money pronouncements, as the flood of interest in "Economics of World Health" suggests. When critics spoke up, they did not shy from valuing lives. Indeed, they said the money values were not high enough.[108] In response to *America's Human Wealth: The Money Value of Human Life*—a book by one of the few agency chiefs in the United States who could rival Charles Ives for size and influence[109]—a reviewer rhapsodized about "one of the comforting messages of modern demography," that "we have a growing cash value to ourselves, our families, our business associates, our State and our nation." Cash values proved that life now mattered as much as property once had. Though being priced as a slave had once been "an insult to human dignity," new valuations increased human dignity, pointing out "a material investment in which both the individual and society have a share."[110] Given frequently cited claims that "the tendency of the age" was to "measure in terms of money," or that "nearly all other values are being capitalized in this modern age," such celebration of monetizing life should come as no surprise.[111] After all, even truth had a cash value in the early twentieth century.[112] Still, the inversion involved here

108. "The Worth of a Man," *New York Times*, 30 December 1926.

109. Edward A. Woods, an Equitable agent writing with Clarence B. Metzger, published *America's Human Wealth: The Money Value of Human Life* (New York: F. S. Crofts, 1927) in a National Association of Life Underwriters series. The book closely paralleled the work being done at Metropolitan—so much so that Lotka, always up for a good priority fight, accused Woods of plagiarism. Lotka and Dublin eventually wrote to Woods, saying they suspected that one of his underlings (or ghostwriters) probably plagiarized without Woods knowing it. Woods admitted no wrongdoing and Metropolitan officials calculated that suing a purveyor of the gospel that "Health Work Pays" would not look good. There the conflict remained, unresolved. See Folder "Metropolitan-Equitable Controversy," Box 10, Dublin Papers.

110. "Two Economic Surveys" *New York Times*, 4 March 1928.

111. C. V. Chapin, "The Value of Human Life," *American Journal of Public Health* 3, no. 2 (1913): 101–105 at 105. Lotka quotes the health officer Chapin to support his own reticence in using insurance data to promote public health in an undated typescript, (from 1931 or later) in Folder 4, Box 31, Lotka Papers. Chapin's essay argued against using money values of human lives, arguing that it would be ineffective rhetorically (that some people wouldn't buy the argument), but not that it was inappropriate. He has sometimes been taken as evidence that monetizing life was unacceptable at the time. Yet Chapin took the strong position he did precisely because using money values of lives to argue for public health improvements was already a common practice, if not one always done perfectly. For an example of modern scholars mildly misreading Chapin, see Dorothy P. Rice and Thomas A. Hodgson, "The Value of Human Life Revisited," *American Journal of Public Health* 72, no. 6 (1982): 536–538. Ironically, the controversy over pricing lives may have been greater in more recent days. See the retort to Rice and Hodgson: Gary Crum, "The Priceless Value of Human Life," *American Journal of Public Health* 72, no. 11 (1982): 1299–1300. The second quote is from Huebner, *Life Insurance: A Textbook*, 14.

112. William James repeatedly refers to the cash value of words, ideas, and experiences in his explanation of pragmatism. See *Pragmatism: A New Name for Some Old Ways of Thinking* (New

is striking: in traditional risk making, the admitted value of a human life justified its dollar valuation, while in the light of the modern conception of death (blended with 1920s enthusiasms for objective numbers), dollar valuations came to argue for the inherent value of life.

So, when Dublin began his *Harper's* essay on valuing lives by saying, "One shudders a bit at the very thought of an economic evaluation of life," he may have been more sensitive than many of his contemporaries. Dublin explained to his readers, "Life in its full implications is not commensurable with money. . . . Life and health are ends in themselves."[113] By 1928, after a few years of Lotka's technical criticisms contrasted with a disconcertingly popular embrace of valuing lives, Dublin not only shuddered, he back-tracked. In a letter to one of Irving Fisher's associates, who was planning a book on valuing lives, Lotka drafted and Dublin signed a letter claiming that big-dollar values of the American population meant to advocate for public health spending had never been that important to Dublin. The newspapers, they claimed disingenuously, "picked it out for particular mention out of its context," spurred on by "their propensity for dealing in striking figures."[114]

When the two published *The Money Value of a Man* in 1930, the chapter on public health included no attempts to value lives to society—Dublin's spectacular totals faded away. The book instead claimed life insurance agents as its central target and the final chapter shifted the focus to that old question: "How much life insurance should be carried?"[115] Charles Ives's program became Dublin and Lotka's.[116] Lives held economic values, they claimed, but mostly in carefully limited contexts—in injury compensation cases and in life insurance.[117] Even so, Isidore S. Falk, formerly a teaching assistant for Dublin in a Yale vital statistics course and soon to be a crucial player in designing and implementing the Social Security system, still read a

---

York: Longmans, Green, 1907), 53, 74, 86, 90. On James's pragmatism and the booming commodity exchanges, see Levy, "Contemplating Delivery."

113. Dublin, *Economics of World Health*, 1.

114. Louis I. Dublin to Royal Meeker, 19 March 1928, in Folder "Economics of World Health," Box 10, Dublin Papers.

115. Dublin and Lotka, *Money Value of a Man*, 138.

116. One important change, however, came in the kind of people targeted. Ives's agents went after the middle and upper-middle classes, especially businessmen. Dublin and Lotka created valuation tables for workers.

117. According to Dublin, *The Money Value of a Man* accidentally "brought order into what had heretofore been an area of uncertainty and even of chaos" in shaping workmen's compensation law. Dublin, *After Eighty Years*, 132.

broader message of social responsibility in Dublin and Lotka's valuations.[118] "Whether society has come wisely or foolishly to hold that life is precious," wrote Falk in a review, "the corollary is that society must (and does) assume the burden of maintaining it when other resources fail or are lacking; and in the political economy, life has a dollar value."[119]

Movement 4

Ives's piece began with an overture to the science behind insurers' production of statistical individuals and their valuing of lives. His second movement imagined that risk makers' science could win the confidence of wary prospects, and the next sought to teach those prospects to think about themselves statistically. But in the variations on Ives's themes, we discerned complications surrounding insurers' practices and doubt in risk-making communities: risk makers disputed privately how to value bodies by risk factors or women in dollars; they cautioned against using paternal values to value a nation and against trusting risk factors too much; and when their numbers indeed escaped into domains outside life insurance, many retreated. Ives's prose composition now ends with a brief final movement, a coda really, in which even he admitted the limits to risk makers' science.

"Go to it hard now," advised Ives, once the prospect's insurance needs had been calculated, "with all the persuasive power you have—(in some cases it is well to use, at this point, the usual moral and emotional appeals)."[120] One of Ives's protégés, Pete Fraser, gave some insight into those "usual moral and emotional appeals." He explained, in his own advice to agents: "Your entire presentation and preliminary talk must be with the idea of sounding him out. His weak spot may be vanity, his selfishness, his love of wife or baby, his fear of the future. But whatever it is, find it, attack it, and you will find as I have found . . . that you need not sell him. He will buy."[121] This was how one closed a scientific sale.

Yet Ives's internal dissonance should not distract us from the larger harmonies in this coda that tie together all three of this piece's variations. Ives, in the end, refused to trust everything to numbers. But that made him *more*

---

118. On Falk, see "Isidore Sydney Falk," *Social Security History,* http://www.ssa.gov/history /ifalk.html; on Falk's teaching for Dublin, see Folder "Yale Vital Statistics Course-Correspondence 1916–23," Box 5, Dublin Papers.

119. Isidore S. Falk, Review of "The Money Value of a Man" by Dublin and Lotka in *Journal of Political Economy* 39, no. 4 (August 1931): 546–548 at 548.

120. Ives, "Amount to Carry," 38.

121. Fraser quoted in Cahn, *Matter of Life and Death,* 178–179.

like the quantifiers and risk makers, rather than *less*. Ives recognized that statistical argument served the sale, but did not constitute it. Actuaries and medical directors believed the numerical method valuable, but only when it served them, when it helped them cheaply predict relative mortality likelihoods. Dublin intentionally thinned Lotka's numbers so that they could serve him rhetorically in justifying public health expenditures. Risk makers most trusted their numbers when they controlled them and understood their production.

When outsiders—encouraged by the modern conception of death's new gospel and excited to have objective facts to riff off—used risk makers' numbers for their own purposes, the quantifiers got nervous. Louis Dublin, as we've seen, abandoned the national human life values (that he had appropriated from Lotka's study) when they became too popular and instead focused on extending Ives's methods for making paternal values to low-income wage earners.[122] Even Irving Fisher—the man responsible for arguing that life insurers' statistical laws should be employed to extend lives—stepped back from advocating the risk-based valuations of individuals that he and LEI had adapted from the life insurance risk makers. When his partner at LEI, the medical director Eugene Fisk, died at the age of sixty-four, Fisher emphasized that age wasn't everything. In the face of LEI critics who delighted in asking why Fisk had not succeeded in extending his own life further, Fisher answered: "Life should be measured in terms of things accomplished rather than in years."[123]

Risk makers and those who took up their tools to lengthen lives appear to have understood, for the most part, that the valuations they made could only be partial. They seem to have recognized that their reductions of complex individuals to numbers were inherently flawed. But those valuations and reductions helped sell insurance, made it possible to process more applications, and even promised to help some Americans live longer, healthier lives. The quantifiers kept on quantifying because they thought they could control the way their numbers were used (they were wrong) and because they could focus on the benefits of their work, while avoiding the costs.

That too was an old life insurance sales strategy. Once a prospect admitted his and his family's calculated need for insurance, Ives warned agents

---

122. This shift challenged industrial insurers' former commitment to insuring all family members individually in the process. It pushed the idea that the male should be head of household and primary provider in working-class households too. Dublin and Lotka, *Money Value of a Man*, 140.

123. Irving Fisher, "In Memoriam: Dr. Eugene Lyman Fisk," *How to Live: A Monthly Journal of Health and Hygiene* 14, no. 8 (August 1931): 2–3 at 2, in Reel 8, Fisher Papers.

that he might still balk at the "cost." For that reason, Ives instructed, the agent must never bring up cost directly. Should the prospect ask about the price of his insurance, the agent could turn the conversation toward scheduling a medical examination. He might even produce a small glass bottle on the spot and direct the prospect's attention from financial outlays toward biological deposits: sir, would you mind?

That is how this piece, inspired by Charles Ives, ends.

SEVEN

Failing the Future

In June 1922, a thirteen-year-old boy living in Jackson, Mississippi, took a job filling out policy applications for his illiterate neighbor, who had quit work as a janitor to try his hand as a life insurance agent.[1] The boy's name was Richard Wright. He had been born on a Mississippi plantation to miserable poverty, a fractured family, brutal punishment, and inconsistent schooling. Still, he could read, write, and "figure" better than Brother Mance, his new employer. So, Wright set off on his "strange job" moving from shack to shack, church to church, selling life insurance to even poorer black families on Delta plantations.[2]

Wright, making an unprecedented five dollars a week, stuffed himself with pork, black-eyed peas, and "for once, all the milk I wanted." Brother Mance gushed over his abilities as Wright filled out applications until his "fingers ached." Wright enjoyed the food, the money, and the attention. But he "hated" the "bare, bleak pool of black life" and the "wall-eyed yokels" whom he awed with his own limited literacy. He brimmed with confusing, conflicting emotions: pride and disgust, satisfaction and deeper hunger.[3]

Brother Mance excelled as a salesman. Every Sunday he filled a country church pulpit and roused each congregation of sharecroppers to the great goal of purchasing a life insurance policy—a small one, undoubtedly just enough to pay for private burial. Wright couldn't tell why people in such poverty gave up what little they had for life insurance, but they did. Wright

1. Constance Webb, *Richard Wright: A Biography* (New York: G. P. Putnam's Sons, 1968), 57–58.
2. Richard Wright, *Black Boy: A Record of Childhood and Youth* (New York: Harper & Row, 1969), 150–151.
3. Wright, *Black Boy*, 150–151.

thought, with a trace of narcissism, that they acted irrationally: "Many of the naïve black families bought their insurance from us because they felt that they were connecting themselves with something that would make their children 'write 'n speak lak dat pretty boy from Jackson.' "[4] They were, he mused, buying dreams of a better future and he was, to some extent, right: they put their trust in life insurance—and thus in traditional risk making— with hopes of securing a brighter life for themselves and their children.

Wright doesn't identify the company whose insurance he sold, but if it served poor blacks on Delta plantations, it must have been a small African American–led company. The big companies who still sold life insurance to African Americans, like Dublin's Metropolitan, seldom ventured into the rural South. And most of the large mainstream companies had effectively stopped selling any life insurance to blacks after the antidiscrimination bills became law in the 1880s and 1890s North. So it fell to small life insurers to sell insurance (and also hope) to poor Delta blacks. Unfortunately, hope remained in short supply.

As we saw in chapter 2, African American reformers beginning in the 1880s had found hope in the promise of post-Emancipation civil rights politics. They looked forward to a future of true equality. Most life insurers, in response, embraced Frederick Hoffman's bleak prophecy of African Americans' extinction. In the early twentieth century, the tradition of egalitarian hope lived on in African American–run life insurance companies. And then, in the context of the modern conception of death, it took (shallow) root in one of the most important mainline companies—Metropolitan— because of the efforts of Louis Dublin.

But in each case, hope housed within corporate risk-making systems failed to bring equality. In fact, despite that hope, these systems perpetuated inequality in the ways they made risks, in their systematic preference for making African Americans into substandard, subprime risks.

At the root of these failures lay life insurers' construction since the Panic of 1873 of an inclusive "white" category for their classificatory systems.[5]

---

4. Wright, *Black Boy*, 151.

5. The best survey of race-based practices in life insurance, by Mary L. Heen, explains that white men stood as the "norm" for life insurers' standard policies in the early twentieth century, but does not consider the effort and intention required to create a unified "white" category in the preceding decades. See Heen, "Ending Jim Crow Life Insurance Rates," 378–379. This chapter not only explains discrimination that harmed African Americans, but also considers how some other forms of discrimination ended with life insurers' extension of the privilege of normality (and, in that sense, whiteness) to new people. For a recent, accessible survey text building on a now voluminous literature on the construction of "whiteness," see Nell Irvin Painter, *The History*

That construction gave their data powerful political resonances. Or, more precisely, it gave life insurance data a particularly pertinent biopolitics, a way of thinking about race at a moment when questions of race loomed large in national debates.[6] In that biopolitics, life insurers designated "white" to be the default standard category encompassing the vast majority of Americans, while "black" or "colored" became, by a cruel data logic, substandard. The implications of white data politics varied, shifting with the hands wielding the data and the problems to which risk makers applied their data, and yet even racial liberals among the risk makers (like Dublin) found themselves limited by the politics of their data.

Risk makers' tools gained influence in the first decades of the twentieth century in large part because Americans more readily accepted being made into statistics when risk making came with the promise of extended and improved lives. But risk-making tools also gained in cultural power generally. Crucially, as more Americans embraced life insurers' numbers in their daily lives, they saw those numbers enter one of the key debates of the age: the fight over immigration restriction and over the racial definition of the nation. This chapter critically examines the rising political significance of risk makers' tools and by-products. Its first section explains how Metropolitan's Louis Dublin mobilized life-span measurements for the anti-restriction side of the immigration debate, even as his opponents wielded data derived from the twentieth century's other prominent form of statistical individuation: intelligence testing. White data politics, this chapter argues, supported Dublin's (ultimately unsuccessful) opposition to the racialization of eugenic thought and its political expression in the stemming of immigration to the United States, especially from outside northern Europe.

But this chapter's investigation of white data politics matters mostly because it allows us to see (in its second section) something that had a greater impact on many Americans' lives: the systematic bias against African

---

of *White People* (New York: Norton, 2010). This chapter's emphasis on political economy and power in explaining who could be "white" and the burdens applied to the nonwhite owes much to Barbara J. Fields's critiques of the whiteness literature, such as in "Whiteness, Racism, and Identity," *International Labor and Working-Class History* 60 (October 2001): 48–56; or "Of Rogues and Geldings," *American Historical Review* 108, no. 5 (2003): 1397–1405.

6. Data can have politics, I argue, in the same limited way that Langdon Winner thought that "artifacts" could have politics. Data, like other artifacts, can carry with them some distinct features of the system or social order that produced them. See Langdon Winner, "Do Artifacts Have Politics?" *Daedalus* 109, no. 1 (1980): 121–136. When Foucault introduced the idea of biopolitics, he was drawing our attention especially to data whose politics pertain to human populations. See Foucault, *History of Sexuality*, 139–141; Hacking, "Biopower and the Avalanche of Printed Numbers," 279–280.

Americans built into life insurers' practices. The people to whom Wright sold life insurance in the early twenties, as well as those to whom he would again sell insurance at the height of the Great Depression (where this chapter ends), never experienced the full benefits of that essential modern commodity. They bought life insurance at higher prices from less secure companies operating out of smaller communities of shared risk. Louis Dublin could use his data to argue for African American equality and he did. Similarly, African American–run life insurers could attempt to charge their policyholders standard rates. But in each case, white data politics trumped intentions. Dublin's arguments did not prevent African Americans from remaining the type for substandard life insurance, and black companies could not ultimately avoid mimicking their mainline corporate peers' discriminatory pricing schemes.

* * *

Risk makers' white data politics found a powerful advocate in Metropolitan's Louis Dublin. Dublin considered himself in the 1910s and early 1920s to be part of the "eugenics movement," as did other believers in the power of risk-making tools, methods, and knowledge to extend lives and improve American society.[7] Like Irving Fisher, who served as the first president for the American Eugenics Society after its formation in 1923, Dublin saw both the similarities of eugenical evaluations to insurers' selection practices and the health implications of controlling reproduction among those with serious impairments.[8] Eugenicists like Dublin and Fisher shared with life insurers the tool of the medical examination, some imagining that the state could make life insurance exams the basis for granting marriage licenses, while the president of the American Medical Association in 1914 shared a "doctor's dream" that a conscientious young man would, before proposing marriage, first ask his doctor to "examine him as carefully as he would were

7. Dublin, *After Eighty Years*, 155. On the history of eugenics in the United States, see Kevles, *In the Name of Eugenics*. Many other valuable studies have followed. For an excellent overview of the field, see the introduction to Alexandra Minna Stern, *Eugenic Nation: Faults and Frontiers of Better Breeding in Modern America* (Berkeley: University of California Press, 2005).

8. Nathaniel Comfort uses Irving Fisher's story to explain how a eugenic impulse and hereditarian ideas entered medical and public health thinking in ways that persist down to today. He also endeavors to keep us from making easy, but too simple, distinctions between benevolent health reformers and villainous eugenicists—the two types, in fact, were often embodied by the same person. See Comfort, *Science of Human Perfection*, chapter 2. Fisher's biographer apologizes for Fisher's eugenical work, but writes around the question of racism, at once noting Fisher's discomfort with miscegenation alongside his confusion over the definition of a race. Allen, *Irving Fisher: A Biography*, 150.

he applying for a large life insurance."[9] Insurer's exams included extensive family histories, which also resembled (at least in intent) the family records gathered by Eugenics Records Office fieldworkers beginning in the 1910s.[10] Public health and eugenics advocates instituted medical exams—streamlined versions of those originally developed by life insurers—for all immigrants in 1892 (administered at new facilities like Ellis Island), and tests of mental deficiency had become available by 1918.[11] Just as life insurers rejected around 10 percent of all applicants, eugenicists spoke of a defective "submerged tenth."[12] Most people fit normally into society (including into insurance societies), both groups believed, but the tenth that did not needed to be kept out or discouraged from reproducing. Insurers believed that knowing about an individual's family could help them predict whether or not he or she would fall in the subnormal 10 percent. Eugenicists shared that same faith in heredity as (part of) fate.

But the question of race exposed the tensions in the marriage of risk making to eugenics. Dublin felt such tensions acutely. Earlier, he had toyed with the concept of "race suicide," the idea that new immigrants reproduced at much higher rates than native-born Americans, and wrote a paper that found its way into the *Congressional Record* at the behest of a restrictionist North Carolina congressman. A related attack on birth control among the upper classes (Dublin advocated "birth release") later made him—to his chagrin—a widely circulated author in Catholic circles.[13] But Dublin grew less interested in race suicide and much more concerned about the dangers of race talk as debates over immigration—especially immigration from southern and eastern Europe—encouraged many within eugenic circles to assert the reality of race and innate racial differences.

9. Victor C. Vaughan, "The Importance of Frequent and Thorough Medical Examinations of the Well," in *Proceedings of the First National Conference on Race Betterment* (Battle Creek, MI: Race Betterment Foundation, 1914), 90–96 at 94. On marriage licenses and the life insurance model, see for one instance: William Alan Chapple, *The Fertility of the Unfit* (Melbourne: Whitcombe & Tombs, 1903), 120.

10. Allen, "Eugenics Record Office," 225–264.

11. John Parascadola, "Doctors at the Gate: PHS at Ellis Island," *Public Health Reports* 113, no. 1 (1998): 83–86. Between 1892 and 1924, twelve million people went through Ellis Island, of which, about 2 percent were denied admission.

12. H. H. Laughlin, "Calculations on the Working Out of a Proposed Program of Sterilization," in *Proceedings of the First National Conference on Race Betterment* (Battle Creek, MI: Race Betterment Foundation, 1914), 478–494 at 480.

13. Dublin, *After Eighty Years*, 136–137. On North Carolina, eugenics, race, and "selectionist" thought, see Gregory P. Downs, "University Men, Social Science, and White Supremacy in North Carolina," *Journal of Southern History* 75, no. 2 (2009): 267–304.

In a presentation to the Second International Congress of Eugenics in 1921, later republished in *Scientific Monthly*, and then again in *Eugenics in Race and State* in 1923, Dublin attempted to use some by-products of risk making—specifically an argument constructed from state data using insurers' analytical tools—to combat criticism of so-called new immigration, without rejecting "race" out of hand.[14] Figure 7.1 displays the results of Dublin's attempt to evaluate New York's "race stocks" from insurers' preferred standpoint—the standpoint of mortality (or longevity). This chart presented two bar graphs, one for men and the other for women, showing the "expectation of life"—the number of years that an individual would expect, on average, to have ahead of him or her—for residents of New York State born in the United States, England/Scotland/Wales, Germany, Ireland, Italy, or Russia.[15] Dublin organized the chart with the native born on top because they represented the control population for the comparison, but otherwise listed birth nationalities in alphabetical order. His visual strategy did not argue for ordering races. But his conclusions did. Interpreting life expectancy as a measure of "natural vigor," Dublin generated a hierarchy: Russians (who he claimed to be mainly Jews)[16], native-born Americans, Italians, the English/Scotch/Welsh, Germans, and with a "particularly low expectation," the Irish.[17]

Another chart, this one from 1923, depended on a different way of thinking about an individual statistically, on a different numerical, objective measure—not longevity, but intelligence—and drew from a different strain of eugenic thinking, one that substituted racial science for risk making. Figure 7.2 presents a ruler graded by two related measures of intelligence (a "combined score" and a "mental age" both of which were essentially arbitrary scales) against which were placed a series of American groups.[18]

14. Dublin published earlier results with similar conclusions in 1916: Louis I. Dublin, "Factors in American Mortality: A Study of Death Rates in the Race Stocks of New York State, 1910," *American Economic Review* 6, no. 3 (1916): 523–548; Louis I. Dublin, "The Mortality of Foreign Race Stocks: A Contribution to the Quantitative Study of the Vigor of the Racial Elements in the Population of the United States," *Scientific Monthly* 14, no. 1 (1922): 94–104; Louis I. Dublin, *The Mortality of Foreign Race Stocks* (New York: Metropolitan Life Insurance Company, 1923), reprinted from *Eugenics in Race and State*, vol. 2.

15. Dublin, "Mortality of Foreign Race Stocks," 98.

16. In his very different take on racial hierarchies, Carl Brigham shared the conclusion that Russian immigrants were primarily Jews. See Carl Brigham, *A Study of American Intelligence* (Princeton, NJ: Princeton University Press, 1923), 190.

17. Notably, Dublin used the male data to produce the hierarchy and not the female data. Dublin, "Mortality of Foreign Race Stocks," 103.

18. Brigham, *Study of American Intelligence*, 124.

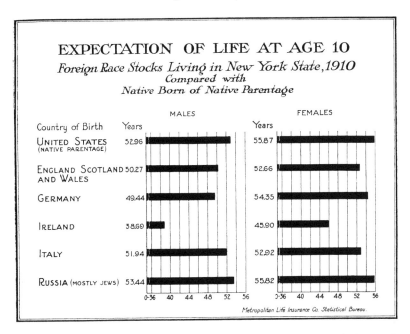

EXPECTATION OF LIFE AT AGE 10
Foreign Race Stocks Living in New York State,1910
Compared with
Native Born of Native Parentage

Country of Birth	MALES Years	FEMALES Years
UNITED STATES (NATIVE PARENTAGE)	52.96	55.87
ENGLAND SCOTLAND AND WALES	50.27	52.66
GERMANY	49.44	54.35
IRELAND	38.69	45.90
ITALY	51.94	52.92
RUSSIA (MOSTLY JEWS)	53.44	55.82

Metropolitan Life Insurance Co. Statistical Bureau.

7.1. Louis Dublin's table of life expectancies for New York "race stocks,"
in Dublin, "Mortality of Foreign Race Stocks," 98.

The psychologist Carl Brigham included the chart in his book *A Study of American Intelligence*, intending it to show the relative intelligence of various foreign-born national groups in the United States as measured by batteries of intelligence tests administered to draftees and officers during the world war. While intelligence tests had enjoyed quite a bit of success already in the twentieth century as tools for sorting schoolchildren or the institutionalized "feeble-minded," their use as tools to aid in categorizing and assigning American draftees catapulted them to a new prominence and almost overnight made them one of the most important ways that Americans came to be understood statistically.[19] Brigham's study drew on intelligence tests administered to nearly 150,000 Americans, a sixth of whom were African Americans and a twelfth of whom were foreign-born.[20]

In its construction and presentation, Brigham's aim of displaying a hierarchy of nationalities came through clearly enough. Like Dublin, Brigham used statistical measures that had been developed for sorting individuals to

19. Carson, *Measure of Merit*, chapters 5 and 6.
20. Brigham, *Study of American Intelligence*, 154.

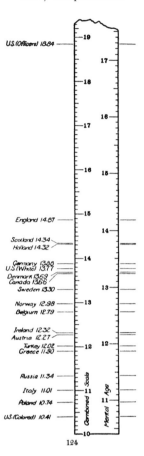

7.2. Carl Brigham's eugenic
hierarchy expressed through
intelligence tests, in Brigham,
Study of American Intelligence, 124.

sort races. The table demonstrated his statistical conclusion that English immigrants were, on average, more intelligent than Germans and the average
native-born white American, who all stood above the Irish, who scored well
above Russians, Italians, and Polish, and finally above the average "colored"
draftee. From this hierarchy, Brigham (in the text) transmuted national origins into "race," drawing on the popular theories of European (white) races
developed by Madison Grant and William Z. Ripley.[21] Eastern and southern

21. Brigham acknowledged both men in *A Study of American Intelligence*, xvii–xviii. See also
Painter, *History of White People*, 278–326.

European immigrants thus became a mix of "Alpine" or "Mediterranean" blood, while northern Europeans became "Nordic" with some (presumably lower quality) "Alpine" blended in. Brigham argued from his results that the Nordic race enjoyed a natural superiority in intelligence, trumping inferior Alpines and Mediterraneans.[22] Brigham made his ruler tell stories of both national and racial difference.

Brigham had also subsumed "defect"—a classic target of eugenicists and insurers—within "race." "American intelligence is declining," warned Brigham, "and will proceed with an accelerating rate as the racial admixture becomes more and more extensive."[23] Brigham's study argued that mental defects would best be screened out by targeting Alpine and Mediterranean "blood," not (only) by regulating the reproduction of the subnormal 10 percent. In a similar manner, the Eugenics Record Office's Harry Laughlin argued before Congress that a disproportionate number of the nation's insane or feebleminded hailed from southern and eastern Europe.[24] Their arguments—blended with concerns about socialist or radical immigrants and with what Dublin called postwar "hypernationalism"—fueled calls by restrictionists to close America's borders.[25]

Dublin, in 1924 the president of the American Statistical Association, argued that the "evidence is hardly sufficient to warrant the sweeping conclusion that certain racial elements in our national life are vastly superior to others."[26] As Dublin had shown: looking at "vigor" instead of "intelligence" could produce a very different hierarchy of races—and vigor or health seemed to Dublin to be more important, especially in light of the modern conception of death. Arguments advanced by Brigham within the racialized strain of eugenics looked, to Dublin, like "organized propaganda for Nordic races" and "simply an effort to give the appearance of respectability and of science to what is fundamentally an expression of unreasoned prejudice."[27] (In 1930, Brigham recanted his conclusions, declaring his racial analysis meaningless. By that point, he seems to have agreed with Dublin.)[28] Dublin

22. Brigham, *Study of American Intelligence,* 168–170.

23. Brigham, *Study of American Intelligence,* 210.

24. Kevles, *In the Name of Eugenics,* 95, 103. Kevles notes on 132 that Laughlin's study drew serious criticism from his contemporary, the Johns Hopkins zoologist Herbert S. Jennings.

25. On war-induced 100 percent Americanism and other factors leading the drive for restriction, see John Higham's classic *Strangers in the Land: Patterns of American Nativism, 1860–1925* (New Brunswick, NJ: Rutgers University Press, 2002), chapters 8–10.

26. Louis I. Dublin, "The Statistician and the Population Problem," *Journal of the American Statistical Association* 20, no. 149 (1925): 1–12 at 11.

27. Dublin, "Statistician and the Population Problem," 3.

28. Kevles, *In the Name of Eugenics,* 130.

took the eugenics movement to task for having abandoned older practices—
practices more consistent with those of insurers.

Dublin warned that the "eugenic movement" had "overstressed the dangers" inherent in differential birth rates—that it worried too much about race suicide—ignoring "the influence of environment and tradition on our conduct and achievement."[29] (Recall that Irving Fisher considered eugenics only part of the solution to America's health crisis, on par with public health reform, and not so important as improved personal hygiene. And Dublin had argued in 1923 that public health, improved well-being, and hygiene had occasioned recent declines in tuberculosis more than had improved heredity.)[30] His address to the American Statistical Association, later adapted for wider publication in the *Atlantic Monthly*, worried even more that eugenicists had undermined "our best ideals of democracy and religion."[31]

He meant that restrictionists had denied the fact around which life insurance, and much of American society, operated: that the "80 per cent" who compose the mass of "plain folks carrying on the world's work"—those left behind when you remove the exceptionally able and the exceptionally impaired—possess no "serious difference in innate ability." In fact, argued Dublin, the leaders of nations had always come from places where they were least expected and would continue to do so if Americans continued to care about creating a society with sufficient opportunity. But the dark side of Dublin's faith in the 80 percent was his prescription for the bottom 10 percent. "If we can be careful to control, or better yet entirely check, the reproduction of the obviously unfit, we are in no danger of racial deterioration," he explained.[32] Racial inferiors did not threaten the American people,

---

29. Dublin, "Statistician and the Population Problem," 11.

30. Dublin admitted that "constitutionalists" interested in hereditary arguments had some strong points (most notably, and awkwardly for racial theorists, in the lower incidence of tuberculosis in Jews and Italians), but he mainly sided with "environmentalists" who thought that "improved wellbeing" and the vast efforts of the "tuberculosis movement" best explained the decline. Louis I. Dublin, *The Causes for the Recent Decline in Tuberculosis and the Outlook for the Future* (New York: Metropolitan Life, 1923). Today, the general argument for the decline credits public health efforts as primary, alongside improved standards of living. See Simon Szreter, "The Importance of Social Intervention in Britain's Mortality Decline c. 1850–1914: A Re-interpretation of the Role of Public Health," *Social History of Medicine* 1, no. 1 (1988): 1–38.

31. Dublin, "Statistician and the Population Problem," 11; Louis I. Dublin, *The Fallacious Propaganda for Birth Control* (New York: Metropolitan Life Insurance Company, 1926) reprinted from the *Atlantic Monthly* (February 1926).

32. Dublin, "Statistician and the Population Problem," 11.

in other words, but the abnormal or defective did. Dublin argued against the racialization of difference in favor of its medicalization.

Dublin's personal and intellectual roots contributed to his distrust of arguments concerning racial inferiority. Born in Lithuania in 1882, Dublin emigrated with his family at the age of four and grew up on New York's Lower East Side, going on to study at City College among soon-to-be famous young cosmopolitans like Morris Cohen, before getting a PhD in biology from Columbia.[33] At Columbia, Dublin studied under the cytologist Edmund B. Wilson and spent enjoyable summers with the soon-to-be-leading eugenicist Charles Davenport. But Dublin felt slighted by Wilson in the end, and the man whom he felt most indebted to at Columbia was the anthropologist Franz Boas, to whom he had turned for help in interpreting articles in Karl Pearson's *Biometrika*. Boas, he later wrote, "welcomed me as though I were his son."[34] The community of secular, working-class, immigrant Jews surrounding Dublin in his early life and his crucial encounter with Boas—known today as a prime intellectual force, with his many students, behind the rejection of fixed, biological "race"—helped inoculate Dublin against theories of racial difference.[35]

But more than Dublin's personal politics shaped his stance. The politics of risk makers' data and statistical classing also argued against drawing invidious racial distinctions among whites. Dublin's Metropolitan had, like all

33. Dublin, *After Eighty Years*, 1–2, 19.

34. Dublin, *After Eighty Years*, 23–33 at 30. Dublin seldom if ever mentioned his ethnic or religious heritage—indeed, he presented himself as secular and cosmopolitan— yet his story of frustrations within Columbia, of E. B. Wilson's advice to him to study medicine as well as biology, and ultimately Wilson's limited help in getting Dublin into an academic position suggest that anti-Semitic prejudice may have played an important role in Dublin's career. Certainly some of Dublin's scientific peers looked down on him because of his Jewish background. Raymond Pearl told a Pearson student and biometrician that Dublin was "the vilest product of low Jew germ-plasm it has been my fortune to meet." Quoted in Edmund Ramsden, "Carving Up Population Science: Eugenics, Demography and the Controversy over the 'Biological Law' of Population Growth," *Social Studies of Science* 32, no. 5/6 (2002): 857–899 at 894.

35. See for a just a few prominent instances, the discussion of Boas and his students in Painter, *History of White People*, 228–244; Kevles, *In the Name of Eugenics*, 134–138; Carl N. Degler, *In Search of Human Nature: The Decline and Revival of Darwinism in American Social Thought* (New York: Oxford University Press, 1991), 59–104, especially 61–63; Elazar Barkan, *The Retreat of Scientific Racism: Changing Concepts of Race in Britain and the United States between the World Wars* (New York: Cambridge University Press, 1992), 77–95; and Joanne Meyerowitz, "'How Common Culture Shapes the Separate Lives': Sexuality, Race, and Mid-Twentieth-Century Social Constructionist Thought," *Journal of American History* 96, no. 4 (March 2010): 1057–1084 at 1062. On Dublin's intellectual milieu and cosmopolitan social thought, see David Hollinger, "Ethnic Diversity, Cosmopolitanism and the Emergence of the American Liberal Intelligentsia," *American Quarterly* 27, no. 2 (1975): 133–151.

the other mainstream life insurers, scrupulously built a big-tent category of "white." There seems to have never been much question of charging higher premiums to particular nationalities as such—even to the very short-lived Irish, although insurers' underwriting practices targeting alcohol consumption probably indirectly penalized German and Irish immigrants into the late nineteenth century.[36] Insurers probably deemed it too dangerous to alienate politically powerful immigrant groups—a fear that seems justified by their defeats in northern states on the question of discrimination by "color." Insurance texts cited intermarriage among whites as one reason that such differences could play but little part in insurance decisions.[37]

Still, erasing differences among white men had taken concerted effort. Regional variations in average mortality rates posed the most significant problem to white unity, as figure 7.3 suggests. Drawn from life insurer mortality investigations in the early twentieth century and translated into map form by the economic geographers Ellsworth Huntington and Frank E. Williams, the figure demonstrates the persistent health disparities between North and South. In the nineteenth century, northern life insurers almost all charged penalties to southern policyholders—a source of considerable tension.[38] Only after the Panic of 1873 did companies begin relaxing their penalties. But the penalties went away long before mortality differences disappeared. Metropolitan's actuary indicated that southern mortality in 1920 still exceeded expectations by 28 percent.[39] Frederick Hoffman's many cemetery trips came in response to just such differences: he hoped to turn an apparently regional difference into many instances of defective locales, to unify the white nation by identifying abnormal sites of high morality. In the end, political and commercial concerns, rather than corporate mortality experience, rid life insurers of regional classifications. So when Louis Dublin

36. Medical examiners received instructions to discuss an applicant's sobriety and temperance (which by the 1880s had come to include significant concerns about opium-eating too), but were not expected to speak to nationality. See, for instance, J. Adams Allen's remarks in *Medical Examinations for Life Insurance* (1886), 21–24.

37. Allen, *Medical Examinations for Life Insurance*, 75.

38. An impolitic actuary for MONY attributed higher southern mortality rates in 1854 to slave society using thinly veiled language: "In some States, the parties insured live simply and frugally and labor steadily and long, while in other States more bountifully favored in some respects, they toil less, live fast and soon accomplish their aim and—end." 16 August 1854 meeting, Minutes of the Insurance Committee (Book 1): 236–237, MONY Papers. MONY commissioned a doctor named James Wynne in 1857 to justify its discriminations to southern agents and doctors. 12 August 1857 meeting, Board of Trustees Minutes (Book 2): 144–146, MONY Papers. See Wynne, *Report on the Vital Statistics of the United States*.

39. J. D. Craig, "Mortality on Colored Lives," *Transactions of the Actuarial Society of America* 21, no. 64 (1920): 452–475 at 459.

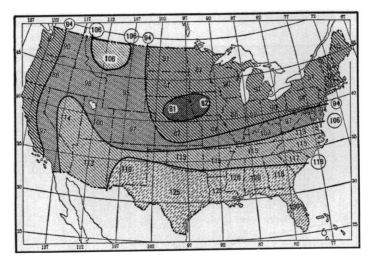

7.3. Relative mortality rates by region of the United States, adapted from life insurance statistics. The numbers indicate percent of mortality compared to that predicted by a recent mortality table. The original data only included state-level figures, but Ellsworth Huntington and Frank E. Williams smoothed it with isolines for this illustration in *Business Geography* (New York: Wiley, 1922), 73.

refused to acknowledge white racial differences, he followed industry precedent, the drift of classing practices, and risk makers' data politics.

Dublin's arguments had little chance of budging a political will bending toward the restriction of immigration. His position, built from one system for thinking statistically about individuals, supported the politically weaker side. Brigham's arguments, built from another, became a tool for the stronger side. Neither set of arguments played a determinative role, but the fact that each side employed cultural resources that began as by-products in statistical sorting—in risk making and intelligence testing—suggests the growing power of such systems and the tools, methods, and knowledge they generated.

In 1924, the Reed-Johnson immigration act became federal law, essentially halting immigration to the United States for nearly half a century. The act allowed only 150,000 overseas immigrants a year subject to quotas for each nation, and excluding almost entirely Asians and Africans. Determined to appear nondiscriminatory in the face of outside criticism from other nations and recent immigrants, the act based its quotas first on the proportion of immigrants from each country in the 1890 census (before the "new" immigration began in earnest) and then on the "national origins" of white Americans in 1920. The concept of "national origins" imagined that each white American could be understood in terms of his or her racial components—a statistical invention (beginning with aggregate immigration and census data) that reified "race" as a heritable entity constituted within individuals. "National origins" gave racial ideas political power. But American statistical systems had data politics resembling life insurers'—although more attentive to nation and race—and those data politics made the calculation with any certainty of the national origins of Americans nearly impossible.[40] Still, it hardly mattered: formerly mighty rivers of immigrants flowing into America dwindled to a series of trickles. And in the process, as historian Mae Ngai has argued, the federal government's classification systems made starker distinctions between white ethnicities and nonwhite races, distinctions that became more and more important as nativist sentiment died down in the post–Reed-Johnson era.[41] White data politics triumphed in the long run, even as Dublin and the anti-restrictionists lost.

Dublin's continued work with white population data led him to rue his defeat even more acutely as he worked with Alfred Lotka to extend risk-making tools toward encompassing more aspects of life than simply death. In 1924, Lotka, only recently hired as Metropolitan's director of mathematical research, developed the theoretical basis for a calculation of the "true rate of natural increase" of the (white) American population. Lotka realized that simply looking at birth rates and death rates, as was the common practice for those setting off Malthusian alarms of runaway population growth, did not provide a complete basis for prophesying about population changes. Birth rates and death rates both depended on the "age structure" of a population. A nation with many young people had higher birth rates and lower death rates, for instance. To take such possibilities into account, Lotka created a tool that calculated the effects of current rates of birth and mortality at every age on *any* population. Insurers used "life tables" that did the same thing, but only with mortality rates. Lotka's invention better deserved the title of life table.[42]

Lotka's new tool demonstrated that the apparent growth rate of the (white) American population massively overstated the "true" growth rate. White Americans constituted a very young population in 1924, both because immigration had brought many young people across the Atlantic and even

40. Mae Ngai, "The Architecture of Race in American Immigration Law: A Reexamination of the Immigration Act of 1924," *Journal of American History* 86, no. 1 (1999): 67–92; Anderson, *American Census*, 142–149.

41. Ngai, "Architecture of Race in American Immigration Law," 67–92.

42. Dublin and Lotka, "On the True Rate of Natural Increase."

more importantly because of a much higher birth rate among Americans in the 1890s. With the collapse of immigration and a falling birth rate, Lotka and Dublin predicted that the long-run growth rate of the United States would be half as much as everyone else expected.[43] While their peers warned of excessive growth, Dublin worried that a "stationary," and therefore aging, population would not be far off, especially if birth control advocates (his wife among them) kept up their campaigns. Dublin, who would be called the "ablest critic of the birth control movement" called on the birth control league to preach birth control to the poor and birth increase to the "antisocial" "selfish" well-off who "escape their responsibility, ultimately to their own detriment and to the injury of the State."[44] Immigration restriction appeared to Dublin a very bad idea indeed, and it made birth control seem even more dangerous. But, once again, few of Dublin's contemporaries came around to his positions.[45]

* * *

After his defeat on immigration restriction, Louis Dublin devoted much intellectual effort to the cause of arguing—with the help of risk-making by-products—that "probably no single American group is experiencing so deep and so intelligent a revival of latent power as is the Negro today."[46] In popular magazines and public forums, Dublin noted that African Americans' 1927 life expectancy of forty-six years was "equal to that of white Americans only thirty years ago," and gaining.[47] Moreover, he explained that declining birth rates among African Americans did not justify the "doleful prophecies"

43. Dublin and Lotka, "On the True Rate of Natural Increase."

44. Dublin, *Fallacious Propaganda for Birth Control*, 8. On Dublin's wife, see *After Eighty Years*, 136. On Dublin as the "ablest critic," see Norman Himes, review of *Population Problems*, in *American Economic Review* 16, no. 3 (1926): 511–514 at 511.

45. Dublin held a minority view on both counts among contemporary economists and demographers, who worried much more about overpopulation at the time. See the other essays in Louis I. Dublin, ed., *Population Problems in the United States and Canada* (Boston: Houghton Mifflin, 1926). A few reviewers praised Dublin's perspective, but most overlooked it. For a typical review citing favorable attitudes toward birth control and restriction, see Hugh Dalton's review in *Economica* 18 (November 1926): 347–349. Notestein reports, however, that the Lotka/Dublin paper on the true rate of natural increase did make its way to the *Congressional Record*. And years later Lotka's tool became crucial to the discipline of demography, at which time it fueled fears of overpopulation. Frank W. Notestein, "Demography in the United States: A Partial Account of the Development of the Field," *Population and Development Review* 8, no. 4 (1982): 651–687 at 655–656.

46. Louis I. Dublin, "Life, Death, and the American Negro," *American Mercury* 12, no. 45 (1927): 37–45 at 44.

47. Dublin, "Life, Death, and the American Negro," 41.

of race "extinction" of the past:[48] having fewer children simply made African Americans look like all other native-born Americans.[49] Dublin interpreted higher rates of African American mortality in terms of lag, and not as evidence of inherent difference. He offered the hope of future equality: a repudiation of Hoffman's earlier forecast and a glimmer of the egalitarian optimism that had fueled the antidiscrimination fight years before.

But the white data politics of corporate life insurance meant that Dublin's arguments could do very little to address the cause of significant racial inequality within traditional risk making. No one understood this better than Lawrence Napoleon Brown, a master's student in the insurance program at the University of Pennsylvania's Wharton business school, who in 1930 wrote a thesis investigating how life insurers at the time viewed African Americans as risks.[50] Brown praised Dublin's positions and drew on his findings. But even as he praised Metropolitan for insuring blacks and for extending its welfare work to black communities, Brown constructed a subversive critique of white data politics and the discriminatory effects of those politics on ordinary folks like those to whom Richard Wright sold insurance.

Brown couched his critique in a reproduction of Dublin's chart of "race stock" life expectancies, presented in figure 7.4—a reproduction lacking Metropolitan's graphical flair, but made profound by one crucial addition. Ordering each "race stock" by increasing life expectancy, Brown inserted "Negro" between "Irish" and "German."[51] He thereby broke Dublin's chart's (and mainline life insurers') assumption of white separateness. African Americans did not exist apart from whites from the standpoint of mortality. They fit *within* a broader spectrum of American mortality experience, not outside of it. Louis Dublin understood this too: "We can readily pick out a number of foreign races in our congested cities with mortality rates in excess of those for the Negro," he wrote in 1928.[52] But Dublin's company, as Brown reported, did not act out this truth. It saw race through the lens of its white/black classificatory binary.

48. The most famous such doleful prophecy, and undoubtedly the one on Dublin's mind, was Hoffman's in *Race Traits and Tendencies of the American Negro* in 1896.

49. Dublin, "Life, Death, and the American Negro," 45.

50. Lawrence N. Brown, "The Insurance of American Negro Lives" (MBA thesis, University of Pennsylvania, 1930).

51. Brown, "Insurance of American Negro Lives," 42.

52. Louis I. Dublin, "Health of the Negro," *Annals of the American Academy of Political and Social Science* 140 (November 1928): 77–85 at 82.

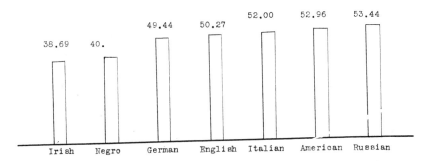

Diagram 4

Comparative Life Expectancy of Racial Stocks

in the United States.

Males at Age 10

7.4. Lawrence N. Brown's hierarchy of "Racial Stocks," ranked by vitality, in Brown, "Insurance of American Negro Lives," 42. Courtesy of University of Pennsylvania Libraries.

Brown—who grew up in Johnson City, Tennessee, and graduated from Lincoln University, a black college in Pennsylvania—began his project already aware that insurers generally classed African Americans as poor risks. He set out to determine why.[53] Masking more radical intentions behind a message of race uplift, Brown claimed his research would provide "a basis for determining what, in the minds of insurance men, makes the Negro an undesirable risk that the undesirabilities may be eliminated to make him a good risk."[54] But as Brown would show, one of the key things that made African Americans inferior, in insurers' eyes, was the fact that insurers classed them apart from whites.

Brown's survey of life insurers' practices disclosed rampant and undisguised discrimination. Nearly three-quarters of all mainline companies had responded to antidiscrimination laws by simply refusing to solicit African

53. Brown's record in the *1926/1927 Graduate School Announcement,* University of Pennsylvania Archives; "Lincoln University Closes with Fifty-one Graduates," *Chicago Defender,* 21 June 1924.

54. Brown, "Insurance of American Negro Lives," 3.

Americans (or by excluding them outright), a handful insured African Americans on an equal rate, and 28 percent insured African Americans on some sort of limited or substandard policy.[55] Companies did not bother to hide any of these practices. Brown found most company attitudes and practices published in the industry trade press, and he received candid responses to letters from more than half of the remaining companies.[56] Indeed, why should they hide their practices, when even regulators in states that barred race discrimination admitted openly that African Americans should pay more for life insurance.[57] Metropolitan had led the way in popularizing that view in the prior two decades and many companies referred to its statistical experience as justification for avoiding African Americans.[58]

In the 1920s, Metropolitan's actuarial and statistical departments possessed data linked to over two million policies covering African Americans, encompassing a fifth of the entire African American population.[59] Metropolitan carried two-thirds of *all* policies on African Americans.[60] Each of those policies bore the distinguishing mark of the applicant's nonwhite "color" or "race," a mark with consequences. That mark pretty much guaranteed that the resulting risk would be labeled substandard. It also facilitated Metropolitan's actuaries in their attempts to justify their discriminations.

But it's worth asking how Metropolitan—which had been singled out for praise by African American leaders for being a major life insurer who agreed to solicit (carefully selected) African Americans even after New York passed its antidiscrimination law—became the model for discriminating firms.[61] Part of the answer comes from regulatory reforms after the Armstrong hearings and part of the answer hails from the continued lengthening of Americans' lives. In 1904, Booker T. Washington had cited Metropolitan

---

55. Brown, "Insurance of American Negro Lives," 31.

56. Brown, "Insurance of American Negro Lives," 23–24.

57. In 1919, an actuary for New York's regulatory body allowed "race" as a permissible reason to rate an applicant "under average." *None* of his audience—regulators all, and many in states with antidiscrimination laws—registered an objection. See C. G. Smith's comments in the discussion following R. E. Ankers, "The Proper Basis of Reserves for Sub-Standard Risks in Life Insurance," *Proceedings of the Annual Session of the National Convention of Insurance Commissioners* 15 (1919): 120–122, 283–288 at 120. A 1917 insurance textbook referred to antidiscrimination laws passed "in the belief that they were thereby enforcing the spirit of the fourteenth amendment," but continued: "In practice such laws can be easily evaded." William Franklin Gephart, *Principles of Insurance*, vol. 1, *Life* (New York: Macmillan, 1917), 105–106.

58. Brown, "Insurance of American Negro Lives," 32.

59. Dublin, "Life, Death, and the American Negro," 42; also in Craig, "Mortality on Colored Lives," 453.

60. Brown, "Insurance of American Negro Lives," 17.

61. James, *Metropolitan Life*, 338.

for its faith in African Americans' innate capacity for equality.[62] But post-Armstrong legislation in 1907 made it illegal for Metropolitan, which was not a mutual company (it was not owned by its policyholders at that time, but would eventually mutualize), to offer policies with dividends and thus inadvertently set the company on the road to discrimination.

Since it could no longer rely on dividends to refund policyholders the amount they were overcharged by out-of-date tables, Metropolitan actuaries took up the task of preparing new tables of mortality and new rate schedules based on a law of mortality (a curve drawn through actual mortality data graphed against age of death) that more closely fit the company's experience.[63] The new tables evidenced racial and sexual differences in life expectancy born out of changing life spans. African Americans, the mortality studies revealed, lagged behind "whites," while women's lives outlasted men's by larger margins than before.[64] One of the strange consequences of improving health conditions is that lives seldom lengthen uniformly across statistical categories, so longer lives create new inequalities.[65] The same changing life spans that suggested the modern conception of death also argued for racial discrimination in risk making.

Metropolitan reformulated its substandard policy table from its experience with African American risks, making African Americans the type case

---

62. His stump speeches included references to a conversation between himself and Metropolitan's chief medical director, who "stated that after twenty years' experience and observation, his company had found that the Negro who was intelligent, who worked regularly at a trade or some industry and owned his home was as safe an insurance risk as a white man in the same station of life." Booker T. Washington, *Negro Education Not a Failure* (Tuskegee, GA: Tuskegee Institute Steam print, 1904).

63. Metropolitan Life, *Epoch in Life Insurance*, vii–viii.

64. James, *Metropolitan Life*, 339. Insurers' responses to lengthening women's lives reveal the space for judgment in setting rates. In 1907, for instance, Metropolitan began charging women more for annuities, creating, according to one company publication, "a much more equitable arrangement, in view of the demonstrated superiority in the longevity of female annuitants." Women did not, however, receive the benefit of lower life insurance premiums from Metropolitan (which was at that point ahead of most other insurers in that it didn't charge women *more*) or from most other companies until the 1950s. Metropolitan Life, *Epoch in Life Insurance* (1924), 87. For Metropolitan's insurance rates for women and similar information for its competitors, see Alfred M. Best Company, *Best's Illustrations of Net Costs, Cash Values, Premium Rates, Policy Conditions of Most Legal Reserve Insurance Companies Operating in the United States* (New York: A. M. Best, 1934 and 1944). Joseph B. Maclean's standard life insurance textbook reported for many years that women, though longer lived, should be charged standard rates. Only the 1962 edition changed that advice, noting a shift around 1956 toward charging women less. See Joseph B. Maclean, *Life Insurance* (New York: McGraw-Hill, 1962), 262–263. Maclean claims on 237 that charging women more for annuities did not become a widely accepted practice until the 1930s.

65. James C. Riley highlights this point in *Rising Life Expectancy: A Global History* (New York: Cambridge University Press, 2001), 30–31.

for substandard or subprime insurance.[66] Those tables would soon (with help from Metropolitan's lobbying) become the New York State standard for regulatory valuations.[67] Metropolitan subsequently discouraged agents from selling standard policies to any "colored person" by refusing a commission on such sales.[68] With this scheme, Metropolitan claimed to avoid the legal prohibition against race discrimination.[69]

In 1920, Metropolitan's actuary, J. D. Craig, quoted from New York's antidiscrimination laws, acknowledging that "legislatures do not look with favor upon an additional premium charge for colored lives." He continued, explaining his company's reasoning in making African Americans the type for subprime policies: those legislatures "probably felt that their 'general condition of health and prospect of longevity' compare favorably with white lives, but where the expectation of life or the prospect of longevity is not the same as for white lives, it would seem to be in accordance with the intent of the law that premiums for colored lives might be based on tables which actually reflect the prospect of longevity for such lives."[70] Craig could speak authoritatively on African American mortality—from one of the best sources for data on African American mortality experience—because Metropolitan's discriminatory scheme meant that it continued to sell life insurance when many of its peer companies did not. As a result, Metropolitan's experience encompassed a fifth of the African American population, drawn from all classes and (nearly all) regions of the United States (the company left blacks in the rural South to the care of Brother Mance, Richard Wright, and their colleagues in smaller burial societies).[71]

---

66. James, *Metropolitan Life*, 339.

67. Metropolitan Life, *Epoch in Life Insurance* (1917), viii.

68. They justified this practice by saying that the withheld commission paid for the extra cost of insurance. But what salesman would sell a standard policy with no commission when he could sell a subprime policy and get paid for it? For a generous interpretation, see James, *Metropolitan Life*, 339.

69. A New York State investigation of Metropolitan's practices conducted at the turn of the twenty-first century found statistical evidence that the company had indeed assigned higher-priced substandard policies to many African Americans on no basis other than race. See New York Insurance Department and Clifford Chance Rogers and Wells, "State of New York Insurance Department Report on Examination of the Metropolitan Life Insurance Company regarding Response to Supplement No. 1 to Circular Letter No. 19 (2000)," Investigative report (2002) at 15. Metropolitan eventually settled with New York State regulators and class-action plaintiffs for their early twentieth-century discriminatory practices for a sum reported to be between $140 and $160 million. "MetLife Is Settling Bias Lawsuit," *New York Times*, 30 August 2002.

70. Craig, "Mortality on Colored Lives," 452.

71. Dublin, "Health of the Negro," 77.

Lawrence Brown's critique pointed out the flaw in such an actuarial justification. J. D. Craig's assertion "that in all classes the mortality on colored lives is materially higher than on white lives" only held because "white" remained a monolith.[72] Craig never had to explain why African Americans, but not Irish or southerners, became the type case for substandard policies. Nor did he reflect on the efforts life insurers had exerted (or the politics and business interests that drove such efforts) to erase mortality differences based on region in order to construct a unified, white standard category. Life insurers' classing practices silently ensured that African Americans were held apart and treated differently, as inferiors.

Brown still looked ahead with optimism, placing his hopes in life extension rather than in the egalitarian civil rights rhetoric that activists like Chappelle had employed at the end of the nineteenth century. Looking forward to the year 2000, he foresaw over fourteen million African Americans who together "expected to realize a higher standard of living in terms of health, wealth, education and self-expression" along with "an improving rate of mortality."[73] Brown may have underestimated the power of white data politics and classing practices, particularly the impact of refusing African Americans entry to segregated risk communities.

White data politics and significant industry segregation forced even African American–run insurers to treat all their policyholders as subprime risks. Years earlier, in the wake of the passage of antidiscrimination laws and subsequent decisions by many companies to abandon selling insurance to blacks, fraternal societies providing small payments for burial costs or to sick members (like that which first employed Richard Wright) had flourished.[74] In the 1890s, black entrepreneurs started even more such

72. Craig, "Mortality on Colored Lives," 453. A similar flaw persists in scholarship today that follows the life insurers' lead by explaining discriminatory practices as being motivated not by racism, but by African Americans' mortality rates—an argument that overlooks the racism built into life insurers' classification systems. See for example, J. Gabriel McGlamery, "Race Based Underwriting and the Death of Burial Insurance," *Connecticut Insurance Law Journal* 15, no. 2 (2009): 531–570 at 537.

73. Of course, Brown's optimism could also have been strategic: he may have been talking of a better future in the hopes of convincing life insurers to reconsider their risk ratings for African Americans. Brown, "Insurance of American Negro Lives," 54.

74. Local lodges and smaller fraternal societies were nearly always quite homogenous in terms of race, class, or ethnicity. African American fraternals were not exceptional in that regard, though black-only insurance companies would be. On the homogeneity of fraternals, see Glenn, "Fraternal Rhetoric," 226. The most important and most copied African American fraternal was Virginia's Grand Fountain of the Grand United Order of True Reformers, founded in 1881 with a business model that resembled commercial insurance closely. See Woodson, "Insurance Business among Negroes," 206, 209–210; Bryson, Jr., "Negro Life Insurance Companies," 9–13.

societies. In the first couple decades of the twentieth century, many of these refashioned themselves as "legal reserve" life insurance companies operating on an actuarially sound basis, creating such prominent firms as North Carolina Mutual Life (founded as North Carolina Mutual and Provident Association) and Atlanta Life Insurance Company (founded as Atlanta Mutual Insurance Association).[75] These firms became crucial economic and social institutions in segregated America that could aggregate capital, support other black enterprises, and fund a wide variety of social "uplift" efforts, including challenges to legal segregation.[76] But their scale and power remained limited. By 1929, thirty-two black life insurers operated in the United States, only a handful of which did business in the North, and only one of which managed to meet the stringent requirements set by New York State for life insurance companies.[77] The $240 million of insurance in force in 1930 in black life insurers accounted for only a sixth of the total held on African Americans and less than 1 percent of the total insurance in force in the United States.[78] Still, at first, these insurers promised African Americans protection and a better future without the presumption of inferiority.

Operating within the same egalitarian frame as the earlier antidiscrimination activists, black insurers got under way employing risk-making systems that charged their policyholders standard (white) industrial insurance premiums. But these companies soon struggled. Unlike larger, better-established life insurers, they lacked experience in selecting good risks and, much more importantly, they made little headway attracting applicants outside the African American community. A firm like Metropolitan could choose not to discriminate against Irish immigrants (or southerners, or any group) and not suffer much: it had a large, diverse risk pool that could absorb some groups with above-average mortality. Black insurers had no room for choice: they attracted African Americans only. In the 1930s, faced with high rates of mortality among their insured, most black insurers raised their rates

75. Excellent histories of both the North Carolina Mutual and the Atlanta Life exist. See Weare, *Black Business in the New South*; and Henderson, *Atlanta Life Insurance Company*.

76. For instance, Earl Dickerson, president of Chicago-based Supreme Life, funded lawsuits to challenge other kinds of discrimination, including restrictive covenants in *Hansberry v. Lee*. See Kenneth W. Mack, "Rethinking Civil Rights Lawyering and Politics in the Era before *Brown*," *Yale Law Journal* 115, no. 2 (2005): 256–354 at 288.

77. Woodson, "Insurance Business among Negroes," 216–220. On an important northern life insurer that began in these years as a funeral society, see Robert E. Weems, *Black Business in the Black Metropolis: The Chicago Metropolitan Assurance Company, 1925–1985* (Bloomington: Indiana University Press, 1996).

78. Brown, "Insurance of American Negro Lives," 14–15. Stalson reports $105 billion in force in 1930. Stalson, *Marketing Life Insurance*, 817.

to the same levels charged by discriminating white companies.[79] "Negro companies," noted an African American insurance researcher, learned to consider "their average risk substandard."[80] Most African Americans—in large part thanks to white data politics—no longer had any choice but to be labeled substandard.

* * *

Richard Wright's first adventure in life insurance had come to an abrupt end when Brother Mance died unexpectedly, taking Wright's five-dollar wage with him.[81] He did not relish returning to insurance sales nearly a decade later after his move north to Chicago. But the Great Depression forced his hand as it stole from him the stable job he'd won in the US Postal Service. Through a distant relative he ended up working for several African American–run burial societies selling funeral policies in 1930.[82] On his weekly rounds, Wright interacted with those Americans who most directly felt the failure of risk making to build a better future.

Wright painted a bleak picture of desperation and corruption. He did ugly work for little money and found himself surrounded by pain, want, and viciousness. Wright "hungered for" and found "relief" from "comely black housewives who, trying desperately to keep up their insurance payments, were willing to make bargains to escape paying a ten-cent premium each week."[83] His fellow agents sometimes did even worse, extorting sex in return for claim payments.[84] The burial society's management, burdened by the ongoing economic disaster, committed its own crimes against policyholders. Wright explained a "swindle" he participated in involving swapping, without the policyholder knowing, a more liberal policy for one with "stricter clauses" that could save the society money (echoing American Popular Life's swindle in the 1870s). Wright did not like it, but he also did not believe he had other choices: "I could quit and starve," he wrote. "But I did not feel that being honest was worth the price of starvation."[85] In his desperation, Wright found far too much company among his bosses, his fellow agents, and so many decent families trying, in a cruel twist of fate, to

79. Bryson, "Negro Life Insurance Companies," 34–39.
80. Bryson, "Negro Life Insurance Companies," 47.
81. Wright, *Black Boy*, 150–152.
82. Richard Wright, *American Hunger* (New York: Harper & Row, 1977), 31.
83. Wright, *American Hunger*, 31–32.
84. Wright, *American Hunger*, 36.
85. Wright, *American Hunger*, 36.

maintain their dignity (and avoid a pauper's burial) by maintaining their insurance at any cost.

In their famous study of Muncie, Indiana, Robert and Helen Lynd had noted that (white)[86] unemployed families cut back on clothing, food, and the telephone, and sought extra work before they gave up an insurance policy. One woman, expressing a popular view, said, "We'll give up everything except our insurance. We just can't let that go."[87] Many African American families clearly felt the same way. But a risk-making system that labeled all African Americans substandard made their commitment to life insurance particularly costly.

Writing his classic work on black insurance, *An Economic Detour*, in 1940, M. S. Stuart talked about the affront to dignity, the "brazen type of American prejudice," exemplified by mainline companies who sold insurance at discriminatory rates to African Americans: "*We will send agents of other races into your homes seeking business and there, across your threshold, around your own fireside, in the presence of loved ones, you may expect the same kind of humiliation.*"[88] Yet the triumph of white data politics in risk making meant that few African Americans avoided some sort of humiliation in their efforts to protect themselves and their families from unknown, uncertain futures. Life insurers' tools for making individuals statistical hurt those individuals, attacking their pride and their wallets.

Even insurers' efforts to improve individuals' futures—to make good on the promise of the modern conception of death and the potential of life insurance for social improvement—could leave African Americans further behind their neighbors. When life insurers began offering group policies sold to employers to cover whole companies, they raised the rate for each "Negro" employed, creating a potential disincentive for employing blacks.[89] As industrial insurers like Metropolitan phased out medical exams, they let white applicants avoid exams for larger policies than black applicants could.[90] Similarly, free nursing visits—the vanguard of insurer-led health

---

86. The Lynds explicitly limited their gaze to white, native-born Muncie residents. See Igo, *Averaged American*, 55.

87. Robert S. and Helen Merrell Lynd, *Middletown: A Study in American Culture* (New York: Harcourt, Brace, 1929), 62–63.

88. Emphasis in original. M. S. Stuart, *An Economic Detour: A History of Insurance in the Lives of American Negroes* (College Park, MD: McGrath, 1969): xxii–xxiii.

89. Brown, "Insurance of American Negro Lives," 27–28.

90. Metropolitan eliminated medical examinations on nearly all industrial applications for $500 or below (for whites age forty or younger), for $300 for whites over forty and for "coloured lives" forty or under, and for $200 for "coloured lives" over forty. Metropolitan Life, *Epoch in Life Insurance* (1924), 59–60.

reform—reached a smaller percentage of black families than white families.[91] Risk makers' tools for making statistical individuals did not aid all individuals' bodies equally. Even before the depression hit, risk makers had often failed to make a better future for African Americans.

For Wright, "the depression deepened" and he "could not sell insurance to hungry Negroes."[92] Walking his insurance "debit," (weekly rounds for collecting premiums) Wright had seen extraordinary poverty and had been exposed to the radical ideas of Chicago's John Reed Club with its brand of American communism. His next big opportunities—the ones that helped him become a celebrated writer of works like *Native Son*—came through the Party.[93] But in the meantime, out of work and out of food, Wright swallowed his considerable pride to wait in a Cook County Bureau of Public Welfare line, "to plead for bread."[94] Like many Americans, he turned to the state for help. The state, in turn, looked to private enterprise—to life insurers—for guidance in constructing a new foundation for "welfare" in America: the Social Security system. With that system, the federal state created a powerful, new institution for risk makers, from which they could extend their reach even further to new corners of American life.

91. Dublin, *Records of Public Health Nursing*, 28.
92. Wright, *American Hunger*, 42.
93. Richard Wright, *Native Son* (New York: Harper's, 1940).
94. Wright, *American Hunger*, 42.

In August 1939, Dorothea Lange, a photographer employed by the Farm Security Administration to document the lives of down-and-out Americans struggling through the Great Depression, stopped in the Pacific Northwest. She snapped a photo (figure 8.1).

Lange's captivating multilayered photo invites decomposition. On its face, it depicts a nation's upheaval, capturing a still moment amidst one young couple's disrupted existence. They stand as types of Depression-era movement and uncertainty. Peeling back a layer, the photo presents two lives: a man's and his wife's. Lange offers these two as economic individuals, a bean picker and her unemployed lumber worker, identified by their occupations, their poverty, and their bad luck. Looking one layer deeper, we realize that Lange composed this photo with an unexpected subject in mind. Her focus lay not on these two enchanting individuals, but on the number tattooed across the lumber worker's arm. It is his Social Security account number.

To conclude this book, we will let Lange's photo and its three layers guide us. Each layer invites us to explore a different answer to this book's title question—how did our days become numbered?—through a partial summary of what has happened so far, followed by more of the story (now centered on the New Deal's risk making) that remains to be told.

Each layer corresponds to one keyword from the title—layer 1 to "our," layer 2 to "days," and layer 3 to "numbered." In the first layer's depiction of upheaval, the photo draws our attention to the ways that crisis and disruption created opportunities for risk-making systems and their by-products to migrate, which they did again during the crisis of the Great Depression. Thinking about such migrations invites us to consider the "our" in this book's title—for with each migration, new lives became risks.

8.1. Dorothea Lange's "Unemployed lumber worker goes with his wife to
the bean harvest. Note social security number tattooed on his arm. Oregon.
See general caption number 46." Courtesy of Library of Congress, Prints and
Photographs Division, FSA/OWI Collection, LC-DIG-fsa-8b15572.

One level deeper, this photo, in showing us lives defined by economic
opportunity and peril, leads us to ask what aspects of our "days" have attracted risk makers' attention and leads us to the conclusion that when
risk-making systems migrated—in part or in whole—they gave risk makers
access to different facets of our lives. In the nineteenth century, making a
life into a risk meant focusing on factors that might bring early death. By
the early twentieth century, health became nearly so important a focus as
death, and with the Great Depression, risk makers pulled individuals' daily
economic existence into their sphere of interest too: more and more of our
days fell under their analytical, statistical gaze.

Finally, at the level in which we realize the photo's alternate subject—
the Social Security number, itself—the photo demands that we think about
what it means to be "numbered," by directing our attention to the numbers created by risk makers and by succeeding generations of risk-making
systems. Moreover, Lange's photo demands that we consider the ways that

some Americans grasped those numbers as their own and the ways the numbers maintain their own hold upon us.

Each section of this concluding chapter begins with Lange's photo, summarizes and analyzes material from the book so far, and then carries the analysis further into the context of risk makers' activities in Depression-era America. The chapter ends with a reflection on the layers of risk making that numbered our days and number them still. Lange's photo will show us the way.

Layer 1: Our

We see a migrant couple in a makeshift camp thrown up by bean pickers on the picking trail.[1] A simple tent, a dirt floor, a pot to cook in, and even a rickety stove (revealed in a companion picture) evoke unsettled living, a home without comforts, easily uprooted. It is a picture, at the broadest level of context—our first layer—of economic destabilization, of crisis, motion, and possibility.[2]

* * *

In this book, moments of economic or industrial upheaval became moments of opportunity and expansion for risk makers, their methods, and their by-products. Panic in 1873 and an ensuing depression spanning much of the seventies laid waste to life insurance corporations—risk makers' chief home at the time. Companies like American Popular Life collapsed and in their implosions became celebrity villains. Those companies that survived the disaster decided that the population of northern, middle-class men, the targets of furious competition over the previous decade, could not be depended on by companies desirous of surviving (sure to be recurrent) downturns. So, Metropolitan joined Prudential in the late 1870s in becoming an industrial insurer and began reaching out to working-class men, women, and children. Around the same time, MONY expanded its territory, more actively cultivating the West and the South—as would most other ordinary

---

1. See general caption no. 46 reprinted in Anne Whiston Spirn, *Daring to Look: Dorothea Lange's Photographs and Reports from the Field* (Chicago: University of Chicago Press, 2008), 154.

2. Lange, for her part, would not necessarily object to our taking apart her pictures this way. She made portraits of individuals, but always individuals in context. She had, after all, an explicit mission to document lives in economic distress in America. On Lange's mastery of portraiture in context, see Linda Gordon, *Dorothea Lange: A Life beyond Limits* (New York: Norton, 2009), 221–227.

insurers in the next two decades, with a few companies even writing insurance on European and South American lives. Life insurance, and through it risk makers, touched many more Americans' lives in the process. The "our" whose days were being numbered became much more inclusive.

By the end of the nineteenth century, life insurers' risk-making systems (taken together) accounted for millions of individuals' pasts and then predicted their futures. Those being numbered often did not approve of the use of statistical methods to forecast their fates, however. African Americans resisted being classed as an inferior race and as poor risks, their destinies determined (in insurers' eyes) by a past of slavery. Fighting back, they called into question risk makers' fatalizing, the idea that statistical pasts could be projected into the future. In 1905, insurers' risk making—their smoothing of past and present experience—looked to investigators like it might be yet another corruption of capitalism foisted on the nation by corporations plagued by power-hungry financiers meddling in democratic politics and benefiting themselves at the expense of the masses. But these challenges—from those made into risks and from the state—failed to dissuade life insurers from dealing in and with statistical individuals. In one case, internal corporate challenges (to the writing networks that allowed life insurers to keep tabs on Americans) even ended up *amplifying* the power of statistics to differentiate and individualize, as actuaries turned away from seeking a "grand average" to join doctors in learning "how fishermen compare with farmers, how physicians compare with clergymen, how brewers compare with manufacturers of soda-water, and the like."[3] The "our" expanded and simultaneously disaggregated, while many resisted.

For life insurers, the 1905 Armstrong hearings in New York, the scandals and revelations surrounding them, and the economic panic that followed shortly after combined to destabilize their industry yet again. This time, risk makers, their know-how, and the materials and knowledge they had created over decades of interpreting individuals through statistics began circulating beyond the confines of traditional life insurance. Doctors putting patients on scales in their offices started referring to insurers' "build" tables and began thinking about build in terms of risks of mortality, while public health nurses joined insurers' correspondent networks, scouring tenements with insurers' statistical guidance and financial support. Public officials and sanitary reformers (and eugenicists too) came to rely on means of valuing lives in dollars and on insurers' risk-making systems more generally to

3. McClintock, "On the Objects to Be Attained," 374.

justify their existence and plead for more support. And Americans across the nation—indeed, individuals around the world—employed risk makers' by-products as tools for improving their own health, thanks to Fisher and Fisk's popular guide, *How to Live*. Now even more Americans quantified their mortality chances or had those chances quantified than ever before. And when risk making promised life extension and not just death predictions, those being numbered showed themselves to be more willing to be interpreted statistically.

* * *

The Great Depression set Americans in motion in extraordinary numbers. It set our bean pickers—in figure 8.1—off following Oregon farm trails. It kicked Richard Wright out of the Post Office and into an insurance office in the throes of failure. It sent Dorothea Lange out West, camera in tow. And it spurred two corporate actuaries to catch trains to Washington, DC. With them, risk making began another significant journey.

Otto Richter hailed from AT&T's Comptroller's Department, Bill Williamson from Travelers Insurance Company's group life division.[4] In 1934, both men accepted offers to run numbers—to do the technical work—for President Franklin Roosevelt's Committee on Economic Security, a working group designed to prepare some sort of answer to the economic insecurity inherent to industrial capitalism, an insecurity the Depression made impossible to ignore.[5]

4. On Richter, see obituaries in *Transactions of Society of Actuaries* 14, no. 38 (1962): 197–198; and *Proceedings of the Casualty Actuarial Society* 48, no. 90 (1961): 242. On Williamson, see his obituary in *Proceedings of the Casualty Actuarial Society* 67, no. 128 (1980): 274; "Statement of W. R. Williamson," Economic Security Act 1009–1020, http://www.ssa.gov/history/pdf/hr35 williamson.pdf; Robert Myers Oral History Taken by Larry Dewitt, 1996, transcript on *Social Security History* website, http://www.ssa.gov/history/myersorl.html; *"Wyco" The Building of a Professional Actuarial and Consulting Organization: The Wyatt Company* (Wyco, 1984), 30.

5. The Committee on Economic Security staff's members contributed to much of the early history of Social Security's origins. See the special issue of *Law and Contemporary Problems* 3, no. 2 (April 1936); Committee on Economic Security, *Social Security in America: The Factual Background of the Social Security Act as Summarized from Staff Reports to the Committee on Economic Security* (Washington, DC: Government Printing Office, 1937), iii–v, 515–558; J. Douglas Brown, *The Genesis of Social Security in America* (Washington, DC: Social Security Administration, 1969); J. Douglas Brown, *The Idea of Social Security* (Washington, DC: Social Security Administration, 1958). Much of the ensuing scholarly work has debated the extent to which Social Security was designed to serve capitalists' interests. For our purposes, it is enough to accept that corporate tools certainly went into designing the Social Security system even as that system differed markedly in the ends it sought to achieve with those tools. For a sampling of the debate, see Jill Quadagno, "Welfare Capitalism and the Social Security Act of 1935," *American Sociological Review* 49, no. 5 (1984): 632–647; Theda Skocpol and Edwin Amenta, "Did Capitalists Shape

Both Richter and Williamson brought expertise to aid the federal government from institutional settings that were themselves relatively recent conquests for risk making. Group insurance emerged on the scene in the 1910s as a privately provided alternative to social insurance for workers, but it proved particularly attractive to insurers because it opened a new scandal-free market for companies stung by the Armstrong scandals or hampered by post-Armstrong regulations.[6] Companies purchased group policies insuring (against death, accident, disability, old age, etc.) their employees (sometimes with employees contributing half the cost). This was one face of welfare capitalism, made possible with insurers' help.[7] Corporate pension plans (run by the companies themselves) offered another face for welfare capitalism, and some, like AT&T, had their own ties to insurance. (AT&T took the pioneering step of employing an in-house actuary to guarantee pension solvency over the long run, beginning in 1927).[8]

When the Depression hit, and kept hitting, group insurance and actuarially run pensions proved resilient. While amounts of life insurance dropped precipitously, and corporate, fraternal, and union-based safety nets buckled under with the downturn, group insurance stumbled only momentarily in

---

Social Security?" *American Sociological Review* 50, no. 4 (1985): 572–575; G. William Domhoff, "The Little-Known Origins of the Social Security Act: How and Why Corporate Moderates Created Old-Age Insurance," in *Social Insurance and Social Justice: Social Security, Medicare, and the Campaign Against Entitlements*, ed. Leah Rogne et al. (New York: Springer, 2009), 47–61. The Social Security Act, and especially old-age insurance, plays a crucial role for historians and sociologists telling the story of social policy in the United States and of the creation of America's welfare state. See, for instance, Theda Skocpol, *Protecting Soldiers and Mothers: The Political Origins of Social Policy in the United States* (Cambridge, MA: Harvard University Press, 1995); Edward D. Berkowitz, *America's Welfare State: From Roosevelt to Reagan* (Baltimore, MD: Johns Hopkins University Press, 1991). On the extent to which private and public forms of insurance and security worked together to create the welfare state, see Jacob S. Hacker, *Divided Welfare State: The Battle over Public and Private Social Benefits in the United States* (New York: Cambridge University Press, 2002); Jennifer Klein, *For All These Rights: Business, Labor, and the Shaping of America's Public-Private Welfare State* (Princeton, NJ: Princeton University Press, 2003).

6. Equitable sold the first group policy to Montgomery Ward in 1912. Other companies joined the field later in the decade after prevailing upon New York legislators to exclude group sales from regulatory ceilings on annual policy sales. See Klein, *For All These Rights*, 18–30.

7. Group life advocates saw it as a way of easing tensions between "capital and labor" and fostering "good will," without being "paternalistic." Francis Bernard Foley, "Group Insurance" (Fellowship diss., Insurance Institute of America, 1934), 8–9. On corporate welfare capitalism, see Brody, "Rise and Decline of Welfare Capitalism." Tone offers a nice summary of subsequent debates over the life and especially "death" of welfare capitalism in Andrea Tone, *The Business of Benevolence: Industrial Paternalism in Progressive America* (Ithaca, NY: Cornell University Press, 1997), 5n6.

8. Nandini Chandar and Paul J. Miranti, Jr., "The Development of Actuarial-Based Pension Accounting at the Bell System, 1913–40," *Accounting History* 12, no. 2 (2007): 205–234 at 212.

1933 and AT&T's pension never posed any danger of failure.[9] As a consequence, neither Richter nor Williamson should be seen as running away from failure in 1934. They ran toward a new opportunity.

Roosevelt's Committee on Economic Security created that opportunity when it decided to make social insurance its tool in battling industrial insecurity. The committee needed a program it could implement quickly: the Civil Works Administration would soon be closing down its work relief programs leaving the unemployed to stew, and Townsendite crusaders (following the activist California physician Francis Townsend) hollered incessantly for large (irresponsible, in committee eyes) pensions to end poverty among the aged. Roosevelt's administration hoped unemployment insurance and old-age insurance—both of which had been employed extensively in various forms around the globe and were well understood in American universities—would hold off more radical reforms while helping Americans win a modicum of economic stability.[10] Committee members decided on the form and structure of the new social insurance program. Richter and Williamson did what they were told for the most part—they did not make policy—but they brought to the table expertise in corporate methods for making risks ready to be adapted to the government's ends.[11]

9. Aggregate group insurance dipped in 1933 before resuming its steady ascent, even as ordinary life insurance totals dropped substantially. Stalson, *Marketing Life Insurance*, 812–15, appendix 23 and 24. In a close study of Depression-era Chicago, Cohen explains how ethnic fraternal societies and banks, whose members could no longer pay premiums—even though they considered it a priority—and whose investments in real estate plummeted, repeatedly failed those who depended on them by cutting benefits or going out of business. The same went for many corporate welfare programs hit by the plunging stock market and diminishing corporate income. Lizabeth Cohen, *Making a New Deal* (Cambridge: Cambridge University Press, 1990), 227–246. AT&T in fact faced a different problem: it assumed the Depression would be shorter than it was and thus kept up reserves for employees who it believed would be only temporarily laid off. It ended up with a scandalously large reserve covering a lucky few who had kept their jobs. Chandar and Miranti, Jr., "Development of Actuarial-Based Pension Accounting at the Bell System, 1913–40," 225–230.

10. Rodgers makes this case in *Atlantic Crossings*, 440. On Roosevelt's tax policy as essentially conservative and Social Security as, at least in part, an effort to keep a more liberal Congress from doing anything rash, see Mark H. Leff, "Taxing the 'Forgotten Man': The Politics of Social Security Finance in the New Deal," *Journal of American History* 70, no. 2 (1983): 359–381.

11. Williamson went on to become the chief actuary (called Actuarial Consultant) for Social Security, but he never made his peace with the form of the program that committee members instituted. For his critiques see W. R. Williamson, "Social Budgeting," *Proceedings of the Casualty Actuarial Society of America* 24, no. 49 (1937): 17–34; W. R. Williamson, "Some Backgrounds to American Social Security," *Proceedings of the Casualty Actuarial Society* 30, no. 60 (November 1943): 5–30; and W. R. Williamson, "Social Insurance," *Journal of the American Association of University Teachers of Insurance* 14, no. 1 (1947): 84–93. On the implications of Williamson's

Layer 2: Days

Lange began the caption to her photo: "Unemployed lumber worker goes with his wife to the bean harvest." He worked once, but no longer, doing dangerous work no doubt, and drawing a wage. His wife, in the background, had to support their family now. She picked beans and her earnings (one dollar per one hundred pounds picked) had to sustain them on the picking trail.[12] Despite the hardship, they appear to have kept up their spirits— although she looks tired, smoking her cigarette, her head on her hand.

Looking at this layer of Lange's photo, at her depiction of two Americans who had bad luck in the economic chance-world of industrial capitalism, we see a different kind of risk making—tied to a different aspect of human life, a different portion of our days—than that which this book has most closely investigated. The first calculable risks of life had to do with the fundamental question of life and death. But in everyday life, other risks—of lost wages, lost work, poverty—could loom just as large.

"Life is safer, but living less secure," explained a 1937 Social Security pamphlet.[13] Public health reforms, safer workplaces, improved hygiene, better food and housing, better care for infants and children, and advancing medical knowledge had combined to lengthen lives and make them healthier. According to Louis Dublin and Alfred Lotka, the expectation of life for white men rose from forty-eight to fifty-nine years in American registration states between 1901 and 1931 and from fifty-one to sixty-three for women.[14] Americans' days had grown more plentiful. But living—making a living—posed its own challenges. And in the Great Depression—in Lange's portrait—(industrial, economic) living looked very much less secure.

* * *

---

discontent, see Larry DeWitt, "Financing Social Security, 1939–1949: A Reexamination of the Financing Policies of this Period," *Social Security Bulletin* 67, no. 4 (2007): 51–69.

12. Spirn, *Daring to Look*, 154.

13. Social Security Board, "Why Social Security?" (Washington, DC: Government Printing Office, 1937). This pamphlet's author is Mary Ross, a former *Survey Graphic* employee who once appealed to Louis Dublin to defend public health against criticisms by Raymond Pearl, who thought medical science deserved more credit for prolonging lives. See Bouk, "Science of Difference," 280–284.

14. These comparisons are tricky. In 1901, only ten US states belonged to the registration area. In 1931, all but a couple states kept good enough mortality records to qualify for the area. Still the basic story of substantial improvements seems sound. Most of those added years of expectation came from decreased mortality among infants and young people. Louis I. Dublin and Alfred J. Lotka, *Length of Life: A Study of the Life Table* (New York: Ronald Press, 1936), 61, 69, 77–79.

When risk making moved, it changed.

Risk making began in life insurers' offices where insurers endeavored to forecast fates and predict deaths. MONY, in an 1845 annual report, bragged of its reliance on "accurate and widely extended observations, in space and time," that made possible "deducing the laws of mortality."[15] On the basis of these laws, life insurance became "an exact science" that could "render sure and safe all computations and contracts based upon it."[16] Yet no sooner had insurers proclaimed the universal truth of mortality laws, than they looked into ways to limit them.[17] When T. S. Lambert advocated throwing out mathematized, statistical laws entirely in the late 1860s and early 1870s, he only carried the individualizing impulse in risk making to its logical conclusion. Most risk makers, however, worked somewhere in the middle of the spectrum running from the statistical to the particular; they described regularities and also asserted differences, defined classes and smoothed experience. Ultimately, they built systems that helped them explain and justify how various Americans' days should be differently numbered. Then, with the help of nationwide networks of informants, they numbered them.

When Frederick Hoffman trudged through southern cemeteries in 1911, he had death and difference in mind. He wanted data that would help his employer—Prudential—better predict the mortality among whites throughout the South (since the company had committed, with most of its peers, to American regional unity and American racial division). Yet Irving Fisher, Lee Frankel, and Louis Dublin were each already finding ways for risk practices to fight death, to improve health. As they—in the wake of the Armstrong hearings' destabilizations in 1905—shepherded risk makers' tools and insights into doctors' offices, settlement houses, hospitals, and public health offices, they oversaw a rethinking of risk making too. Risk makers would no longer only predict the future; they would (intentionally) reshape the future.[18] Those who began numbering Americans' days in the early twentieth century with Fisher's "modern conception of death" in mind viewed their numbering as inherently temporary, as counting days that were bound to multiply. Hoffman tried to follow the other progressive risk makers

---

15. Mutual Life Insurance Company of New York, *Annual Report* (1845): 14.

16. Mutual Life Insurance Company of New York, *Annual Report* (1845): 17.

17. They also demonstrated their (lack of) faith in these laws by adding a 30 percent cushion atop whatever premium rate their tables called for.

18. Of course, insofar as insurers' predictions affected terms for granting insurance and thus affected the economic and social opportunities open to individuals, such predictions had always also reshaped the future. But, for the most part, risk practitioners had not intended for their predictions to either change or fulfill the future they predicted.

(although he had his limits, just as risk making had its limits) and ended up in the Loomis Sanatorium, the Henry Street Settlement (with its nurse service), and the Johns Hopkins Hospital too: learning about health and living as much as, or more than, about dying. He visited his last cemetery (on Prudential's dime, at least) in 1916.

So as Americans' days grew more plentiful, risk makers sliced each day more finely. They colonized "life" and multiplied the ways they conceived of risk. Men and women discovered themselves plagued with previously unknown "impairments"—some of which had only just been invented by doctors cooperating with actuaries. At the same time, entire populations became fodder for projections and predictions about their births, developments, and deaths. In 1925, Lotka and Dublin created a new kind of life table that could forecast the growth rate of the entire nation. They foresaw a stagnating, aging America that public policy makers, state bureaucrats, and corporate managers all needed to prepare for. Risk makers described existing hazards in new ways, while also identifying new uncertainties, framing new problems. They invented new forms of risk and analyzed men and women's bodies (and the metaphorical body of the population) in light of them.

When risk making changed, older practices seldom disappeared. Instead, risk making proliferated in form and variety, taking different shapes in different institutions, and, sometimes, different shapes within the same institutions. The modern conception of death did not so much displace risk prediction as form a new layer of practice, technique, and knowledge atop the existing one—a layer that touched more American lives in different ways. New-York Life, for instance, continued to refine its risk-making treadmill and invented the numerical rating system to price each and every one of its applicants individually: Americans insured by New-York Life had their days numbered more precisely than ever before.[19] From New-York Life's fine-tuning of risk prediction, the Life Extension Institute developed a tool for scaring Americans into reforming their lives: losing weight, seeing a doctor, or—with no apparent irony—cultivating a worry-free lifestyle. Americans' days became things to be forecast and also things to be extended, at the same time.

* * *

19. Contemporary actuaries disagreed as to whether these precise valuations deserved to be called not just precise, but accurate.

When risk makers moved into corporations and into government in the early twentieth century, they sliced up Americans' days a bit more. This time, risk makers moved beyond biological life (health or death) to more fully address the economic uncertainties of life in industrial America. Louis Dublin and Alfred Lotka—in the process of trying to value the life of an American worker in 1926—had discovered just how little one could tell from existing statistics about an individual's economic life. A key problem they faced was figuring the life course of wages—how wages, on average, moved with the age of the worker. In attempting to create a wage table or wage curve, Dublin and Lotka began the process of making economic lives susceptible to risk makers' methods. The life table—risk makers' paradigmatic tool— correlated death rates and life expectancies to age. To create economic risks, Dublin and Lotka needed to show that wages varied by age in regular, lawlike ways too.

But no one else had bothered tying wages to ages on a large scale. As a preliminary step, they turned to their fellow corporate statisticians—a new breed of corporate employee in the early twentieth century. E. E. Lincoln provided Lotka with data from Western Electric's payrolls, which Lotka blended with public New York data to prepare a rough wage by age curve as part of an article on the value of a wage earner.[20] Using tracing paper, Lotka drew a gently graduated curve by hand through Western Electric's data—one more example of smoothing. But when publication time came, Lincoln, worried that employees might be upset about the data he shared, asked Dublin and Lotka to say as little as possible about his cooperation.[21]

Otto Richter and Bill Williamson soon made similar discoveries about the paucity of data on economic lives. Williamson found very little sound data on unemployment ("so incomplete" he wrote in 1937, "that it is necessary to resort to estimates of the number out of work"),[22] and what data he did find produced only "intricate and, so far, baffling frequency distributions."[23] Richter doubted the numbers he had to work with so much that he denied their precision entirely. None of his predictions, he warned,

20. Metropolitan Life, "Value of a Man as a Wage-Earner."

21. See Edmond E. Lincoln to Louis I. Dublin, 1 July 1926, in Folder 4, Box 31, Lotka Papers.

22. Committee on Economic Security, *Social Security in America*, 55. Unemployment had been recognized as a large hole in government statistics for some time. The politics of counting the unemployed during the Depression (neither Hoover's nor Roosevelt's administrations cared to broadcast a precise unemployment figure) made the data at once important and obscure. The first reasonably sound unemployment numbers only came out in 1937. See Anderson, *American Census*, 176–177, 183–185; Stapleford, *Cost of Living in America*, 148, 168.

23. Williamson, "Social Budgeting," 31.

were likely to turn out absolutely true. He argued only that his calculations showed the *comparative* costs of various plans accurately.[24] Richter and Williamson must have agreed with Isaac Rubinow—the influential advocate of social insurance (and a socialist to boot)[25]—who wrote in 1934: "America is, one might say, a highly statistical country," and yet: "in all matters dealing with human life as distinct from business, our statistical information is very fragmentary at best."[26] Indeed, even in business matters, human lives remained mysterious.

Still, Richter needed to predict Americans' futures within his admittedly hazy calculations and that meant both modeling the vital (that is, bodily or biological) transformations of the entire population while attempting to account for Americans' changing economic lives. Risk makers already had ready solutions at hand for the vital components. Alfred Lotka's population table—his "life" table admitting a fuller sense of life—allowed researchers from 1925 on to predict the ("true" "natural") growth rate of a population from its age and sex structures (how many men and women at each age) and its mortality and fertility rates at each age. His contemporaries at the Scripps Foundation for Research in Population Problems, Warren S. Thompson and P. K. Whelpton, used similar methods to construct population predictions for the entire United States, usually broken up by race and including a range of possible population outcomes (low, medium, high).[27]

24. Richter noted another reason to treat his numbers with a grain of salt as predictions: "For it is a practical certainty that the terms and provisions of the plan will be revised many times and in many ways before the period covered by these estimates has elapsed." Otto C. Richter, "Actuarial Basis of Cost Estimates of Federal Old-Age Insurance," *Law and Contemporary Problems* 3, no. 2 (April 1936): 212–220 at 220. Of course, Richter's caveats did not keep the numerical cat in the bag—once a number escapes into the wild, it often goes feral. Richter's models produced an astounding figure of $47 million as the eventual size (by 1980) of the Social Security reserve fund. Even Williamson (who should have known better, but was no fan of creating a giant reserve) cited the number in congressional testimony as if it were the actual amount that would one day accrue. Williamson, "Statement of W. R. Williamson," Economic Security Act 1009–1020 at 1012.

25. Roosevelt apparently read Rubinow's *The Quest for Security* (New York: Henry Holt, 1934), or at least inscribed a copy: "For the Author—Dr. I. M. Rubinow. This reversal of the usual process is because of the interest I have had in reading your book." "Isaac M. Rubinow," *Social Security History*, http://www.ssa.gov/history/rubinow.html. See also Philip Elkin, "The Father of American Social Insurance: I. M. Rubinow," *Review of Insurance Studies* 3, no. 3 (1956): 105–112. On Rubinow as a socialist, see Rodgers, *Atlantic Crossings*, 242–245.

26. Rubinow, *Quest for Security*, 41.

27. On Thompson and Whelpton's methods and various predictions over a range of years, see "The Population Forecasts of the Scripps Foundation," *Population Index* 14, no. 3 (July 1948): 188–195. On both men's centrality to demography in the mid-twentieth century, see Notestein, "Demography in the United States."

Lotka, Thompson, and Whelpton could sketch a picture of what Americans' (vital) days looked like in the aggregate fifty years into the future. Richter began with actual numbers from the 1930 census, adopted Thompson and Whelpton's "medium" projection for the population in 1980, drew (probably by hand) a logistic curve (an s-shaped "smooth curve of the type characteristic of population-growth") connecting the present to the future, and derived all relevant vital statistics from that curve.[28]

Richter's model described a nation that would reach a maximum population in 1975 of 150 million people. It would grow little from immigration. It would grow more industrial (which would mean that more individuals would be working for wages in occupations that qualified for old-age insurance). And it would grow older, from 5.4 percent of the population over age sixty-five in 1930 to 11.3 percent in 1980. Richter announced "the unescapable trend toward an increasing number and proportion of the aged in the population."[29] Two years later, the authors of the landmark *Problems of a Changing Population*, sponsored by the National Resources Committee, declared the United States to be on the road toward a "mature population."[30] Richter and his contemporaries projected an older, stagnant, industrialized American risk community by 1980. This vision informed all of Richter's calculations.

Risk makers' existing tools and statistics had less to say about Americans' economic lives. Census data gave little insight into the course of Americans' careers. Because no government body regularly collected income information for most Americans, no comprehensive data existed showing the average course of wages in America.[31] Richter could have turned to estimates like those that Dublin and Lotka had constructed, but instead he assumed an average wage of $1,100 for 1930 to 1980, because such an assumption eased calculations and because his data could not be trusted anyway. Similarly, Richter did not know when Americans usually retired with any accuracy (which mattered in debates over whether old-age insurance should kick in

---

28. Richter, "Actuarial Basis of Cost Estimates of Federal Old-Age Insurance," 214. On the logistic curve's development and use, see Kingsland, *Modeling Nature*, 64–76.

29. Richter, "Actuarial Basis of Cost Estimates of Federal Old-Age Insurance," 214.

30. *The Problems of a Changing Population: Report of the Committee on Population Problems to the National Resources Committee May 1938* (Washington, DC: Government Printing Office, 1938), 7.

31. The federal income tax touched very few workers before World War II. Less than 5 percent of the public paid income taxes at all in the 1930s. That number rose to 11 percent in 1941 and then to 36 percent (or about three-quarters of the labor force) in 1945. Leff, "Taxing the 'Forgotten Man,'" 378–379; Boorstin, *Americans*, 205–213.

automatically at sixty-five or only when the worker actually retired).[32] He had to guess. Risk makers struggled to understand—and thus to predict—Americans' economic days.

That would begin to change when Social Security began numbering Americans' days with dollar signs. Committee for Economic Security staff insisted—alongside President Roosevelt—that the old-age insurance program tie eventual benefits to individuals' entire history of wages, to their so-called contributions. The Social Security Act—signed by Roosevelt on 14 August 1935—called for percentage taxes, beginning at 1 percent, on all wages up to $3,000 every year for eligible workers (notably excluding agricultural laborers, domestic servants, government employees, and the self-employed). Against these regressive taxes (which would hit the lowest paid hardest), it set a progressive, redistributive benefit structure that fixed a worker's eventual monthly income at 1/2 percent of the first $3,000 in lifetime wages earned, 1/12 percent of the next $42,000, and 1/24 percent of the next $45,000.[33] Taxes required no records of wages, but benefits were meant to sustain "a man's standard of living as reflected in the wage level at which he has worked" and thus depended on keeping accurate records of each individual's entire employment history.[34]

Contributions had been employed in other nations' social insurance systems to avoid subjecting recipients to invasive need tests, and Roosevelt appears to have thought, as Mark Leff has argued, that those who contributed to the system would make sure the government ran it safely, conservatively, for the long term.[35] Contributory group insurance and pensions had also become the preferred method among leading corporations.[36] But no existing contributory system deemed it necessary to record an individual's entire economic history to figure his or her benefits. Other nations simply fit individuals into broader classes, while corporate pensions, like Richter's AT&T plan, based old-age payments on limited historical data: years of service

---

32. Richter, "Actuarial Basis of Cost Estimates of Federal Old-Age Insurance," 212–220.

33. David F. Cavers, "Federal Old-Age Insurance: Benefit Payments and Tax Collection," *Law and Contemporary Problems* 3, no. 2 (April 1936): 199–211 at 202–203.

34. Committee on Economic Security, *Social Security in America*, 214.

35. Barbara Nachtrieb Armstrong, a Berkeley law professor on the committee staff, made the case for contributions from international experience in "Old-Age Security Abroad: The Background of Titles II and VIII of the Social Security Act," *Law and Contemporary Problems* 3, no. 2 (April 1936): 175–185. Leff, "Taxing the 'Forgotten Man,'" 377.

36. Industrial Relations Section, Princeton University, *Memorandum: Group Insurance* (Ann Arbor, MI: Edwards Brothers, 1927) [mimeograph], 12.

or an average of recent wages.[37] Williamson blamed Social Security's odd contribution scheme on the "popularity of life insurance, with its banking devices."[38] He had a point. Committee staff built "equity" ideas from life insurance into the legislation. For instance, workers who failed to accumulate enough wages to qualify for old-age payments or the beneficiaries of those who died before receiving benefits would receive their own (but not their employer's) Social Security tax payments back, with interest.[39] Committee staff argued that workers should pay Social Security taxes by citing the "virtual equity" that would come from guaranteed benefits, even as they claimed that "by contributing . . . the individual worker would establish an earned right to a benefit related to the contribution made."[40] In a neat tautology premised on the "banking" element of commercial life insurance, entitlements and "contributions" created one another. They, in turn, opened a window onto Americans' livings, in dollars and cents.

Social Security's architects—without quite meaning to—made economic lives visible in the aggregate and individual economic risks as calculable as vital risks. Statisticians in the coming decades would mine Social Security contribution records—a new and very valuable by-product of the Social Security system—for all sorts of data. Interested in administering benefits for widows and widowers, for instance, they investigated the relative ages of husbands and wives and described remarriage in America. They also took advantage of the long-term nature of Social Security data, which followed individuals and groups over much of their lives, to secure more accurate data on that great unknown "old age" while acknowledging that mortality levels remained in flux and so created new "generation mortality tables" specific to every cohort of Americans.[41] Most importantly, they created crucial data sets detailing Americans' working lives generally: their wages, mobility, distribution, and uncertainties over time.[42] Risk makers' increased capacity for

37. Rodgers, *Atlantic Crossings*, 445; Chandar and Miranti, Jr., "Development of Actuarial-Based Pension Accounting at the Bell System, 1913–40," 210.

38. Williamson, "Social Budgeting," 33.

39. Committee on Economic Security, *Social Security in America*, 212. This emulated one form of group annuity, which returned contributions with interest, but minus a surrender charge, to employees and employers in the case of early withdrawal. Birchard E. Wyatt, *Private Group Retirement Plans* (Washington, DC: Graphic Arts Press, 1936), 60–64.

40. Committee on Economic Security, *Social Security in America*, 204–205.

41. Robert J. Myers, "Beneficiary Statistics under the Old-Age and Survivors Insurance Program and Some Possible Demographic Studies Based on These Data," *Journal of the American Statistical Association* 44, no. 247 (1949): 388–396 at 393–395.

42. The "continuous work history sample," begun in 1941 as a 4 percent sample (later 1 percent) launched the longest continuous longitudinal sample in the federal government

parsing days in economic terms did not do away with older risk-making methods and aims: the vital and economic lived on, one layered upon the other in Social Security records.

Layer 3: Numbered

In the second line of her caption, Lange instructed her viewer: "Note social security number tattooed on his arm." Across the unemployed lumber worker's bicep—we can just make it out—runs a tattoo reading "SSA" and under that: "535–07–5248."[43] Lange took three photos by the migrants' tent and in each she made this number her subject as much as the man. What we are looking at here, peering one layer deeper into this image, is a new kind of numbered American. Indeed, a numbered American who bears (proudly) a number inscribed on his body. But how could this American (and not just his days) be numbered in this way?

* * *

Risk makers, for the most part, never set out to number lives (that is, to fill our lives with numbers derived from them or assigned to them). They numbered days with more intention, but even then they only predicted individual fates—which is to say, made risks—because it helped them sell life insurance. Making risks and numbering days were epiphenomenal to life insurance, but became ends in themselves within their corporate homes. As risk makers did their jobs, they assembled population statistics, filled in ledgers, built card-filing systems, wrote books or articles, designed reports, and maintained networks of distributed informants: these were all by-products of risk making which combined in various ways to make up risk-making systems. From those by-products, from that system, emerged numbered lives.

This book has told the story of evolving risk-making systems. As risk makers moved, encompassed a wider "our," and parsed Americans' days more finely, they invented new risk-making systems to be hosted in new spaces for the processing and translating into numbers of more and more lives.

---

and spurred much later research on Americans' working lives. Joseph W. Duncan and William C. Shelton, *Revolution in United States Government Statistics, 1926–1976* (Washington, DC: Government Printing Office, 1978), 188.

43. "SSA" must have stood for either Social Security Act or Social Security Account. The Social Security Administration, with which we associate the acronym today, did not yet exist.

Midcentury Americans often lamented the state of vital statistics in their nation. The federal government and the states, for the most part, responded to such laments slowly and haltingly. In the meantime, life insurers built their own systems for tracking Americans, recording data about lives and rating them as "risks." New-York Life bragged of its extraordinary "machinery" at the turn of the twentieth century, of a home office that brought together inspection reports, medical examinations, applicants' self-descriptions, and the intelligence gathered by other life insurers in order to—in a single day, after thirty-six steps—transform all those writings about a Tennessee man into a "new baby," a life insurance policy to be "christened '4,133,587.' "[44] Risk makers' systems for numbering days ended by assigning a serial number to each new risk.

Risk-making systems proved over decades to be flexible, mobile, and durable. Take, for example, one key component of risk-making systems: medical examinations. Life insurance medical directors feared in the late nineteenth century that examinations and the networks of correspondents maintained to conduct them might be in peril. Their concerns resulted, through the collective action of directors from many companies and the added help of actuaries, in numerical rating and the Medical Information Bureau card system—innovations that increased medical department efficiency and quieted angry agents for a while. But life insurers' urge to limit medical examinations did not go away. By the first few decades of the twentieth century, competitive pressures led many life insurers to reconsider. They began to forgo medical examinations in the case of smaller life insurance policies.[45]

But medical examiners and risk makers by that time hardly needed to worry. They already had new clients anxious for their services (in the case of the examiners) and new opportunities for gathering individual medical data (in the case of the risk makers), thanks especially to the key role medical examinations played in plans to use risk-making tools to extend lives. The Life Extension Institute provided preventative medical exams to insurance policyholders of all classes, but also to corporate employees and to private individuals, and it helped make the annual medical checkup a common American custom.[46] It also gathered and analyzed data pertaining

44. New-York Life, *Temple of Humanity*, 27–38, especially 37.

45. Industrial insurers led the way in scaling back on medical examinations, but many ordinary insurers followed in the 1920s and 1930s. Metropolitan Life, *Epoch in Life Insurance* (1924), 59–60; Buley, *American Life Convention*, 2:698.

46. Hirshbein, "Masculinity, Work, and the Fountain of Youth," 96.

to the health of all its clients. Industrial policyholders who fell ill underwent something similar to an insurance medical examination after Louis Dublin's reforms of nursing record systems, bringing their health data within the reach of a risk-making system, often for the first time. When group insurers did away with medical examinations (and did away with collecting any data about individuals other than age, sex, race, and occupation)[47]—another seeming blow to the medical examiner and the risk-making system's capacity for gathering individual data—they did so knowing that most of their corporate clients in the 1920s had already built their own medical examination infrastructures (and group insurers advised those that had not on how to do so).[48]

We might have expected the decline of medical examinations in life insurers' risk-making systems to be a detriment to medical examination on the whole, but the opposite appears to have been true. Even as insurers phased out medical examinations, the proliferation of risk-making techniques meant that the number of Americans subject to a doctor's risk-inflected evaluation was probably increasing. No wonder that Hooper Holmes, the pioneering credit reporting agency, eventually decided to become a different kind of risk-making system: today it's a leading provider of individual health information and home health exams.[49]

Risk-making systems rated and valued "human machines," and were in their own way human machines—hybrids of organization and information technology. In the nineteenth century, life insurers appropriated (and

47. Group insurers used a mass census to gather individual data, which determined the aggregate rate charged employers. But each employee paid the same rate regardless of his or her individual characteristics. Foley, "Group Insurance," 17. Rates also depended on other employer-controlled factors, like manufacturing techniques, factory safety measures, on-site medical services, and even the availability of clean water. See Equitable Life Assurance Society of the United States, "Application for Group Insurance," in 80–28, Box 1, RG4, Equitable Papers.

48. On medical examinations and corporate employment, see Angela Nugent, "Fit for Work: The Introduction of Physical Examinations in Industry," *Bulletin of the History of Medicine* 57, no. 4 (1983): 578–595; Nate Holdren, "Incentivizing Safety and Discrimination: Employment Risks under Workmen's Compensation in the Early Twentieth Century United States," *Enterprise and Society* 15, no. 1 (2014): 31–67. Railroads had been early adopters of medical examinations in hiring, and railway workers knew well that examinations could be misused to pry into workers' lives and justify firing the aged, injured, or potentially radical. See White, *Railroaded*, 320–321. On insurers advising corporations on how to build their own medical examination infrastructures, see, for instance, Equitable Life Assurance Society of the United States, "Group Department—Physical Examination," in 80–28, Box 1, RG4, Equitable Papers; and Harold A. Ley, *Employees Welfare Work That Pays: How Employers and Employees Are Going Fifty-Fifty on a New Life Extension and Insurance Service* (New York: Life Extension Institute, 1929), 4–5.

49. Mozzone, "Hooper Holmes, Inc.," 266. See also Hooper Holmes's corporate website at http://www.hooperholmes.com/.

disciplined) the traditional professions to serve as nation-spanning observational communities recording and reporting details about their fellow citizens' lives. Actuaries and medical directors decided, from objective data shaped by subjective judgments, what characteristics mattered to risk ratings and what did not. They made complicated, culturally contingent decisions: for instance, that race mattered, but only some kinds of "race," and that the same data could justify discriminating by sex in one case and not discriminating in the next.[50] Armies of clerks (men and women) processed intelligence reports from around the nation, writing a "risk" onto a set of cards (some of which were machine-punched, others typed) to be kept locally or shared with other insurers through the secretive Medical Information Bureau. Punched cards became fodder for new machines that sorted and tabulated them, making them at once easier to manage individually and easier to aggregate into a whole.[51] Risk-making systems built new statistical communities and statistical individuals as they transformed lives into risks—one card at a time.

* * *

In December 1936, hived off from the Christmas bustle that usually occupied all the attention of the Post Office in Chicago, nine hundred typists ("girls" hired as temporary stenographers) sat at typewriters (rented for seventy cents a week) resting atop makeshift plywood tables, all busily numbering their fellow Americans.[52] The Social Security Board lacked the infrastructure, staff, and organization necessary to register well over twenty million individuals in the space of a few months. So it turned to the Post Office. A new risk-making system was taking shape.

On 16 November (after the presidential election that year, not coincidentally), postal carriers made a census of employers on their routes and determined the number of employees in each. On the twenty-fourth, they delivered bundles of SS-5 forms, the individual application for a Social Security account, to be distributed by employers—not unlike the common

50. Sex demonstrates powerfully the role of judgment in translating objective statistics into company policy. Life insurers began charging women more for annuities (because they lived longer) in the 1930s, but it would be another thirty years before standard industry practice advised that women should be charged less for life insurance (because they lived longer). See Maclean, *Life Insurance*, 237, 262–263.

51. For the definitive account of life insurers and their tabulating technologies, see Yates, *Structuring the Information Age*.

52. Charles McKinley and Robert W. Frase, *Launching Social Security: A Capture-and-Record Account 1935–1937* (Madison: University of Wisconsin Press, 1970), 369.

practice in group insurance. SS-5s required only name, date, and place of birth; sex; "color"; mother's maiden name and father's given name; and home and employment addresses—no medical examinations, no personal or family histories. The Social Security official responsible for the forms explained to critics that the system required "color" (originally "race," although state unemployment administrators objected to the term) because of actuarial realities of "important differences in life expectancies among different races," a nod to the persistent power of white data politics in risk calculations.[53] A handful of employers took it upon themselves to surreptitiously expand the applications, like the New Jersey firm whose fake supplementary "Social Security System. Employee History Record" inquired into workers' nationality, religion, education, and union affiliations.[54]

All forms eventually crossed the desks of typists like those in Chicago. Social Security's risk-making system mass-produced punch-card coded risks (or, as I sometimes think of them, machine-readable men)[55] on an unprecedented scale. Board statisticians anticipated that 26 million accounts would be necessary to cover all eligible workers immediately, with that number growing by 2.5 million every year to a projected peak of 40–50 million.[56] To make risks at such an extraordinary scale required that names become numbers.

Social Security officials sometimes pretended that something other than the logic of punch cards and federal statistical bureaucracy drove the creation of account numbers. A popular story of the time, repeated by Social

---

53. McKinley and Frase, *Launching Social Security*, 326, 343, 368.

54. Concerned about employer intrusions (or about workers' distrust of their bosses), the Social Security Board also allowed all applicants to submit their forms directly to a post office, to a postal carrier, or to their union. Birchard E. Wyatt and William H. Wandel, *The Social Security Act in Operation: A Practical Guide to the Federal and Federal-State Social Security Programs* (Washington, DC: Graphic Arts Press, 1937), 54, 57.

55. I use "men" intentionally. Even as the system did not discriminate directly with respect to sex or race, its exclusions (of farmworkers and domestic servants especially) fell disproportionately on women and African Americans. But more than that, the program imagined a breadwinning male supporting a family to be its primary recipient. On the gendered and racialized imaginaries of Social Security, see Alice Kessler-Harris, "In the Nation's Image: The Gendered Limits of Social Citizenship in the Depression Era," *Journal of American History* 86, no. 3 (1999): 1251–1279. Julian Zelizer argues that to fully understand the sexual and racial exclusions of the old-age insurance system, the financial and (antistatist) political limitations of the state (rather than the failures of liberalism) need to be understood. See Julian E. Zelizer, "Where Is the Money Coming From?" *Journal of Policy History* 9, no. 4 (1997): 399–424 at 417n4.

56. By early 1937, over 23.6 million numbers had been assigned. Wyatt and Wandel, *Social Security Act in Operation*, 44–45, 370. The number had already surpassed forty million by August 1938, just three years in. Arthur J. Altmeyer, "Three Years' Progress toward Social Security," *Social Security Bulletin* 1, no. 8 (August 1938): 1–7 at 1.

Security insiders, explained the necessity for account numbers in terms of the "Fred Smith" problem. All of the Fred Smiths of New York City, the story went, had come together in the 1930s to form a clearinghouse for improperly directed inquiries which swelled with the increasing volume and density of Fred Smiths. Numerical identification, explained one former insurance official turned Social Security advisor, sidestepped such name duplication complexities.[57] Yet the debates over creating account numbers tell a different story. The bigger Fred Smith problem derived from technological, rather than social, concerns. It had to do with the greater expense of printing letters rather than numbers.

Early proposals for Social Security account identifiers included three letters, but each died as government statisticians' tempers flared over practical issues of cost. Alphanumeric Social Security numbers would not work with many state registration systems and unemployment schemes that rented expensive tabulators equipped to deal with numbers only. Even more expensive alphanumeric tabulators were out of the question for most statisticians.[58] If statisticians could not abide a three-letter identifier, there was never any chance that they would use whole names like Fred Smith.

But tempers flared over other proposed registration numbering details too. Particularly provocative were concerns around worker liberties and fears that a new numbering system would complicate states' existing unemployment compensation efforts. Outside the 1935 meeting of the American Statistical Association, government statisticians holed up in a private room, refusing to leave until all were satisfied. Plans for the account number to include birth years faced stout criticism from those convinced that workers should have the power to dissemble before age-discriminating employers (while being free to tell the state the truth).[59] The final number contained nine digits: the first three identified a region of the United States, the next two a group number used for mechanical tabulating purposes, and the final four an individual serial identifier.

All those tempers flared because many saw a propitious, perhaps once in a lifetime, moment to remake American statistics around this new risk-making system. Stuart Rice, chairman of the Central Statistical Board—itself a utopian effort to bring order out of bureaucratic confusion—saw in Social Security the opportunity to join the functioning death registration area and the nearly complete birth registration area with a worker registration system

57. Wyatt and Wandel, *Social Security Act in Operation*, 45.
58. McKinley and Frase, *Launching Social Security*, 320–322.
59. McKinley and Frase, *Launching Social Security*, 323–324.

reaching "half the population at the time it begins work." What a waste, he thought, to overlook this chance for "the ultimate acceptance of universal registration." But the same dreams of universal registration made statisticians nervous about their ambitions. They abandoned the word "registering," which sounded too regimented, and presumably too German, in favor of "assigning" account numbers to American workers.[60]

So, in December 1936, handwritten SS-5 forms became typed carbons, each preprinted with a nine-digit Social Security account number. Those carbons traveled to a 120,000-square-foot Baltimore warehouse, outfitted to record tens of millions of Americans' economic lives. Provisioned by International Business Machines (IBM) with punch-card systems capable of sorting, duplicating, tabulating, checking, and printing or summarizing, the warehouse became a source of amazement (and good publicity).[61] The press dubbed it "the world's biggest accounting job"[62] or "the biggest bookkeeping job in history."[63] Over two thousand employees, with IBM technicians waiting in the wings to help, set to work creating "master cards" anchoring individuals' many disparate wage records and other bureaucratic details together.[64] Nearly a third of the staff processed dummy cards, learning how to operate the machines and awaiting the moment that the Treasury Department got its act together and started transferring wage data.[65] An enormous risk-making system—a human machine—waited, ready to number Americans' days.

* * *

Reporters and photographers tracked the first American assigned a number (055–09–0001)[66] to his Princeton Club's bar in Manhattan. "John David Sweeney Jr., blue-eyed, sturdy, unmarried, 23" reported *Time*, worked as a clerk for his father's electrical supply company, lived in his father's

60. McKinley and Frase, *Launching Social Security*, 322, 329.

61. Wyatt and Wandel, *Social Security Act in Operation*, 59, 118–119; McKinley and Frase, *Launching Social Security*, 364, 376n22.

62. "Three Candles," *Time*, 22 August 1938, 24.

63. Altmeyer, "Three Years' Progress toward Social Security," 1.

64. McKinley and Frase, *Launching Social Security*, 364.

65. The first wage records were supposed to arrive in July 1937, but Treasury fell behind and the record system was still waiting for data to arrive in January 1938. McKinley and Frase, *Launching Social Security*, 375.

66. "The First Social Security Number and the Lowest Number," Social Security Administration website, http://www.ssa.gov/history/ssn/firstcard.html.

fifteen-room house in New Rochelle, New York, and had recently voted against Roosevelt and for Alf Landon.[67] But Sweeney approved of Social Security.[68]

So did Mabel Shea—"a sedate, graying personage with tortoise-shell glasses and schoolteacherish appearance"—much to the chagrin of her conservative columnist employer, who had tried to make her weekly thirty-five-cent payroll deductions (from a thirty-five-dollar-a-week secretary's wages) into a case against Social Security. Shea refused to cooperate.[69] According to public opinion polls of the time, most Americans, regardless of class, region, or age, shared Sweeney's and Shea's approval of Social Security.[70] Such support for old-age benefits is all the more surprising since most Americans' own payments stood, as Sweeney put it, "a long way off."[71]

About the only things that most Americans who filled out an SS-5 got in return for their efforts in the 1930s were their Social Security numbers. They greeted these new possessions ambivalently. Some worried about their bureaucratic rechristening. A 1937 political cartoon featured a stern-looking Uncle Sam asking a man, coming hat in hand, "What did you say your name is?" The man replied, obediently, with an account number.[72] Others took the extraordinary step of inscribing their numbers on their bodies in permanent ink. "Holder of social-security number" joined "women interested in fashionable patches, sea sailors, circus freaks, and stevedores, and truckmen" as classes of individuals likely to possess tattoos, according to the *New York Times* in 1937.[73] Men tattooed their arms or feet, women

67. "Pensioners," *Time*, 14 December 1936.

68. "The First Social Security Number and the Lowest Number," Social Security Administration website, http://www.ssa.gov/history/ssn/firstcard.html.

69. "Secretary and Program," *Time*, 22 November 1937, 23–24.

70. Americans do not appear to have differentiated between need-based old-age assistance and old-age insurance. Two-thirds of those surveyed supported old-age benefits generally in 1936. By 1944, nearly 96 percent did the same. Michael E. Schiltz, *Public Attitudes toward Social Security 1935–1965* (Washington, DC: Government Printing Office, 1970), 35–37.

71. "Pensioners," *Time*, 14 December 1936. Even for those much older than Sweeney, payments would have to wait: the first annuities would not be paid until the 1940s. Some 1930s retirees received lump-sum payments refunding their contributions in the meantime and many more received need-based assistance also brought into being by the Social Security Act.

72. Clifford Kennedy Berryman, "The New Numbers Game" reproduced in Steven Lubar, "'Do Not Fold, Spindle, or Mutilate,': A Cultural History of the Punch Card," *Journal of American Culture* 14, no. 4 (1992): 43–55 at 50.

73. "Decorative Art," *New York Times*, 20 June 1937.

their shoulders or thighs.[74] The tattoos served as memory aids, a means of identification, and often something more.

In the case of Lange's unemployed lumber worker, we find a man who might have opted for a tattoo because it could identify him in the case of an accident. After all, he did dangerous work. Or, as he paid money into the Social Security system, he may have wanted to ensure he'd one day be able to redeem his annuity, that he would not forget his number.[75] Neither explanation, however, encompasses his evident pride in his number, his willingness to show off for Lange (who undoubtedly encouraged him, maybe posed him).

Was he proud to have been, as the *New York Times* had put it, a "holder" of a Social Security number? Possessing such a number meant a promise of future income, protection for himself, and—after the just-completed 1939 amendments that included expanded support for families—protection for his wife and future children against his old age or death.[76] Maybe he felt pride in the New Deal, in a government committed to him and his family, and advertised that pride on his bicep.[77] Or perhaps, his tattoo displayed a workingman's pride in his gender and class identities. Possessing such a number signaled his status as an industrial worker.[78] Such status mattered in a bean-picking camp where most pickers had been excluded from

74. "Sidelights of the Week: Security Tattoo," *New York Times*, 17 January 1937; Amelia Klem, "A Life of Her Own Choosing: Anna Gibbons' Fifty Years as a Tattooed Lady," *Wisconsin Magazine of History* 89, no. 3 (2006): 28–39 at 36.

75. This is actually more speculative than we would today imagine. Half of all those surveyed who paid Social Security taxes and had a number did not believe they would eventually get a pension. Schiltz, *Public Attitudes toward Social Security*, 35.

76. With the amendments, a man who retired with a wife over sixty-five received 50 percent more. If he died, his wife received a 75 percent annuity and minor children received 50 percent annuities. "Congress Writes 45,000,000 Life Insurance Policies," *Life*, 7 August 1939, 51–57. The portrait's form glorifies the man as the breadwinner here, but it also ties his capacity as a breadwinner to his Social Security account, in line with the prevailing idea of Social Security planners and propagandists that it primarily support the male wage earner. Kessler-Harris, "In the Nation's Image."

77. David Peeler reads the tattoo this way, as an indication of a man "confident of the nation's course" while also seeing the couple's stylish flair as a display of fortitude in the face of hardship. David P. Peeler, *Hope among Us Yet: Social Criticism and Social Solace in Depression America* (Athens: University of Georgia Press, 1987), 103–104.

78. Conservative critics in actuarial circles attacked Social Security precisely for being "class legislation" that benefited workers, while ignoring other individuals and taxing employers. See, for instance, Williamson, "Social Insurance"; Clarence W. Hobbs, "Social Insurance and the Constitution," *Proceedings of the Casualty Actuarial Society* 22, no. 44 (1935): 32–49.

eligibility for Social Security. Possessing an account set this unemployed lumber worker apart, probably even from his bean-picking wife.[79]

* * *

There are some things we do not need to speculate about, thanks to the number in Lange's picture. The federal government maintains a Social Security Death Master File (now accessible online and especially useful to credit reporters on guard against individuals applying for credit with pre-owned Social Security numbers) that ties each number to a date of birth, a date of death, and a name. With the number 535–07–5248 and ten dollars, the historian can gain some solid facts.

This unemployed worker was born on 2 July 1912: he had just turned twenty-seven when Lange found him.[80] Census records suggest his wife was about the same age. According to tables prepared by Metropolitan, he had 611 in 1,000 chances of living long enough to draw a Social Security check eventually, while she (had she been eligible) enjoyed better odds of 680 in 1,100.[81] He did not stay unemployed long—he worked thirty-two weeks in 1939, hauling produce for some part of that time, and made $400 for his efforts—while she picked fruit for twenty weeks, bringing in $150 total.[82] The family's gross pay of $550 paled next to Otto Richter's assumption of a $1,000 wage, and it looked much worse next to the $2,500 in Dublin and Lotka's calculations. But then, before Social Security, Richter, Dublin, and Lotka had little economic data from real American lives to go on.

The crisis of the Great Depression changed that. It brought a new risk-making system of unprecedented size into existence. As Social Security expanded to encompass excluded professions—farmworkers, domestic servants, the self-employed—work registration became nearly as extensive as birth and death registration. By the 1970s, 90 percent of all employment fell under Social Security's gaze—although those who did not work for formal wages remained outside the system.[83] Social Security's machinery layered atop existing risk-making systems that recorded Americans' vital

79. As Linda Gordon notes, migrant farmworkers (especially "Okies") like those in these and similar camps suffered from exclusion from most federal and state welfare programs, as well as the hostility of many farmers. Gordon, *Dorothea Lange*, 226–227.

80. Social Security Death Master File.

81. Based on a table in Metropolitan Life Insurance Company, "The Chances of Reaching Age 65," *Statistical Bulletin* 18, no. 3 (March 1937): 2–5 at 3.

82. See entries for Thomas and Vivian Cave, 1940 US Federal Census, Shasta, Klamath County, Oregon, enumeration district 18–66B, page 6b, roll T627_3366, www.ancestry.com.

83. Duncan and Shelton, *Revolution in United States Government Statistics*, 187.

details in doctors' offices, state registries, corporate headquarters, and in their own homes. It transformed individuals into statistical citizens, into state-produced risks.

Thanks to Social Security's risk-making system, we now know the identities of Lange's subjects: Thomas and Vivian Cave. But we should pause at this awesome power accessible even to, well, a historian in the early twenty-first century. Risk-writing machines made individuals—citizens—legible to the state, to insurers, to employers, to anyone with a few bucks in his pocket.[84] Controlled by institutions with power (in most cases), these machines conferred on their creators even more power: the capacity to track individuals, to acquire valuable information about them for analysis or exchange. Thomas Cave inked his account number on his own body (we think) and smiled about it, but soon enough tattooed serial numbers—like those inscribed by Nazi bureaucrats on victims bound for death camps in the Holocaust—took on tragic resonances. And as the Social Security number became tied to other state bureaucracies (as it had been designed to do) and other risk-making systems (which it had not been designed to do), the implications (personal and financial) of wearing his Social Security number on his sleeve must have impelled Cave to rethink his smile and his tattoo.[85]

* * *

Thomas Cave, with a face fit for a movie star and that photogenic tattoo adorning his bicep, doesn't make a good candidate to be named a typical American. Still, Cave serves well as an example of what it was like to be a statistical individual in the early to mid-twentieth century. Indeed, as a statistical individual—an American whose days were becoming numbered—Cave may indeed have been typical.

Cave and his contemporaries had lived through the first great proliferation of national risk-making systems. In his lifetime, risk makers—spurred to migrate, search for more subjects, and change their methods by economic and corporate crises—had made statistical individuals from millions of Americans by promising them better futures. They promised Cave and his cohort futures improved in the traditional ways (by the protections of insurance, offered by corporations or by the state) or in the spirit of the modern

84. On legibility, and the dangerous implications of legibility, see James C. Scott, *Seeing like a State: How Certain Schemes to Improve the Human Condition Have Failed* (New Haven, CT: Yale University Press, 1999).

85. On Social Security numbers as required by a widening array of government programs and agencies until recent efforts to scale back their use, see Carolyn Puckett, "The Story of the Social Security Number," *Social Security Bulletin* 69, no. 2 (2009): 55–74 at 67–69.

conception of death (by the adaptation of risk-making methods and materials for preventative health care). They made these promises without having resolved the fundamental tension between classing and smoothing that characterized and plagued risk making. They made risks from Cave and his cohort without solving difficult problems pertaining to the roles of race or sex in risk making, without revealing the ways they traded impairment information about individuals with one another, and with questions still unanswered about the propriety of fatalizing or the corrupting power of smoothing.

Cave's cohort did not give up on critiquing risk makers. But many did embrace the by-products of risk making in all manner of ways, despite their ambivalence about being made into risks, despite often having little understanding of where those by-products came from. Interwar Americans readily took up build tables tied to mortality readings and dollar valuations derived from insurance sales, to take two prominent examples, using such by-products of risk making to guide their day-to-day choices and shape social or political debates, even as risk makers worried about how little most people appreciated the limitations built into such tools and figures. In this way, Thomas Cave's decision to appropriate his Social Security number—yet another risk-making by-product—as a statement of his politics or class identity does not seem so odd. That he probably had very little idea how Social Security's risk-making system worked—that it reduced complex men and women to data that could fit on a few punch cards, that it created centralized proto-databases capable of recording much of Americans' economic lives, that the number it created would encourage even greater centralization of data about Americans—makes him all the more typical.

Yet what makes Cave so wonderfully exemplary of what it meant to become a risk in mid-twentieth-century America is also what makes him appear so odd: the fact that he decided to inscribe his Social Security number *on his body*. Cave inadvertently summed up the changes that had been wrought in risk making since 1873, and since 1905 especially. Risk-making systems, initially intended as a solution to the economic problems caused by death and tied to the prediction of death, had come to be seen as a solution to the biological problems of living as well. Traditionally, risk makers wielded economic power: they determined who gained access to what Charles Ives called an "essential commodity" and on what terms; they determined who had access to the dignity that came with proper burial and who would gain access to an investment market made rational and smooth by corporate actuaries. In the early twentieth century, risk makers' power took on biological or biopolitical shades as well. Americans' bodies (and

Americans' interactions with their bodies) changed because of the tools, knowledge, and actions of risk makers.

So it is only fitting that Thomas Cave took a by-product of a new state-administered, yet traditional, risk-making system and applied it to his body. In the juxtaposition of a symbol of economic identity to a site of vital power—in the layering of a risk makers' number atop human flesh—we find the story of this book encapsulated, more neatly than any author could have hoped for. From Lange's portrait of Thomas Cave's bicep emerges a pithy answer to our question. From that image we read how our days became numbered, in layers.

For those Americans pulled into the new Social Security system, two things happened right away: they began paying new taxes and they received a paper card with a number on it. With that card in hand, a few went so far as the tattoo studio; many more stopped at Woolworth's and bought a wallet. In 1938, the E. H. Ferree Company in Lockport, New York, began shipping wallets nationwide that accommodated the new cards. It advertised this new use of a wallet by placing sample specimen cards in each one sold.[1] Wallets were becoming receptacles for the paper (and soon plastic) residuals of risk making, a reminder of the extent to which Americans' days were becoming more thoroughly numbered.

If we look into the wallet of an American living today among the remnants of what Lizabeth Cohen has called our "consumers' republic," we find many more cards there too.[2] Nearly all have something to do with the continued expansion of corporate and state bureaucracies, the scaffolding of modern life. We find drivers' licenses, student IDs, or library cards that grant us privileges and also identify us, and perhaps selective service cards that indicate a presumed—if long dormant—obligation. Most of the rest of the cards in wallets derive in some way from the making of risks.

We find health insurance cards, of course. The risk making behind them descends directly from life insurers' systems. Commercial health insurers with their "scientific" (that is, risk-based) pricing have out-competed or transformed earlier Blue Cross hospital plans and Blue Shield physician

---

1. Social Security Administration, "Social Security Numbers: Social Security Cards Issued by Woolworth," http://www.ssa.gov/history/ssn/misused.html.

2. Lizabeth Cohen, *A Consumers' Republic: The Politics of Mass Consumption in Postwar America* (New York: Vintage, 2003).

plans that began offering health insurance in the 1930s without making individual risks. Even health insurance that comes through group plans—as so much of it now does—needs risk making: up to half of all insured individuals in most group plans become the basis for an individual risk in an insurer's file.[3] And every individual generates his or her own claim history, thus becoming a thoroughly defined risk in that way.

We find credit cards too. They are the impetus for some of the most widespread and important risk-making operations in modern life. In the 1960s big national banks began offering credit cards at scale and turned to new credit scoring firms to help them approve applicants speedily using impersonal scoring techniques that could be readily justified. Raters and lenders sought efficiency, but even more importantly they sought protection amidst noisy controversies over credit discrimination by race and gender. For their part, ordinary Americans submitted to increased surveillance in the hope that they would be treated more fairly.[4] Following the Equal Credit Opportunity Act of 1974, Fair, Isaac and Company (FICO) introduced new credit scorecards, distant descendants of New-York Life's numerical method transformed by sophisticated techniques borrowed from the young field of academic mathematical statistics.[5] The now ubiquitous FICO score became widely available to the dominant credit bureaus between 1989 and 1991—life risks and credit risks could finally be reduced to a single number in most cases.[6]

Less obviously risk driven are our shopper discount or loyalty cards issued by supermarkets or big-box stores. But they too owe their existence to risk-making systems. The earliest—and generally feeble—attempts to make consumer risks came from department store credit agencies in the 1930s

3. This was Deborah A. Stone's estimate in "Struggle for the Soul of Health Insurance," 305–306. On the early history of voluntary health insurance in America, see Christy Ford Chapin, "Ensuring America's Health: Publicly Constructing the Private Health Insurance Industry" (PhD diss., University of Virginia, 2011), chapter 1; and Ronald L. Numbers, "The Third Party: Health Insurance in America," *Sickness and Health in America*, ed. Judith W. Leavitt and Numbers (Madison: University of Wisconsin: 1997), 269–283.

4. I rely on Louis Hyman for this sketch of the history of credit cards and credit scoring. See Hyman, *Debtor Nation: The History of America in Red Ink* (Princeton, NJ: Princeton University Press, 2011), especially 205–219.

5. On credit scoring techniques, see Raymond Anderson, *The Credit Scoring Toolkit: Theory and Practice for Retail Credit Risk Management and Decision Automation* (New York: Oxford University Press, 2007). Anderson does not mention any insurance influence, instead drawing a straight line from R. A. Fisher's work on discriminant analysis in the 1930s to later scoring methods, but this history seems calculated to advance a particular statistical approach to scoring today.

6. See FICO's company timeline, available online at http://www.fico.com/en/about-us/history/.

that tracked customer data in order to predict future purchases and target marketing through a method known as "customer control."[7] Then in the 1980s and 1990s airline frequent-flyer programs drawing on evolving computer database technologies demonstrated the potential that lay in making risks from consumers. Bookstores' reader loyalty cards subsequently allowed merchants to track purchases and predict what might interest each reader next. Supermarkets' cards allowed wholesalers to target coupons while stores repackaged customers' purchasing data as a valuable (and saleable) commodity.[8] Now in the age of e-commerce and Big Data, of Amazon and Facebook, marketing analytics departments can track customers and make predictions (of their risk to buy!) even without a special card.[9] Social networks make no cards for our wallets, but they now boast both huge market capitalizations and caches of personal data that—because they open the door to a probabilistic prediction of each individual's risk to do, well, just about anything—account for a significant portion of companies' assets.[10]

Risk makers' cards inhabit our wallets, stacked one atop the other in leather pockets mirroring the multiple layers of traditional, predictive risk-making systems in which we are imbricated. Such cards make visible what we can otherwise so easily overlook: that our days are becoming numbered on a greater scale than ever before.

* * *

Today, the cards in our wallets are only the beginning. Risk makers play important roles in deciding who can afford a home and where.[11] They decide

---

7. See Josh Lauer, "Making the Ledgers Talk: Customer Control and the Origins of Retail Data Mining, 1920–1940," in *The Rise of Marketing and Market Research*, ed. Harmut Berghoff, Philip Scranton, and Uwe Spiekermann (New York: Palgrave Macmillan, 2012), 153–169.

8. My discussion of 1980s and 1990s developments depends on Joseph Turow, *Breaking up America: Advertisers and the New Media World* (Chicago: University of Chicago Press, 1997), chapter 6, especially at 138–144. On the developing market for data (often from public sources) useful to corporations or political groups for forecasting and targeting, see Paul Starr and Ross Corson, "Who Will Have the Numbers? The Rise of the Statistical Services Industry and the Politics of Public Data," in *The Politics of Numbers*, ed. William Alonso and Paul Starr (New York: Russell Sage Foundation, 1987), 415–447.

9. Charles Duhigg, "How Companies Learn Your Secrets," *New York Times Magazine*, 16 February 2012.

10. Somini Sengupta, "Facebook's Prospects May Rest on Trove of Data," *New York Times*, 14 May 2012.

11. On risk rating and its spread through federal mortgage programs, see Jennifer Light, "Discriminating Appraisals: Cartography, Computation, and Access to Federal Mortgage Insurance in the 1930s," *Technology and Culture* 52, no. 3 (2011): 485–522; and Kenneth T. Jackson, *Crabgrass Frontier: The Suburbanization of the United States* (New York: Oxford University Press, 1985), 190–218.

who gets through the security line first at the airport.[12] They predict who can safely be paroled and who should be policed.[13] They can even track cars' movements and drivers' behaviors, setting individualized rates for auto insurance in the process.[14]

From time to time, we—like Americans over a century ago—resist the risk makers and the power they wield. Some of the most serious challenges to the premises and practices of risk making came in the 1970s when credit scoring, automotive risk rating, and the discriminations enabled by both raised the sustained, organized ire of women's and African Americans' advocacy groups.[15] More recently, critics have contested the utility and justice of predictive policing and risk profiling.[16] And every so often, small furors erupt over stories of individuals apparently unfairly denied insurance, which draw even more scowls when it is a child who has been rejected. At such moments, individualist ideas of fairness fuel condemnations of risk makers' ongoing project to fill the nation with statistical individuals.

Yet such controversies drive what we might call the dialectic of numbering (where conflicts over traditional, predictive risk making spur the creation of new forms of risk making supposed to change fates, such that more risks get made from more people in new ways). The original proponents of the modern conception of death invented this dialectic when they adopted and adapted risk making at a moment when traditional risk makers in their insurance offices were reeling. They promised to use risk-making practices

12. The Transportation Security Administration put into effect in 2011 an "intelligence-driven, risk-based analysis" system drawing on a new "PreCheck" database that asks travelers to trade their fingerprints and a background check for the right to get through airport security lines more quickly. http://www.tsa.gov/tsa-precheck/.

13. Bernard E. Harcourt, *Against Prediction: Profiling, Policing, and Punishing in an Actuarial Age* (Chicago: University of Chicago Press, 2006), especially part 1.

14. Telematics, the darlings of many Big Data enthusiasts, are tracking devices in insured automobiles that record speeds attained, times stopped short, or time on the road. From such data, auto insurers hope to better assess the risk that a driver will get in an accident (and thus more finely class drivers and price policies), while placing even less emphasis on sensitive demographic categories—like race or sex. Bill Franks, *Taming the Big Data Tidal Wave: Finding Opportunities in Huge Data Streams with Advanced Analytics* (Hoboken, NJ: Wiley, 2012), 54–55.

15. On the proliferation of actuarial thinking generally and broad cultural responses to that proliferation with a particular emphasis on discrimination debates, see Caley Dawn Horan, "Actuarial Age: Insurance and the Emergence of Neoliberalism in the Postwar United States" (PhD diss., University of Minnesota, 2011), chapters 3 and 4.

16. Bernard E. Harcourt, "A Reader's Companion to *Against Prediction*: A Reply to Ariela Gross, Yoram Margalioth, and Yoav Sapir on Economic Modeling, Selective Incapacitation, Governmentality, and Race," *Law and Social Inquiry* 33, no. 1 (2008): 265–283 at 273. See page 280 on race and problems with profiling: "Race is the first place where we see the poison, but it is a poison that affects everyone else."

to help people in new ways—and they did. But they never actually resolved the fundamental challenges inherent in approaching individuals through statistics, in contemplating the part by way of the whole. Those challenges migrated alongside new risk-making systems into doctors' offices, corporate welfare programs, the United States government, and beyond. Now, as traditional risk makers have continued to expand and face scrutiny, we see the same pattern repeating. With each expansion of unpopular predictive risk systems come new tools and databases that can be used (and are often embraced) to improve lives or control fates. Risk makers' efforts to predict the future and to change the future do not oppose one another so much as they work together to make ever more risks.

The allure of life extension continues to be one of the most important creators of new opportunities for risk making. In the 1970s at the same time that critics decried race and sex discrimination in insurance, the "risk factor" became a primary tool for health campaigners emphasizing personal responsibility in battles against cancer and heart disease.[17] At that juncture, the modern conception of death won a foothold in Americans' subjectivities as well as a firmer grasp on their bodies. Not only did ordinary people manage their health with risk makers' charts or submit themselves for risk-derived medical examinations, but many also started thinking about themselves as risks.

Then came the genomic age. The potential to read individuals' genetic secrets and tease out propensities for later health troubles raised serious questions that traditional risk makers had to confront (such as whether insurers had a right to individual genetic data).[18] That same potential has also inspired new forays into control through risk making. Start-up biotechs, for instance, offer personalized medical risk analyses to individuals eager to protect their health and to see themselves through the lens of genetic risk. Such companies gather DNA samples for analysis and use their growing genetic collections to fine-tune their risk making.[19] In them, we find the Life Extension Institute's business model alive and well.

---

17. Rothstein, *Public Health and the Risk Factor*, 359–367. Robert A. Aronowitz puts the date for the widespread use of the risk factor back a decade to the 1960s. Aronowitz, *Making Sense of Illness: Science, Society, and Disease* (New York: Cambridge University Press, 1998), chapter 5.

18. Mark A. Rothstein, ed., *Genetics and Life Insurance: Medical Underwriting and Social Policy* (Cambridge, MA: MIT Press, 2004); Shobita Parthasarathy, "Regulating Risk: Defining Genetic Privacy in the United States and Britain," *Science, Technology, and Human Values* 29, no. 3 (2004): 332–352.

19. On one such company, 23andMe, see Lisa Miller, "The Google of Spit," *New York Magazine*, 22 April 2014.

Yet critics warn that risk makers' methods for improving bodies can be less effective than advertised. "Build" in the guise of body mass index (BMI), for instance, regularly comes under assault, especially when used by the state to label individuals.[20] So too, preventative medical risk assessments—for breast cancer most prominently—stir up controversies between passionate believers in screening (trained very well by LEI and its successors) and those who assert that too much screening can do more harm than good because of the high rate of false positives leading to unnecessary, expensive treatments and worry.[21] Critics can sometimes convince an institution to decrease screening, but slowing the dialectic of numbering on a larger scale has proven very difficult.[22] More risks keep getting made.

To be sure, we can improve our lives through some of this risk making. For while we usually have little problem thinking about ourselves as individuals, we often need more help to see ourselves through the widening lens of population statistics, help that risk numbers provide. Members of the Quantified Self community seek exactly such insights through their self-tracking.[23] Indeed, a risk analysis, at its best—when we fully understand it and exert some measure of control over it—gives us a new way to examine our lives and new questions to ask.

But individuals can have a difficult time exerting control over the risks they have spawned, whether such risks were made in hopes of predicting fates or changing them. Many risks end up in corporations, like those that predict our purchases (as consumers) or pay us (as employees) to submit to biometric screenings and wear tiny computing devices that track our activity.

20. Arkansas officials in 2003 sent out "report cards" stating children's personal risks of obesity based on their BMIs. Despite public controversies surrounding this action, other states have followed Arkansas's lead. See Lauren Vogel, "The Skinny on BMI Report Cards," *Canadian Medical Association Journal* 183, no. 12 (2011): E787–E788.

21. Robert A. Aronowitz argues that Americans too readily discount the costs of screening and make decisions as if the detection of risk factors for breast cancer were equivalent to the discovery of actual symptoms. Aronowitz, *Unnatural History: Breast Cancer and American Society* (New York: Cambridge University Press, 2007), 269–270.

22. Breast cancer again exemplifies the point. In 2009, the US Preventative Services Task Force called on most women to wait until age fifty for regular mammograms. At least one recent study finds that mammogram use—after much public outcry against the recommendation—did not subsequently decrease for those aged forty to forty-nine. See Lydia E. Pace, Yulei He, and Nancy L. Keating, "Trends in Mammography Screening Rates after Publication of the 2009 US Preventative Services Task Force Recommendations," *Cancer* 119, no. 14 (2013): 2518–2523.

23. Not all Quantified Self tracking contributes to a risk analysis, however. Much self-tracking aims only to make visible gradual changes in individuals' bodies and behaviors that might not otherwise be evident. On Quantifed Self, see Gary Wolf, "The Data-Driven Life," *New York Times Magazine,* 28 April 2010.

Or they reside with governments who underwrite our mortgages or engage in mass surveillance of our bodies as well as our words.[24] We are seldom, it is crucial to remember, the sole drivers of our data-driven existences.

* * *

I began this book by talking about the strange and fascinating books that life insurers generated or inspired in the period beginning with the Panic of 1873 and ending with the Great Depression. But just as the industry did not stop making risks and just as a wide variety of new institutions have since decided to make risks (and print accompanying cards to go in wallets), peculiar and powerful books grounded in the techniques pioneered by life insurers have continued to emerge.

Talk of "risk" binds many of them together. Some provide instructions to a new breed of corporate or state "risk managers," a profession built on the faith that lives present still more opportunities for risks to be made and that more than lives can be made into risks, a profession convinced that the making of risks can lead to safer, more profitable, more secure endeavors, a profession whose fate comes tied inexorably to the financialization of society.[25] Other books, written by critics of nuclear energy or rampant chemical use, debate whether we have recently witnessed a sudden explosion in new risks and now belong to a "risk society."[26]

24. Prominent scandals over the collection of telephone metadata by the National Security Administration underline public concern about improper state uses of data. Extended critiques of Big Data in state and corporate hands include Jaron Lanier, *Who Owns the Future?* (New York: Simon & Schuster, 2013); and Daniel J. Solove, *Nothing to Hide: The False Tradeoff between Privacy and Security* (New Haven, CT: Yale University Press, 2011).

25. A cursory sampling of books on risk management published in the last few years reveals titles pertaining to topics so various as credit, pharmaceuticals, finance, supply chains, business, nuclear energy, and suicide. For an introduction to such books, and to risk management generally, see Michael Power, *Organized Uncertainty: Designing a World of Risk Management* (New York: Oxford University Press, 2007); Bridget Hutter and Michael Power, eds., *Organizational Encounters with Risk* (New York: Cambridge University Press, 2005); and Scott Gabriel Knowles, *The Disaster Experts: Mastering Risk in Modern America* (Philadelphia: University of Pennsylvania Press, 2011). On financialization and its causes, see Greta R. Krippner, *Capitalizing on Crisis: The Political Origins of the Rise of Finance* (Cambridge, MA: Harvard University Press, 2012).

26. While the world may well be getting riskier, the amplitude of the risk explosion that such authors point to undoubtedly also owes much to the enthusiasm with which professionalizing risk managers have made new quantified risks from older hazards. The key work here is Ulrich Beck, *Risk Society: Towards a New Modernity* (London: Sage, 1992). Beck summarizes his case in "Politics of Risk Society," *The Politics of Risk Society* ed. Jane Franklin (Cambridge, UK: Polity Press, 1998), 9–22. See also Richard V. Ericson and Aaron Doyle, "Risk and Morality," in *Risk and Morality*.

Books in the first set look ahead optimistically toward mastering the future by calculation; those in the second worry that calculation cannot live up to its promises. Both help "risk" make the jump from commodity to social reality. A third class of books, responding to the others and also motivated by debates over risk perception originally spurred on by nuclear controversies, decenters the calculative rationality of risk that grew up out of probability. These books think more expansively about the various ways that societies have conceived of the dangers or risks that surround them.[27]

I am most sympathetic to this third set of books. So in writing this one, I have endeavored to contextualize and historicize the powerful, generalizable systems for surveillance and calculation that risk managers rely on and social theorists think with. I have written a history of the systems that helped to inspire those first two sets of books.

I hope that such a history can aid everyone engaged in discussions about risk. Bearing on such momentous topics as business organization, nuclear power, chemical manufacturing, governance, finance, and war and peace, our risk discussions deserve as firm a foundation as can be poured for them.

But I hope even more fervently that this history will encourage some readers to dream not only of better ways to make risks, but also of alternatives to risk making. We should, I think, return to the psalmists' prayer: teach us, again, to number our days. There must be more ways to wisdom.

---

27. The foundational work in this vein is Mary Douglas and Aaron Wildavsky, *Risk and Culture: An Essay on the Selection of Technological and Environmental Dangers* (Berkeley: University of California Press, 1983). On risk perception and the nuclear energy context, see Gigerenzer et al., *Empire of Chance*, 265–269.

ACKNOWLEDGMENTS

I began writing this book after I stumbled upon strange old books in Princeton University's Firestone Library. When I later went looking for other curious books to develop my story, I had the good fortune to be able to turn to Ann Ackerson and the Colgate University Library and to harness the magic of Urs Schoepflin, Ruth Kessentini, and the entire Max Plank Institute for the History of Science's (MPIWG) library team.

When I moved out of the library and into the archives, many archivists and corporate officers helped me navigate archival holdings and generously granted me permission to publish excerpts from the materials they protect and organize. My deep gratitude to: Jonathan Coss and Discretion Winter at AXA; Daniel B. May at MetLife; Daniel J. Linke, Adriane Hanson, Christopher Shannon, and Jessie Thompson at Princeton's Seeley G. Mudd Manuscript Library; Jocelyn K. Wilk, Jennifer B. Lee, and Jane Siegel at Columbia's Rare Books and Manuscripts Library; Stephen Greenberg and Crystal Smith at National Library of Medicine and Jeff Knab at Light Inc.; Kristen McDonald and Claryn Spies at Yale University Library's Manuscripts and Archives; and Dolly von Hollen and Doris Singleton at Prudential Financial. I thank Rose Johnson Tate for helping me find out more about her cousin, Lawrence N. Brown, and Dr. Francis J. Rigney, Jr. for his permission to publish from his grandfather Frederick L. Hoffman's works. I have also enjoyed the special support of college and university archivists who agreed to plumb their alumni files to help me reconstruct the lives of doctors, actuaries, and corporate officers important to this story. My thanks to Nancy R. Miller at University of Pennsylvania, Linda Hall at Williams College, Marlaine DesChamps at Union College, Sushan Chin at NYU Health Sciences, Stephen Novak at Columbia University Medical Center, Katelyn Lamontagne at Hobart and William Smith Colleges, Martha Tenney at Barnard College, and especially to Sarah Keen at Colgate University for teaching me about alumni files in the first place.

As I have crafted my own book, I have been fortunate enough to share company with other writers who encouraged me, challenged me, and inspired me. Noah Dauber and Liz Marlowe kept me writing even while teaching, as did my quietly typing neighbors at the Colgate research retreats Georgia Frank organized. The members of the MPIWG Data Reading group got me thinking and writing in new and exciting ways. On balmy summer nights in Berlin, Megan McNamee, Alma Steingart, and Oriana Walker sat inside with me racing to finish their projects as I raced to wind up mine. Henry Cowles, Helen Curry, Joanna Radin, Lukas Rieppel, and Lee Vinsel started conversations as we wrote together for the blog *AmericanScience* that have continued on in this book. Tim Alborn, Caley Horan, Sharon Murphy, and Megan Wolff reveled with me in the intricacies of insurance and risk—and taught me much in the process.

My thanks go to Princeton University's Department of History, the Program for the History of Science and its weekly seminar, American Studies, and the Modern America Workshop for tending this project in its earliest stages. This book matured as it faced enthusiastic questions and insistent but kind critiques offered on both sides of the Atlantic. For providing opportunities for it to be challenged and criticized in formal settings, I thank: Elena Aronova, Tom Baker, Lindy Baldwin, Bruce Carruthers, Gregoire Chamayou, Jefferson Cowie, Raine Daston, Will Deringer, Dan Epstein, Wendy Espeland, Ann Fabian, Jennifer Green, Joel Hewett, Louis Hyman, Xan Karn, John Krige, Stephen Mihm, Rob Nemes, Christine von Oertzen, Julia Ott, Mike Pettit, Lukas Rieppel, Seth Rockman, Phil Scranton, David Sepkoski, Bob Turner, and Cyrus Veeser.

A close friend attended a seminar with me once in which I presented part of this book. He marveled afterward that a room filled with brilliant scholars had engaged so earnestly with my manuscript in order to help me improve it. I marvel too, and I express my deep gratitude to the participants who made the following conferences, seminars, and workshops marvelous: Brown University's Nineteenth-Century US History Workshop and Science and Capitalism Lecture Series; Colgate's History Department reading group, Social Sciences Division seminar, and PPE reading group; Columbia University's Calculating Capitalism conference; Cornell University's History of American Capitalism lunch workshop; Georgia Tech's History, Technology, and Society speaker series; Hagley Museum and Library's Crisis and Consequence conference; Huntington Library's Capitalizing on Finance conference; MPIWG's Department II colloquium, Historicizing Big Data working group, and Machines of Memory conference; the New School's Power and the History of Capitalism Conference; Penn's Insurance and Society Study Group; the Wissenschaftskolleg's History of Quantification Group; and York University's Science Studies seminar.

Many others have given generously of their time outside formal settings to read parts of this book in progress; share their work; offer encouragement, assistance, or advice; and point me in useful directions. I thank especially Dave Bailey, Jenny Bangham, Etienne Benson, D. Graham Burnett, John Carson, Alan Cooper, Henry Cowles, Raine Daston, Will Deringer, Emmanuel Didier, Daniella Doron, Faye Dudden, Cara Fallon, Yulia Frumer, Courtney Fullilove, Dan Hirschman, Nate Holdren, Caley Horan, Sarah Igo, Chin Jou, Judy Kaplan, Greta Krippner, Josh Lauer, Rob Nemes, Dael Norwood, Jason Oakes, Dominique Pestre, João Rangel de Almeida, Meghan Roberts, Dan Rosenberg, Robin Sloan, Tom Stapleford, Alma Steingart, Steve Usselman, Dora Vargha, Oriana Walker, Matt Wisnioski, and Viviana Zelizer. I thank three anonymous reviewers for the University of Chicago Press for comments that sometimes pleased and sometimes pushed, but ultimately helped. Barbara Welke's encouragements and critiques shaped the manuscript at a series of key moments. Christopher Phillips helped exceedingly by reading the entire manuscript in the final stages. Jon Levy read the whole book when I felt lost and showed me a way forward. Ted Porter encouraged this project from early on and never stopped. Finally, I thank Dan Rodgers, who not only read many iterations of this book with care and unsurpassed insight, but somehow knew the perfect way to help a young historian develop his own voice.

Princeton University, the Max Planck Institute for the History of Science (MPIWG), and Colgate University offered me stimulating intellectual communities and crucial financial support for my research and writing. Princeton's Porter Ogden Jacobus fellowship and an Andrew Mellon fellowship from the Woodrow Wilson Foundation supported my earliest research and writing. My colleagues at Colgate welcomed me and supported me through my first full-time teaching job as well as my introduction to fatherhood. Graham Hodges, Rob Nemes, David Robinson, and Andy Rotter have all served as mentors. When we hired Heather Roller, I knew we had landed an inspiring scholar, but not that I had found a close intellectual companion too. Throughout my five years at Colgate, the Research Council chaired by Lynn Staley and Ken Belanger has supported this book through discretionary grants, publication grants, and student wage grants. When my colleagues at Colgate graciously excused me to write for a year in Berlin in 2012–2013, Raine Daston and the MPIWG offered me not only a postdoctoral stipend, but a desk (and a vibrant community) in the single best space in the world to write about the history of science. I wrote most of this book there.

In my time at Colgate, I have had the honor of teaching a wide array of talented students who have shaped this book's eventual form. My research assistants Christina Crowley, Allie Hester, Chris Johnson, and Brittany Schwab proved diligent diggers, smart summarizers, and rigorous readers, who often made me smile with their witty notes. Students in my history workshop sections and my course on the history of numbers in America have read and commented on portions of this book in draft. I thank them for their patience and curiosity as I worked out many of the book's themes and ideas (as well as its title) while teaching them.

I first dreamed of publishing a book with Chicago while reading for my PhD general exams. Stephen Mihm, in a characteristically generous act, went out of his way to help me realize that dream. At Chicago, Sophie Wereley, Russ Damian, Tadd Adcox, and Carrie Adams attended to my book and to me. Robert Devens shepherded us into Chicago's fold with his keen eye and kind praise. Karen Darling welcomed us upon Robert's departure. Under her steady, inspired editorial guidance, this book has kept getting better. For that I will long be grateful. In copyediting, Jennifer Rappaport lessened ambiguities and straightened out my usage; and she rescued many sentences tortured by dashes.

It has been my joy to have Elizabeth Bouk as my partner this entire time in living the life of the artist. While our arts are different, seeing her courage and passion has given me courage and passion. Together we introduced William James Dromgold Bouk to this world. I thank you, my dear William, for thousands of hugs and for giving me so many reasons to be hopeful about life and also critical of institutions that fail to cultivate all of the varieties of human experience. I know you are already a statistical individual, but I hope that this book can help ensure that you will always be allowed to be so much more.

I offer my profound thanks to Ladd and Cathy Dromgold, Allison and Jared Adams, Michelle Dromgold, and Josh, Cherilyn, Sam, Annabella, and Joe Bouk. I could not have written this book without their love, support, and friendship. Finally, Gail and Ted Bouk have given me every form of aid imaginable. For countless drafts proofread, pizzas purchased, babysitting hours logged, and encouragements offered, and for never doubting that writing was a worthwhile way to spend a life, I dedicate this book to them.

BIBLIOGRAPHY

MANUSCRIPT COLLECTIONS
AXA Archives, AXA Equitable Financial Services, New York, NY
    Equitable Life Assurance Society of the United States Papers (Equitable Papers)
    Mutual Life Insurance Company of New York Papers (MONY Papers)
Columbia University in the City of New York, New York, NY
    Frederick L. Hoffman Papers, Rare Books and Manuscript Library (Hoffman Papers)
MetLife Archives, MetLife, Inc., Long Island City, NY
    Dublin Speeches and Writings, RG/03
    Medical Examination Materials, Box M11, RG/03
National Library of Medicine (NLM) Historical Collections, Bethesda, MD
    Louis I. Dublin Papers, MS C 316 (Dublin Papers)
    Prudential/Hoffman Public Health and Medical Reports by State, MS B 250
Princeton University Library, Princeton, NJ
    Alfred J. Lotka Papers, Public Policy Papers, Department of Rare Books and Special
    Collections (Lotka Papers)
Yale University Library, New Haven, CT
    Irving Fisher Papers, Manuscripts and Archives (Fisher Papers)

COLLEGE AND UNIVERSITY ARCHIVES (ALUMNI FILES)
Columbia University Medical Center Archives, New York, NY
Hobart and William Smith Colleges Archives, Geneva, NY
Union College Special Collections, Union College, Schenectady, NY
University of Pennsylvania Archives, University of Pennsylvania, Philadelphia, PA

PRIMARY SOURCE DATABASES
America's Historical Newspapers, http://infoweb.newsbank.com
    *Aberdeen Daily News; Boston Journal; Boston Evening Transcript; Cleveland Gazette; The
    Freeman* (IN); *Huntsville Gazette; Minneapolis Journal; New York Age; New York Freeman;
    New York Tribune; Springfield Republican; Wilkes-Barre Times; Worcester Daily Spy*
Ancestry.com, http://www.ancestry.com
Chronicling America: Historic American Newspapers, http://chroniclingamerica.loc.gov
    Salt Lake Herald; Los Angeles Herald

Columbia Historical Corporate (Annual) Reports, http://library.columbia.edu/locations
/business/corpreports.html
    MONY, 1865 and 1869
Ebscohost Academic Search Premier, http://www.ebscohost.com/academic/academic
-search-premier
    Life, Time
HathiTrust Digital Library, http://www.hathitrust.org
New York Times Digital Archive, http://www.nytimes.com
ProQuest Historical Annual Reports, http://www.proquest.com
    Connecticut Mutual Life, 1871; Equitable Life Assurance Society of the United States,
    1870; Metropolitan Life, 1918; MONY, 1845, 1875, and 1903; New-York Life, 1908;
    Prudential Insurance Company of America, 1914
ProQuest Historical Newspapers, http://www.proquest.com
    *Atlanta Constitution; Chicago Defender; Chicago Tribune; Los Angeles Times; Wall Street
    Journal; Washington Post*
Social Security Death Master File, http://www.ssdmf.com

BOOKS, DISSERTATIONS, AND JOURNAL ARTICLES

Abbott, Lawrence F. *The Story of NYLIC: A History of the Origin and Development of the New
    York Life Insurance Company from 1845 to 1929.* New York: New York Life Insurance
    Company, 1930.
Actuarial Society of America. "Abstract from the Minutes of the Annual Meeting." *Trans-
    actions of the Actuarial Society of America* 6, no. 24 (1900): 449.
———. "Abstract from the Minutes of the Annual Meeting." *Transactions of the Actuarial
    Society of America* 7, no. 25 (1901): 64–69.
———. *Experience of Thirty-Four Life Companies upon Ninety-Eight Special Classes of Risks.*
    New York: Actuarial Society of America, 1903.
Adams, Henry. *The Education of Henry Adams.* New York: Modern Library, 1999 [1918].
Alborn, Timothy. "A Calculating Profession: Victorian Actuaries among the Statisticians."
    Science in Context 7, no. 3 (1994): 433–468.
———. "The First Fund Managers: Life Insurance Bonuses in Victorian Britain." *Victorian
    Studies* 45, no. 1 (2002): 65–92.
———. "Insurance against Germ Theory: Commerce and Conservatism in Late-Victorian
    Medicine." *Bulletin of the History of Medicine* 75, no. 3 (2001): 406–445.
———. *Regulated Lives: Life Insurance and British Society 1800–1914.* Toronto: University
    of Toronto Press, 2009.
Alborn, Timothy, and Sharon Ann Murphy, eds. *Anglo-American Life Insurance, 1800–1914.*
    London: Pickering & Chatto, 2013.
Allen, Garland E. "The Eugenics Record Office at Cold Spring Harbor, 1910–1940: An
    Essay in Institutional History." *Osiris* 2 (1986): 225–264.
Allen, J. Adams. *Medical Examinations for Life Insurance.* Chicago: Clarke, 1866 and Chicago:
    J. H. and C. M. Goodsell, 1869 and New York: Spectator Company, 1886.
Allen, Robert Loring. *Irving Fisher: A Biography.* Cambridge, MA: Blackwell, 1993.
Altmeyer, Arthur J. "Three Years' Progress toward Social Security." *Social Security Bulletin* 1,
    no. 8 (August 1938): 1–7.
American Popular Life Insurance Company. *Sources of Longevity, Its Indications and Practical
    Applications.* New York: Wm. Wood, 1869.
Amrhein, George L. "The Liberalization of the Life Insurance Contract." PhD diss.,
    University of Pennsylvania, 1933.

Anderson, Benedict. *Imagined Communities: Reflections on the Origin and Spread of Nationalism.* Rev. ed. New York: Verso, 1991.

Anderson, Margo. *The American Census: A Social History.* New Haven, CT: Yale University Press, 1988.

Anderson, Raymond. *The Credit Scoring Toolkit: Theory and Practice for Retail Credit Risk Management and Decision Automation.* New York: Oxford University Press, 2007.

Ankers, R. E. "The Proper Basis of Reserves for Sub-Standard Risks in Life Insurance." *Proceedings of the Annual Session of the National Convention of Insurance Commissioners* 15 (1919): 120–122, 283–288.

Armstrong, Barbara Nachtrieb. "Old-Age Security Abroad: The Background of Titles II and VIII of the Social Security Act." *Law and Contemporary Problems* 3, no. 2 (April 1936): 175–185.

Aronowitz, Robert A. *Making Sense of Illness: Science, Society, and Disease.* New York: Cambridge University Press, 1998.

———. *Unnatural History: Breast Cancer and American Society.* New York: Cambridge University Press, 2007.

Ashworth, William J. "The Calculating Eye: Baily, Herschel, Babbage and the Business of Astronomy." *British Journal for the History of Science* 27, no. 4 (1994): 409–441.

Association of Life Insurance Medical Directors of America (ALIMDA) and Actuarial Society of America (ASA). *Medico-Actuarial Mortality Investigation.* Vols. 1 and 2. New York, 1912 and 1913.

Babbage, Charles. *On the Economy of Machinery and Manufactures.* London: John Murray, 1846.

Baker, Tom. "Containing the Promise of Insurance: Adverse Selection and Risk Classification." *Connecticut Insurance Law Journal* 9, no. 2 (2012): 371–396.

Balka, Ellen, and Susan Leigh Star. "Mapping the Body across Diverse Information Systems: Shadow Bodies and How They Make Us Human." Paper presented at the *4S annual meeting,* Cleveland, OH, 2 November 2011.

Barkan, Elazar. *The Retreat of Scientific Racism: Changing Concepts of Race in Britain and the United States between the World Wars.* New York: Cambridge University Press, 1992.

Basye, Walter. *History and Operation of Fraternal Insurance.* Rochester, NY: Fraternal Monitor, 1919.

Beard, Patricia. *After the Ball: Gilded Age Secrets, Boardroom Betrayals, and the Party That Ignited the Great Wall Street Scandal of 1905.* New York: HarperCollins, 2003.

Beck, Ulrich. "Politics of Risk Society." In *The Politics of Risk Society.* Edited by Jane Franklin, 9–22. Cambridge, UK: Polity Press, 1998.

———. *Risk Society: Towards a New Modernity.* London: Sage, 1992.

Beckert, Sven. *Monied Metropolis: New York City and the Consolidation of the American Bourgeoisie, 1850–1896.* New York: Cambridge University Press, 2003.

Beito, David T. *From Mutual Aid to the Welfare State: Fraternal Societies and Social Services, 1890–1967.* Chapel Hill: University of North Carolina Press, 2003.

Bellamy, Edward. *Looking Backward 2000–1887.* Boston: Ticknor, 1888.

Berkowitz, Edward D. *America's Welfare State: From Roosevelt to Reagan.* Baltimore, MD: Johns Hopkins University Press, 1991.

Bernstein, Peter. *Against the Gods: The Remarkable Story of Risk.* New York: Wiley, 1998.

Best Company, Alfred M. *Best's Illustrations of Net Costs, Cash Values, Premium Rates, Policy Conditions of Most Legal Reserve Insurance Companies Operating in the United States.* New York: A. M. Best, 1934 and 1944.

Blight, David W. *Race and Reunion: The Civil War in American Memory*. Cambridge, MA: Belknap Press of Harvard University Press, 2001.

Boorstin, Daniel J. *The Americans: The Democratic Experience*. New York: Vintage, 1973.

Bouk, Daniel B. "The Science of Difference: Developing Tools for Discrimination in the American Life Insurance Industry, 1830–1930." PhD diss., Princeton University, 2009.

Bowker, Geoffrey C., and Susan Leigh Star. *Sorting Things Out: Classification and Its Consequences*. Cambridge, MA: MIT Press, 1999.

Bridgman, A. M., ed. *A Souvenir of Massachusetts Legislators, 1898*. Vol. 7. Stoughton, MA: A. M. Bridgman, 1898.

Brigham, Carl. *A Study of American Intelligence*. Princeton, NJ: Princeton University Press, 1923.

Brody, David. "The Rise and Decline of Welfare Capitalism." In *Workers in Industrial America: Essays on the Twentieth Century Struggle*. New York: Oxford University Press, 1993.

Broussard, Albert S. *African-American Odyssey: The Stewarts, 1853–1963*. Lawrence: University Press of Kansas, 1998.

Brown, J. Douglas. *The Genesis of Social Security in America*. Washington, DC: Social Security Administration, 1969.

————. *The Idea of Social Security*. Washington, DC: Social Security Administration, 1958.

Brown, Lawrence N. "The Insurance of American Negro Lives." MBA thesis, University of Pennsylvania, 1930.

Bryson, Jr., Winfred Octavus. "Negro Life Insurance Companies: A Comparative Analysis of the Operating and Financial Experience of Negro Legal Reserve Life Insurance Companies." PhD diss., University of Pennsylvania, 1947.

Buley, R. Carlyle. *The American Life Convention 1906–1952: A Study in the History of Life Insurance*. 2 vols. New York: Appleton-Century-Crofts, 1953.

Cahn, William. *A Matter of Life and Death: The Connecticut Mutual Story*. New York: Random House, 1970.

Cain, James M. *Cain x 3*. New York: Alfred A. Knopf, 1969.

Calkins, Gary N. Review of "Race Traits and Tendencies of the American Negro," by Frederick L. Hoffman. *Political Science Quarterly* 11, no. 4 (1896): 754–757.

Cannon, Susan Faye. "Humboldtian Science." In *Science in Culture: The Early Victorian Period*. New York: Dawson and Science History, 1978.

Carruthers, Bruce G., and Wendy Nelson Espeland. "Accounting for Rationality: Double-Entry Bookkeeping and the Rhetoric of Economic Rationality." *American Journal of Sociology* 97, no. 1 (1991): 31–69.

Carson, John. *The Measure of Merit: Talents, Intelligence, and Inequality in the French and American Republics, 1750–1940*. Princeton, NJ: Princeton University Press, 2007.

Cassedy, James H. *American Medicine and Statistical Thinking, 1800–1860*. Cambridge, MA: Harvard University Press, 1984.

————. "The Registration Area and American Vital Statistics." *Bulletin of the History of Medicine* 39 (May–June 1965): 221–231.

Cavers, David F. "Federal Old-Age Insurance: Benefit Payments and Tax Collection." *Law and Contemporary Problems* 3, no. 2 (April 1936): 199–211.

Chandar, Nandini, and Paul J. Miranti, Jr. "The Development of Actuarial-Based Pension Accounting at the Bell System, 1913–40." *Accounting History* 12, no. 2 (2007): 205–234.

Chandler, Jr., Alfred D. *The Visible Hand: The Managerial Revolution in American Business*. Cambridge, MA: Belknap Press of Harvard University Press, 1977.

Chandler, Jr., S. C. *A Comparative Atlas and Graphical History of American Life Insurance Embracing a Period of Twenty Years Previous to January 1, 1880.* New York: Spectator Company, 1880.

Chapin, Christy Ford. "Ensuring America's Health: Publicly Constructing the Private Health Insurance Industry." PhD diss., University of Virginia, 2011.

Chapin, C. V. "The Value of Human Life." *American Journal of Public Health* 3, no. 2 (1913): 101–105.

Chapple, William Alan. *The Fertility of the Unfit.* Melbourne: Whitcombe & Tombs, 1903.

Charap, Mitchell H. "The Periodic Health Examination: Genesis of a Myth." *Annals of Internal Medicine* 95, no. 6 (1981): 733–735.

"Circular of the Atlantic Mutual Life Insurance Company." *North American Journal of Homeopathy* 15 (November 1866): 291–295.

Clark, Geoffrey, Gregory Anderson, Christian Thomann, and J.-Matthias Graf von der Schulenburg, eds. *The Appeal of Insurance.* Toronto: University of Toronto Press, 2010.

Clement, Luo. *The Ancient Science of Numbers: The Practical Application of Its Principles in the Attainment of Health, Success, and Happiness.* New York: Roger Brothers, 1908.

Clews, Henry. *Fifty Years in Wall Street.* New York: Irving Publishing Company, 1908.

Clough, Shepard Bancroft. *A Century of American Life Insurance: A History of the Mutual Life Insurance Company of New York, 1843–1943.* New York: Columbia University Press, 1946.

Cohen, Lizabeth. *A Consumers' Republic: The Politics of Mass Consumption in Postwar America.* New York: Vintage, 2003.

———. *Making a New Deal.* Cambridge: Cambridge University Press, 1990.

Cohen, Patricia Cline. *A Calculating People: The Spread of Numeracy in Early America.* Chicago: University of Chicago Press, 1982.

Coleman, William. "Cognitive Basis of the Discipline: Claude Bernard on Physiology." *Isis* 76, no. 1 (1985): 49–70.

Comfort, Nathaniel. *The Science of Human Perfection: How Genes Became the Heart of American Medicine.* New Haven, CT: Yale University Press, 2012.

Committee on Economic Security. *Social Security in America: The Factual Background of the Social Security Act as Summarized from Staff Reports to the Committee on Economic Security.* Washington, DC: Government Printing Office, 1937.

"Correspondence." *Medical Examiner* 5, no. 55 (1895): 221.

Court, Andrew T. "Measuring Joint Causation." *Journal of the American Statistical Association* 25, no. 171 (1930): 245–254.

Cowell, Henry, and Sidney Cowell. *Charles Ives and His Music.* New York: Oxford University Press, 1969.

Craig, J. D. "Mortality on Colored Lives." *Transactions of the Actuarial Society of America* 21, no. 64 (1920): 452–475.

Cronon, William. *Nature's Metropolis: Chicago and the Great West.* New York: Norton, 1991.

Crum, Gary. "The Priceless Value of Human Life." *American Journal of Public Health* 72, no. 11 (1982): 1299–1300.

Czerniawski, Amanda M. "From Average to Ideal: The Evolution of the Height and Weight Table in the United States, 1836–1943." *Social Science History* 31, no. 2 (2007): 273–296.

Dalton, Hugh. Review of *Population Problems in the United States and Canada.* Edited by Louis I. Dublin. *Economica* 18 (November 1926): 347–349.

Dalton, John C. *History of the College of Physicians and Surgeons in the City of New York; Medical Department of Columbia College.* New York: College of Physicians and Surgeons, 1888.

———. *A History of Columbia University 1754–1904.* New York: Columbia University Press, 1904.

Daston, Lorraine. *Classical Probability in the Enlightenment.* Princeton, NJ: Princeton University Press, 1988.

———. "On Scientific Observation." *Isis* 99, no. 1 (2008): 91–110.

Daston, Lorraine, and Elizabeth Lunbeck, eds. *Histories of Scientific Observation.* Chicago: University of Chicago Press, 2011.

Davies, John D. *Phrenology: Fad and Science.* New Haven, CT: Yale University Press, 1955.

Davis, Audrey B. "Life Insurance and the Physical Examination: A Chapter in the Rise of American Medical Technology." *Bulletin of the History of Medicine* 55, no. 3 (1981): 392–406.

Davis, William T. *Bench and Bar of Massachusetts.* Vol. 1. Boston: Boston History Company, 1895.

Dawson, Miles Menander. Review of "Race Traits and Tendencies of the American Negro," by Frederick L. Hoffman. *Publications of the American Statistical Association* 5, no. 35/36 (1896): 142–148.

Degler, Carl N. *In Search of Human Nature: The Decline and Revival of Darwinism in American Social Thought.* New York: Oxford University Press, 1991.

Desrosières, Alain. *The Politics of Large Numbers: A History of Statistical Reasoning.* Cambridge, MA: Harvard University Press, 1998.

DeWitt, Larry. "Financing Social Security, 1939–1949: A Reexamination of the Financing Policies of this Period." *Social Security Bulletin* 67, no. 4 (2007): 51–69.

Didier, Emmanuel. "Cunning Observation: US Agricultural Statistics in the Time of Laissez-Faire." *History of Political Economy* 44, suppl. 1 (2012): 27–45.

———. *En quoi consiste l'Amérique? Les statistiques, le New Deal et la Démocratie.* Paris: La Découverte, 2009.

Domhoff, G. William. "The Little-Known Origins of the Social Security Act: How and Why Corporate Moderates Created Old-Age Insurance." In *Social Insurance and Social Justice: Social Security, Medicare, and the Campaign against Entitlements.* Edited by Leah Rogne et al., 47–61. New York: Springer, 2009.

Douglas, Mary, and Aaron Wildavsky. *Risk as Culture: An Essay on the Selection of Technological and Environmental Dangers.* Berkeley: University of California Press, 1983.

Downs, Gregory P. "University Men, Social Science, and White Supremacy in North Carolina." *Journal of Southern History* 75, no. 2 (2009): 267–304.

Dublin, Louis I. "Address by Louis I. Dublin, Ph.D." *Addresses Delivered by August S. Knight, M.D. and Dr. Louis I. Dublin at the Annual Banquet of the Officers and Hygiene Reference Board of the Life Extension Institute.* New York: Life Extension Institute, 1923.

———. *After Eighty Years: The Impact of Life Insurance on the Public Health.* Gainesville: University of Florida Press, 1966.

———. *The Causes for the Recent Decline in Tuberculosis and the Outlook for the Future.* New York: Metropolitan Life, 1923.

———. *Economics of World Health.* New York: Metropolitan Life Insurance Company, 1926. Reprinted from *Harper's Monthly Magazine* (November 1926).

———. *The Effect of Life Conservation on the Mortality of the Metropolitan Life Insurance Company: A Summary of the Experience, Industrial Department, 1914, for Superintendents,*

Medical Examiners and Visiting Nurses. New York: Metropolitan Life Insurance Company, 1916.

———. "Factors in American Mortality: A Study in Death Rates in the Race Stocks of New York State, 1910." *American Economic Review* 6, no. 3 (1916): 523–548.

———. *The Fallacious Propaganda for Birth Control*. New York: Metropolitan Life Insurance Company, 1926. Reprinted from *Atlantic Monthly* (February 1926).

———. *A Family of Thirty Million: The Story of the Metropolitan Life Insurance Company*. New York: Metropolitan Life, 1943.

———. "Health of the Negro." *Annals of the American Academy of Political and Social Science* 140 (November 1928): 77–85.

———. *Health Work Pays*. New York: Metropolitan Life Insurance Company, 1925. Reprinted from *Survey Graphic* (November 1925).

———. "The Improvement and Extension of the Registration Area." *Publications of the American Statistical Association* 14, no. 110 (1915): 578–582.

———. *The Insurability of Women: An Address Delivered before the Medical Section of the American Life Convention in St. Paul, Minnesota*. New York: Metropolitan Life Insurance Company, 1913.

———. "Life, Death, and the American Negro." *American Mercury* 12, no. 45 (1927): 37–45.

———. "The Mortality of Foreign Race Stocks: A Contribution to the Quantitative Study of the Vigor of the Racial Elements in the Population of the United States." *Scientific Monthly* 14, no. 1 (1922): 94–104.

———. *The Mortality of Foreign Race Stocks*. New York: Metropolitan Life Insurance Company, 1923. Reprinted from *Eugenics in Race and State*, vol. 2.

———. ed. *Population Problems in the United States and Canada*. Boston: Houghton Mifflin, 1926.

———. *Records of Public Health Nursing and Their Service in Case Work, Administration and Research*. New York: Metropolitan Life Insurance Company, 1922. Originally published in *The Public Health Nurse*.

———. *The Registration of Vital Statistics and Good Business: An Address Delivered before the Annual Conference of Health Officers of the State of Indiana, Indianapolis, May 13, 1913*. New York, Metropolitan Life Insurance Company, 1913.

———. *The Reporting of Disease: The Next Step in Life Conservation*. New York: Association of Life Insurance Presidents, 1914.

———. "The Statistician and the Population Problem." *Journal of the American Statistical Association* 20, no. 149 (1925): 1–12.

———. "Teaching Nurses in Training the Uses and Value of Sickness Statistics." *American Journal of Nursing* 17, no. 12 (September 1917): 1157–1165.

———. "To Be or Not to Be?" *Harper's Monthly Magazine* 161, no. 964 (1930): 486–494.

———. *Vital Statistics in Relation to Life Insurance: Paper Read before Subsection B, Section VIII, "Public Health and Medical Science," of the Second Pan-American Scientific Congress, Washington, December 30, 1915*. New York: Metropolitan Life Insurance Company, 1916.

Dublin, Louis I., and Bessie Bunzel. *To Be or Not to Be: A Study of Suicide*. New York: Harrison Smith & Robert Haas, 1933.

Dublin, Louis I., and John C. Gebhart. *Do Height and Weight Tables Identify Undernourished Children?* New York: New York Association for Improving the Condition of the Poor, 1923.

Dublin, Louis I., and Edwin W. Kopf. "The Improvement of Statistics of Cause of Death through Supplementary Inquiries to Physicians." *Publications of the American Statistical Society* 15, no. 114 (1916): 175–191.

Dublin, Louis I., and Alfred J. Lotka. *Length of Life: A Study of the Life Table.* New York: Ronald Press, 1936.

———. *The Money Value of a Man.* New York: Ronald Press, 1930.

———. "On the True Rate of Natural Increase." *Journal of the American Statistical Association* 20, no. 151 (1925): 305–339.

Du Bois, W. E. B. Review of "Race Traits and Tendencies of the American Negro," by Frederick L. Hoffman. *Annals of the American Academy of Political and Social Science* 9 (January 1897): 127–133.

Duncan, Joseph W., and William C. Shelton. *Revolution in United States Government Statistics, 1926–1976.* Washington, DC: Government Printing Office, 1978.

Dwight, E. W. "Latent Powers of Life Insurance Companies for the Detection and Prevention of Diseases." *Proceedings of the Annual Meeting of the Association of Life Insurance Presidents* 3 (1910): 102–108.

"Editorial." *Medical Examiner* 9, no. 7 (July 1899): 197–204.

"Editorial Notes." *Medical Examiner* 5, no. 12 (1895): 251–252.

Elkin, Philip. "The Father of American Social Insurance: I. M. Rubinow." *Review of Insurance Studies* 3, no. 3 (1956): 105–112.

Emerson, Ralph Waldo. *Conduct of Life.* Boston: Ticknor & Fields, 1860.

———. *Essays.* Boston: Riverside Press Cambridge, 1903 [1865].

Ericson, Richard V., and Aaron Doyle, eds. *Risk and Morality.* Toronto: University of Toronto Press, 2003.

Espeland, Wendy Nelson, and Mitchell L. Stevens. "A Sociology of Quantification." *European Journal of Sociology* 49, no. 3 (2008): 401–436.

Eyler, John M. *Victorian Social Medicine: The Ideas and Methods of William Farr.* Baltimore, MD: Johns Hopkins University Press, 1979.

Fabian, Ann. *The Skull Collectors: Race, Science, and America's Unburied Dead.* Chicago: University of Chicago Press, 2010.

Falk, Isidore S. Review of "The Money Value of a Man" by Dublin and Lotka. *Journal of Political Economy* 39, no. 4 (August 1931): 546–548.

Fields, Barbara J. "Of Rogues and Geldings." *American Historical Review* 108, no. 5 (2003): 1397–1405.

———. "Whiteness, Racism, and Identity." *International Labor and Working-Class History* 60 (October 2001): 48–56.

Fish, Henry Clay. *The Agent's Manual of Life Assurance.* New York: Equitable Life Assurance Society, 1867.

Fisher, Irving. *Bulletin 30 of the Committee of One Hundred on National Health, Being a Report on National Vitality, Its Wastes and Conservation.* Washington, DC: Government Printing Office, 1909.

———. *Economic Aspect of Lengthening Human Life.* New York: Association of Life Insurance Presidents, 5 February 1909.

———. *How to Live Long.* New York: Metropolitan Life Insurance Company, 1916.

Fisher, Irving, and Eugene Lyman Fisk. *How to Live: Rules for Healthful Living Based on Modern Science.* New York: Funk & Wagnalls, 1920.

Fisk, Eugene Lyman. "Life Insurance and Life Conservation." *Scientific Monthly* 4, no. 4 (1917): 330–342.

Fiske, Thomas S. "Emory McClintock." *Bulletin of the American Mathematical Society* 23, no. 8 (May 1917): 353–357.

Fiss, Andrew. "Professing Mathematics: Science and Education in Nineteenth-Century America." PhD diss., Indiana University, 2011.

Flanzraich, Gerri. "The Library Bureau and Office Technology." *Libraries and Culture* 28, no. 4 (Fall 1993): 403–429.

Fleishman, S. L. *The Fallacies of the Assessment Plan of Life Insurance.* Philadelphia: Edward Stern, 1897.

Foley, Francis Bernard. "Group Insurance." Fellowship diss., Insurance Institute of America, 1934.

Foucault, Michel. *Discipline and Punish: The Birth of the Prison.* Translated by Alan Sheridan. New York: Random House, 1995.

———. *History of Sexuality, Volume 1.* Translated by Robert Hurley. New York: Vintage, 1990.

Fox, Renee C. "The Medicalization and Demedicalization of American Society." *Daedalus* 106, no. 1 (1977): 9–22.

Frängsmyr, Tore, J. L. Heilbron, and Robin E. Rider, eds. *The Quantifying Spirit in the 18th Century.* Berkeley: University of California Press, 1990.

Frankel, Lee K. *The Influence of Private Life Insurance Companies on Tuberculosis: Reprint of a Paper Prepared for the Eleventh International Tuberculosis Congress Held in Berlin, Germany, October 22–26, 1913.* New York: Metropolitan Life Insurance Company, 1913.

———. *Insurance Companies and Public Health Activities.* New York: Metropolitan Life Insurance Company, 1913.

Frankel, Lee K., and Miles M. Dawson. *Workingmen's Insurance in Europe.* New York: Charities Publication Committee, 1911.

Frankel, Lee K., and Louis I. Dublin. "Visiting Nursing and Life Insurance." *Publications of the American Statistical Association* 16, no. 122 (June 1918): 58–112.

Franks, Bill. *Taming the Big Data Tidal Wave: Finding Opportunities in Huge Data Streams with Advanced Analytics.* Hoboken, NJ: Wiley, 2012.

Friedman, Walter A. *Fortune Tellers: The Story of America's First Economic Forecasters.* Princeton, NJ: Princeton University Press, 2013.

Geison, Gerald. *Michael Foster and the Cambridge School of Physiology.* Princeton, NJ: Princeton University Press, 1978.

"General Discussion—Movement to Prolong Human Life." *Proceedings of the Annual Meeting of the Association of Life Insurance Presidents* 3 (1910): 108–111, 114–118.

Gephart, William Franklin. *Principles of Insurance.* Vol. 1, *Life.* New York, Macmillan: 1917.

Gerber, David A. "A Politics of Limited Options: Northern Black Politics and the Problem of Change and Continuity in Race Relations Historiography." *Journal of Social History* 14, no. 2 (Winter 1980): 235–255.

Gigerenzer, Gerd, Zeno Swijtink, Theodore Porter, Lorraine Daston, John Beatty, and Lorenz Krüger. *The Empire of Chance: How Probability Changed Science and Everyday Life.* Cambridge: Cambridge University Press, 1989.

Gitelman, Lisa. *Paper Knowledge: Toward a Media History of Documents.* Durham, NC: Duke University Press, 2014.

Glad, Betty. "Hughes, Charles Evans." *American National Biography Online* (2000), http://0-www.anb.org.library.colgate.edu/articles/11/11-00439.html.

Glenn, Brian J. "Fraternal Rhetoric and the Development of the U.S. Welfare State." *Studies in American Political Development* 15, no. 2 (2001): 220–233.

————. "The Shifting Rhetoric of Insurance Denial." *Law and Society Review* 34, no. 3 (2000): 779–808.

Gloninger, D. S. *The Medical Examiner's Manual*. Philadelphia: Claxton, Remson, & Haffelfinger, 1869.

Goodheart, Lawrence B. *Abolitionist, Actuary, Atheist: Elizur Wright and the Reform Impulse*. Kent, OH: Kent State University Press, 1990.

Gordon, Linda. *Dorothea Lange: A Life beyond Limits*. New York: Norton, 2009.

Gore, John K. "Should Life Companies Discriminate against Women?" *Transactions of the Actuarial Society of America* 6, no. 24 (1900): 380–388.

Griscom, John H. "Physical Indications of Longevity in Man." In American Popular Life Insurance Company, *Sources of Longevity, Its Indications and Practical Applications*. New York: Wm. Wood, 1869.

————. "To the United States National Medical Association." *Transactions of the American Medical Association* 1 (1848): 339–340.

Hacker, Jacob S. *Divided Welfare State: The Battle over Public and Private Social Benefits in the United States*. New York: Cambridge University Press, 2002.

Hacking, Ian. "Biopower and the Avalanche of Printed Numbers." *Humanities in Society* 5, no. 3 and 4 (1982): 279–295.

————. "The Looping Effects of Human Kinds." In *Causal Cognition: An Interdisciplinary Approach*. Edited by D. Sperber, D. Premack, and A. Premack, 351–383. Oxford: Oxford University Press, 1995.

————. "Making Up People." *London Review of Books*, 17 August 2006, 7–11.

————. "Risk and Dirt." In *Risk and Morality*. Edited by Richard V. Ericson and Aaron Doyle, 22–47. Toronto: University of Toronto Press, 2003.

————. *The Taming of Chance*. New York: Cambridge University Press, 1990.

"Hahnemann Life Insurance Co." *Hahnemannian Monthly* 2, no. 11 (June 1867): 506–508.

Haller Jr., John S. *The History of American Homeopathy: The Academic Years, 1820–1935*. New York: Pharmaceutical Products Press, 2005.

————. *Outcasts from Evolution: Scientific Attitudes of Racial Inferiority 1859–1900*. Carbondale: Southern Illinois University Press, 1996.

————. "Race, Mortality, and Life Insurance: Negro Statistics in the Late Nineteenth Century." *Journal of the History of Medicine and Allied Sciences* 25, no. 3 (1970): 247–261.

Haggerty, Kevin D., and Richard V. Ericson. "The Surveillant Assemblage." *British Journal of Sociology* 51, no. 4 (2000): 605–622.

Hamilton, Diane. "The Cost of Caring: The Metropolitan Life Insurance Company's Visiting Nurse Service, 1909–1953." *Bulletin of the History of Medicine* 63, no. 3 (1989): 414–434.

Hankins, Thomas L. "A 'Large and Graceful Sinuosity': John Herschel's Graphical Method." *Isis* 97, no. 4 (2006): 605–633.

Harcourt, Bernard E. *Against Prediction: Profiling, Policing, and Punishing in an Actuarial Age*. Chicago: University of Chicago Press, 2006.

————. "A Reader's Companion to *Against Prediction*: A Reply to Ariela Gross, Yoram Margalioth, and Yoav Sapir on Economic Modeling, Selective Incapacitation, Governmentality, and Race." *Law and Social Inquiry* 33, no. 1 (2008): 265–283.

Hardy, Charles O. *Risk and Risk-Bearing*. Chicago: University of Chicago Press, 1923.

Haskell, Thomas. *The Emergence of Professional Social Science: The American Social Science Association and the Nineteenth-Century Crisis of Authority*. Baltimore, MD: Johns Hopkins University Press, 2000.

Haswell, Charles H. *Reminiscences of New York by an Octogenarian (1816 to 1860).* New York: Harper and Brothers, 1896.

Heen, Mary L. "Ending Jim Crow Life Insurance Rates." *Northwestern Journal of Law and Social Policy* 4, no. 2 (2009): 360–399.

Henderson, Alexa Benson. *Atlanta Life Insurance Company: Guardian of Black Economic Dignity.* Tuscaloosa: University of Alabama Press, 1990.

Higham, John. *Strangers in the Land: Patterns of American Nativism, 1860–1925.* New Brunswick, NJ: Rutgers University Press, 2002.

Himes, Norman. Review of *Population Problems. American Economic Review* 16, no. 3 (1926): 511–514.

Hirshbein, Laura Davidow. "Masculinity, Work, and the Fountain of Youth: Irving Fisher and the Life Extension Institute, 1914–31." *Canadian Bulletin of Medical History* 16, no. 1 (1999): 89–124.

Hirschl, Andrew J. *The Law of Fraternities and Societies: With Special Reference to Their Insurance Feature.* St. Louis, MO: William H. Stevenson, 1883.

Hobbs, Clarence W. "Social Insurance and the Constitution." *Proceedings of the Casualty Actuarial Society* 22, no. 44 (1935): 32–49.

Hodes, Martha. "Fractions and Fictions in the United States Census of 1890." In *Haunted by Empire.* Edited by Ann Laura Stoler, 240–270. Durham, NC: Duke University Press, 2006.

Hoff, Derek S. *The State and the Stork: The Population Debate and Policy Making in US History.* Chicago: University of Chicago Press, 2012.

Hoffman, Beatrix. "Scientific Racism, Insurance, and Opposition to the Welfare State: Frederick L. Hoffman's Transatlantic Journey." *Journal of the Gilded Age and Progressive Era* 2, no. 2 (2003): 150–190.

———. *The Wages of Sickness: The Politics of Health Insurance in Progressive America.* Chapel Hill: University of North Carolina Press, 2001.

Hoffman, Frederick L. *History of the Prudential Insurance Company of America.* Newark, NJ: Prudential Press, 1900.

———. *The Mortality from Cancer throughout the World.* Newark, NJ: Prudential Press, 1915.

———. "Practical Statistics of Public Health Nursing and Community Sickness Experience." *American Journal of Nursing* 14, no. 1 (1914): 948–960.

———. "Race Traits and Tendencies of the American Negro." *Publications of the American Economic Association* 11, no. 1/3 (1896): 1–329.

———. "Some Fallacies of Compulsory Health Insurance." *Scientific Monthly* 4, no. 4 (1917): 306–316.

———. *The Statistical Experience Data of the Johns Hopkins Hospital, Baltimore, MD., 1892–1911.* Baltimore, MD: Johns Hopkins University Press, 1913.

———. "Vital Statistics of the Negro." *Arena* 24 (April 1892): 529–542.

Hofstadter, Richard. *Age of Reform.* New York: Vintage, 1960.

Holden, Edgar. "The Object of the Association." *Abstract of the Proceedings of the Association of Life Insurance Medical Directors of America* 6 (1895): 50–57.

Holdren, Nate. "Incentivizing Safety and Discrimination: Employment Risks under Workmen's Compensation in the Early Twentieth Century United States." *Enterprise and Society* 15, no. 1 (2014): 31–67.

Hollinger, David. "Ethnic Diversity, Cosmopolitanism and the Emergence of the American Liberal Intelligentsia." *American Quarterly* 27, no. 2 (1975): 133–151.

Holmes, Charles B. *Elsieville: A Tale of Yesterday.* New York: Charles B. Holmes, 1903.

"Holmes' Mercantile Agency" (classified ad). *American Monthly Review of Reviews* 34 (December 1906), classified advertising section: 157.

"Holmes Mercantile Agency" (advertisement). *Spectator* 63, no. 1 (1899): v.

"Holmes Mercantile Agency" (advertisement). *Spectator* 73, no. 8 (25 August 1904): vi.

"Holmes Mercantile Agency." *Spectator* 74, no. 24 (15 June 1905): 331.

Homans, Sheppard. "Life Insurance: Sheppard Homans on Life Insurance. Paper Read before the American Social Science Association. Revised by the Author." *Insurance Times* 2 (1869): 798.

———. "On the Equitable Distribution of Surplus." *Journal of the Institute of Actuaries* 11 (October 1863): 121–129.

———. *Report Exhibiting the Experience of the Mutual Life Insurance Company of New-York for Fifteen Years Ending February First, 1858.* New York, November 1859.

"Hooper-Holmes Information Bureau." *Spectator* 78, no. 6 (7 February 1907): 80.

Hoppit, Julian. "Political Arithmetic in Eighteenth-Century England." *Economic History Review* 49, no. 3 (1996): 516–540.

Horan, Caley Dawn. "Actuarial Age: Insurance and the Emergence of Neoliberalism in the Postwar United States." PhD diss., University of Minnesota, 2011.

Hounshell, David, and John Kenly Smith. *Science and Corporate Strategy: DuPont R&D, 1902–1980.* New York: Cambridge University Press, 1988.

Howells, William Dean. *A Hazard of New Fortunes.* New York: Harper & Brothers, 1889.

"How to Figure the Extinction of a Race." *Nation* 64, no. 1657(1 April 1897): 246–248.

Huebner, Solomon S. *Life Insurance: A Textbook.* New York: D. Appleton, 1919.

Hunter, Arthur. "Insurance on Sub-Standard Lives." *Annals of the American Academy of Political and Social Science* 70 (March 1917): 38–53.

———. "Selection of Risks from the Actuarial Standpoint." *Transactions of the Actuarial Society of America* 12, no. 45 (1911): 1–17, 281–298.

Huntington, Ellsworth, and Frank E. Williams. *Business Geography.* New York: Wiley, 1922.

Hutcheson, William A. "In Memoriam: Emory McClintock." *Transactions of the Actuarial Society of America* 17, no. 56 (1916): 373–381.

Hutter, Bridget, and Michael Power, eds. *Organizational Encounters with Risk.* New York: Cambridge University Press, 2005.

Hyman, Louis. *Debtor Nation: The History of America in Red Ink.* Princeton, NJ: Princeton University Press, 2011.

Ide, George E. "Report of the Life Extension Committee of the Association of Life Insurance Presidents." *Proceedings of the Annual Meeting of the Association of Life Insurance Presidents* 3 (1910): 82–85.

Igo, Sarah E. *The Averaged American: Surveys, Citizens, and the Making of a Mass Public.* Cambridge, MA: Harvard University Press, 2008.

"In and about New York." *Spectator* 78, no. 6 (7 February 1907): 69.

Industrial Relations Section, Princeton University. *Memorandum: Group Insurance.* Ann Arbor, MI: Edwards Brothers, 1927.

"Information Bureau Adds to its Facilities." *Standard* 60, no. 7 (16 February 1907): 195.

"Insurance Bureau" (advertisement). *Spectator* 63, no. 1 (1899): v.

Isaac, Joel. *Working Knowledge: Making the Human Sciences from Parsons to Kuhn.* Cambridge, MA: Harvard University Press, 2012.

Ives, Charles E. "The Amount to Carry—Measuring the Prospect." *Eastern Underwriter* (17 September 1920): 35–38.

Jackson, Kenneth T. *Crabgrass Frontier: The Suburbanization of the United States.* New York: Oxford University Press, 1985.

James, Marquis. *Metropolitan Life: A Study in Business Growth*. New York: Viking Press, 1947.

James, William. *Pragmatism: A New Name for Some Old Ways of Thinking*. New York: Longmans, Green, 1907.

Jenkins, F. W. "Report of Health Committee of Association of Life Insurance Presidents." *Proceedings of the Annual Meeting of the Association of Life Insurance Presidents* 6 (1912): 78–85, 139.

Joffe, S. A. "Concerning the American Experience Table of Mortality." *Transactions of the Actuarial Society of America* 14, no. 49 (1913): 24–37.

Johnson, Walter. *Soul by Soul: Life inside the Antebellum Slave Market*. Cambridge, MA: Harvard University Press, 1999.

Kafka, Ben. "Paperwork: The State of the Discipline." *Book History* 12 (2009): 340–353.

Kasson, John F. *Houdini, Tarzan, and the Perfect Man: The White Male Body and the Challenge of Modernity in America*. New York: Hill & Wang, 2002.

Keller, Morton. *The Life Insurance Enterprise: A Study in the Limits of Corporate Power*. Cambridge, MA: Belknap Press of the Harvard University Press, 1963.

Kessler-Harris, Alice. "In the Nation's Image: The Gendered Limits of Social Citizenship in the Depression Era." *Journal of American History* 86, no. 3 (1999): 1251–1279.

Kevles, Daniel J. *In the Name of Eugenics: Genetics and the Uses of Human Heredity*. Cambridge, MA: Harvard University Press, 1995.

Kilgour, D. E. "Life Insurance without Medical Examination." *Transactions of the Actuarial Society of America* 22, no. 65 (1921): 120–139, 145–175.

Kindleberger, Charles P. *Manias, Panics, and Crashes: A History of Financial Crisis*. New York: Wiley, 2000.

Kingsland, Sharon E. *Modeling Nature: Episodes in the History of Population Ecology*. Chicago: University of Chicago Press, 1995.

Klein, Jennifer. *For All These Rights: Business, Labor, and the Shaping of America's Public-Private Welfare State*. Princeton, NJ: Princeton University Press, 2003.

Klein, Judy L. *Statistical Visions in Time: A History of Time Series Analysis, 1662–1938*. New York: Cambridge University Press, 1997.

Klem, Amelia. "A Life of Her Own Choosing: Anna Gibbons' Fifty Years as a Tattooed Lady." *Wisconsin Magazine of History* 89, no. 3 (2006): 28–39 at 36.

Knapp, Moses. *Lectures on the Science of Life Insurance Addressed to Families, Societies, Trades, Professions—Considerate Persons of All Classes*. 2nd ed. Philadelphia: E. S. Jones, 1853.

Knowles, Scott Gabriel. *The Disaster Experts: Mastering Risk in Modern America*. Philadelphia: University of Pennsylvania Press, 2011.

Kohler, Robert. *Labscapes and Landscapes: Exploring the Lab-Field Border in Biology*. Chicago: University of Chicago Press, 2002.

Koven, Seth and Sonya Michel, eds. *Mothers of a New World: Maternalist Politics and the Origins of Welfare States*. New York: Routledge, 1993.

———. "Womenly Duties: Maternalist Politics and the Origins of Welfare States in France, Germany, Great Britain, and the United States, 1880–1920." *American Historical Review* 95, no. 4 (1990): 1076–1108.

Krajewski, Markus. *Paper Machines: About Cards and Catalogs, 1548–1929*. Translated by Peter Krapp. Cambridge, MA: MIT Press, 2011.

Krippner, Greta R. *Capitalizing on Crisis: The Political Origins of the Rise of Finance*. Cambridge, MA: Harvard University Press, 2012.

Lambert, Thomas Scott. *Popular Anatomy and Physiology, Adapted to the Use of Students and General Readers*. Portland, ME: Sanborn & Carter, 1851.

———. *Practical Anatomy, Physiology, and Pathology; Hygiene and Therapeutics.* Portland, ME: Sanborn & Carter, 1851.

Lamoreaux, Naomi R. *The Great Merger Movement in American Business, 1895–1904.* Cambridge: Cambridge University Press, 1985.

———. "Rethinking the Transition to Capitalism in the Early American Northeast." *Journal of American History* 90, no. 2 (2003): 437–461.

Lanier, Jaron. *Who Owns the Future?* New York: Simon & Schuster, 2013.

Latour, Bruno. *Science in Action: How to Follow Scientists and Engineers through Society.* Cambridge, MA: Harvard University Press, 1987.

Lauer, Josh. "The Good Consumer: Credit Reporting and the Invention of Financial Identity in the United States, 1840–1940." PhD diss., University of Pennsylvania, 2008.

———. "From Rumor to Written Record: Credit Reporting and the Invention of Financial Identity in Nineteenth-Century America." *Technology and Culture* 49, no. 2 (2008): 301–324.

———. "Making the Ledgers Talk: Customer Control and the Origins of Retail Data Mining, 1920–1940." In *The Rise of Marketing and Market Research.* Edited by Harmut Berghoff, Philip Scranton, and Uwe Spiekermann, 153–169. New York: Palgrave Macmillan, 2012.

Laughlin, H. H. "Calculations on the Working Out of a Proposed Program of Sterilization." In *Proceedings of the First National Conference on Race Betterment,* 478–494. Battle Creek, MI: Race Betterment Foundation, 1914.

Lawson, Thomas W. *Frenzied Finance: The Crime of the Amalgamated.* Vol. 1. New York: Ridgway-Thayer, 1905.

Lears, Jackson. *Rebirth of a Nation: The Making of Modern America, 1877–1920.* New York: Harper, 2009.

———. *Something for Nothing: Luck in America.* New York: Viking, 2003.

Leff, Mark H. "Taxing the 'Forgotten Man': The Politics of Social Security Finance in the New Deal." *Journal of American History* 70, no. 2 (1983): 359–381.

Lehrman II, William George. "Organizational Form and Failure in the Life Insurance Industry." PhD diss., Princeton University, 1989.

Lemann, Nicholas. *The Big Test: The Secret History of the American Meritocracy.* New York: Farrar, Straus & Giroux, 1999.

Lengwiler, Martin. "Double Standards: The History of Standardizing Humans in Modern Life Insurance." In *Standards and Their Stories: How Quantifying, Classifying, and Formalizing Practices Shape Everyday Life.* Edited by Martha Lampland and Susan Leigh Star, 95–113. Ithaca, NY: Cornell University Press, 2009.

Levine, Deborah. "Managing American Bodies: Diet, Nutrition, and Obesity in America 1840–1920." PhD diss., Harvard University, 2008.

Levy, Jonathan Ira. "Contemplating Delivery: Futures Trading and the Problem of Commodity Exchange in the United States, 1875–1905." *American Historical Review* 111, no. 2 (2006): 307–335.

———. *Freaks of Fortune: The Emerging World of Capitalism and Risk in America.* Cambridge, MA: Harvard University Press, 2012.

Lewin, C. G. *Pensions and Insurance before 1800: A Social History.* East Linton, UK: Tuckwell, 2003.

Lewis, John B., and Charles C. Bombaugh. *Stratagems and Conspiracies to Defraud Life Insurance Companies: An Authentic Record of Memorable Cases.* Baltimore, MD: James H. McClellan, 1896.

Lewis, Michael. *Moneyball: The Art of Winning an Unfair Game.* New York: Norton, 2003.

Ley, Harold A. *Employees Welfare Work That Pays: How Employers and Employees Are Going Fifty-Fifty on a New Life Extension and Insurance Service.* New York: Life Extension Institute, 1929.

Life Extension Institute. *Addresses at the Banquet of the Life Extension Institute.* New York: Life Extension Institute, 1921.

"Life, Fire and Miscellaneous Notes," *Spectator* 73, no. 3 (21 July 1904): 32.

"Life Insurance Notes." *Spectator* 63, no. 19 (1899): 211.

"Life Insurance—Wright's Tables." *Hunt's Merchants' Magazine and Commercial Review* 31, no. 6 (December 1854): 734–736.

Light, Jennifer. "Discriminating Appraisals: Cartography, Computation, and Access to Federal Mortgage Insurance in the 1930s." *Technology and Culture* 52, no. 3 (2011): 485–522.

Linker, Beth. *War's Waste: Rehabilitation in World War I America.* Chicago: University of Chicago Press, 2011.

Lippmann, Walter. *Drift and Mastery.* New York: Mitchell Kennerley, 1914.

Lotka, Alfred J. *Elements of Physical Biology.* Baltimore: Williams & Wilkins, 1925.

———. "An Objective Standard of Value Derived from the Principle of Evolution." *Journal of the Washington Academy of Sciences* 4, no. 14 (1914): 409–418; 4, no. 15 (1914): 447–457; and 4, no. 17 (1914) 499–500.

Lubar, Steven. " 'Do Not Fold, Spindle, or Mutilate': A Cultural History of the Punch Card." *Journal of American Culture* 14, no. 4 (1992): 43–55.

Lynd, Robert S., and Helen Merrell Lynd. *Middletown: A Study in American Culture.* New York: Harcourt, Brace, 1929.

Mack, Kenneth W. "Law, Society, Identity, and the Making of the Jim Crow South: Travel and Segregation on Tennessee Railroads, 1875–1905." *Law and Social Inquiry* 24, no. 377 (1999): 377–409.

———. "Rethinking Civil Rights Lawyering and Politics in the Era before *Brown.*" *Yale Law Journal* 115, no. 2 (2005): 256–354.

MacKenzie, Donald. *An Engine, Not a Camera.* Cambridge, MA: MIT Press, 2008.

MacKenzie, Donald, and Yuval Millo, "Constructing a Market, Performing Theory: The Historical Sociology of a Financial Derivatives Exchange." *American Journal of Sociology* 109, no. 1 (July 2003): 107–145.

Maclean, Joseph B. *Life Insurance.* New York: McGraw-Hill, 1962.

Massachusetts, Commonwealth of. *Acts and Resolves Passed by the General Court of Massachusetts in the Year 1884.* Boston: Wright & Potter, 1884.

———. *Forty-First Report to the Legislature of Massachusetts Relating to the Registry and Return of Births, Marriages, and Deaths in the Commonwealth.* Boston: Wright & Potter, 1883.

Masur, Louis P. " 'Age of the First Person Singular': The Vocabulary of the Self in New England, 1780–1850." *Journal of American Studies* 25, no. 2 (1991): 189–211.

Mayer-Schönberger, Viktor, and Kenneth Cukier. *Big Data: A Revolution That Will Transform How We Live, Work, and Think.* Boston: Houghton Mifflin Harcourt, 2013.

McCall, John A. *A Review of Life Insurance from the Date of the First National Convention of Insurance Officials. 1871–1897.* Milwaukee, 1898.

McCay, Charles F. "American Tables of Mortality." *Journal of the Institute of Actuaries* 16 (October 1870): 20–33.

McClintock, Emory. "On the Objects to Be Attained in Future Investigations of Mortality and Death Loss." *Transactions of the Actuarial Society of America* 6, no. 24 (1900): 373–379.

McFall, Liz. "A 'good, average man': Calculation and the Limits of Statistics in Enrolling Insurance Customers." *Sociological Review* 59, no. 4 (2011): 661–684.

McGee, W. J. Review of "Race Traits and Tendencies of the American Negro," by Frederick L. Hoffman. *Science* 5, no. 106 (1897): 65–68.

McGlamery, J. Gabriel. "Race Based Underwriting and the Death of Burial Insurance." *Connecticut Insurance Law Journal* 15, no. 2 (2009): 531–570.

McKinley, Charles, and Robert W. Frase. *Launching Social Security: A Capture-and-Record Account 1935–1937*. Madison: University of Wisconsin Press, 1970.

Meagher, Thomas F. *The Commercial Agency "System" of the United States and Canada Exposed*. New York, 1876.

"Medical Examiners' Blanks." *Medical Examiner* 5, no. 46 (1895): 138.

"Medical Examiners Organizing." *Medical Examiner* 5, no. 46 (1895): 138.

"Medical Examination Fees." *Medical Examiner* 5, no. 12 (1895): 259–260.

Meech, Levi W. *System and Tables of Life Insurance. A Treatise Developed from the Experience and Records of Thirty American Life Offices, under the Direction of a Committee of Actuaries.* New York: Spectator Company, 1898.

Melville, Herman. *Moby-Dick, or The White Whale*. Boston: St. Botolph Society, 1892 [1851].

Menand, Louis. *The Metaphysical Club: A Story of Ideas in America*. New York: Farrar, Straus & Giroux, 2001.

Metropolitan Life Insurance Company. "The Chances of Reaching Age 65," *Statistical Bulletin* 18, no. 3 (March 1937): 2–5.

———. *An Epoch in Life Insurance: A Third of a Century of Achievement*. New York: Metropolitan Life Insurance Company, 1924.

———. *An Epoch in Life Insurance: Twenty-Five Years of Administration of the Metropolitan Life Insurance Company*. New York: Metropolitan Life, 1917.

———. "Height and Weight Standards in Child Nutrition Work." *Statistical Bulletin* 2, no. 3 (March 1921): 5–6.

———. "The Metropolitan Life Insurance Co." *American Health* 2, no. 1 (1909): iv–v.

———. "Stimulants, Sedatives or Food." *Harper's Monthly Magazine* 158 (May 1929).

———. "The Value of a Man as a Wage-Earner." *Statistical Bulletin* 7, no. 6 (1926): 1–4.

———. "The Value of Human Life." *Statistical Bulletin* 6, no. 11 (1925): 4–5.

Meyerowitz, Joanne. " 'How Common Culture Shapes the Separate Lives': Sexuality, Race, and Mid-Twentieth-Century Social Constructionist Thought." *Journal of American History* 96, no. 4 (March 2010): 1057–1084.

Mihm, Stephen. *A Nation of Counterfeiters: Capitalists, Con Men, and the Making of the United States*. Cambridge, MA: Harvard University Press, 2009.

Miller, Kelly. *A Review of Hoffman's Race Traits and Tendencies of the American Negro*. Washington, DC: American Negro Academy, 1897.

Miller, Lisa. "The Google of Spit." *New York Magazine*, 22 April 2014.

Mills, Asher S. "Hahnemann Life Insurance Company." *Ohio Medical and Surgical Reporter* 1, no. 5. (September 1867): 151–152.

Mohun, Arwen P. "On the Frontier of *The Empire of Chance*: Statistics, Accidents, and Risk in Industrializing America." *Science in Context* 18, no. 3 (2005): 337–357.

———. *Risk: Negotiating Safety in American Society*. Baltimore, MD: Johns Hopkins University Press, 2013.

Moorhead, E. J. *Our Yesterdays: The History of the Actuarial Profession in North America 1809–1979*. Schaumburg, IL: Society of Actuaries, 1989.

Mozzone, Terri. "Hooper Holmes, Inc." In *International Directory of Company Histories.* Edited by Thomas Derdak, 22:264–267. New York: St. James Press, 1998.

Murphy, Sharon Ann. *Investing in Life: Insurance in Antebellum America.* Baltimore, MD: Johns Hopkins University Press, 2010.

Mutual Life Insurance Company of New York. *Preliminary Report of the Mortuary Experience of the Mutual Life Insurance Company of New-York.* New York: Mutual Life of New York, 1875.

Myers, Robert J. "Beneficiary Statistics under the Old-Age and Survivors Insurance Program and Some Possible Demographic Studies Based on these Data." *Journal of the American Statistical Association* 44, no. 247 (1949): 388–396.

New York Insurance Department and Clifford Chance Rogers and Wells. "State of New York Insurance Department Report on Examination of the Metropolitan Life Insurance Company regarding Response to Supplement No. 1 to Circular Letter No. 19 (2000)." Investigative report (2002).

New York Legislature. *Laws of the State of New York, Passed at the Ninety-First Session of the Legislature, Begun January Seventh, and Ended May Sixth, 1868, in the City of Albany.* Albany: Van Benthuysen & Sons, 1868.

———. *Report of the Joint Committee of the Senate and Assembly of the State of New York: Appointed to Investigate the Affairs of Life Insurance Companies.* New York: M. B. Brown, 1906.

———. *Testimony Taken before the Joint Committee of the Senate and Assembly of the State of New York to Investigate and Examine into the Business and Affairs of Life Insurance Companies Doing Business in the State of New York.* 10 vols. New York: Brandow Printing Company, 1905.

———. *Testimony Taken before the Joint Committee of the Senate and Assembly of the State of New York to Investigate and Examine into the Business and Affairs of Life Insurance Companies Doing Business in the State of New York.* 7 vols. New York: J. B. Lyon, 1906.

New-York Life Insurance Company. *A Temple of Humanity.* New York: New-York Life, 1909.

Ngai, Mae. "The Architecture of Race in American Immigration Law: A Reexamination of the Immigration Act of 1924." *Journal of American History* 86, no. 1 (1999): 67–92.

Nichols, Walter S. "Dr. Emory McClintock as a Great Creative Mathematician.—The Calculus of Enlargement." *Transactions of the Actuarial Society of America* 17, no. 56 (1916): 290–302.

———. "The Value of Medical Examinations in Industrial Insurance." *Transactions of the Actuarial Society of America* 3, no. 10 (1893): 225–231 and 3, no. 11 (1894): 410–418.

Norris, James D. *R. G. Dun and Co. 1841–1900: The Development of Credit-Reporting in the Nineteenth Century.* Westport, CT: Greenwood Press, 1978.

North, Douglass. "Life Insurance and Investment Banking." *Journal of Economic History* 14, no. 3 (1954): 209–227.

Notestein, Frank W. "Demography in the United States: A Partial Account of the Development of the Field." *Population and Development Review* 8, no. 4 (1982): 651–687.

Numbers, Ronald L. "The Third Party: Health Insurance in America." *Sickness and Health in America.* Edited by Judith W. Leavitt and Numbers, 269–283. Madison: University of Wisconsin, 1997.

Nugent, Angela. "Fit for Work: The Introduction of Physical Examinations in Industry." *Bulletin of the History of Medicine* 57, no. 4 (1983): 578–595.

O'Connor, Alice. *Social Science for What? Philanthropy and the Social Question in a World Turned Rightside Up.* New York: Russell Sage Foundation, 2007.

Ogborn, Maurice Edward. *Equitable Assurances: The Story of Life Assurance in the Experience of the Equitable Life Assurance Society 1762–1962.* London: George Allen and Unwin, 1962.

Olegario, Rowena. *A Culture of Credit: Embedding Trust and Transparency in American Business.* Cambridge, MA: Harvard University Press, 2006.

"Otto Richter." *Proceedings of the Casualty Actuarial Society* 48, no. 90 (1961): 242.

"Otto Richter." *Transactions of Society of Actuaries* 14, no. 38 (1962): 197–198.

Pace, Lydia E., Yulei He, and Nancy L. Keating. "Trends in Mammography Screening Rates after Publication of the 2009 US Preventative Services Task Force Recommendations." *Cancer* 119, no. 14 (2013): 2518–2523.

Painter, Nell Irvin. *The History of White People.* New York: Norton, 2010.

Palmer, G. S. "Partial Report upon a Uniform System of Registration of Births, Marriages, and Deaths, and the Causes of Death." *Transactions of the American Medical Association* 9 (1856): 775–777.

Parascadola, John. "Doctors at the Gate: PHS at Ellis Island." *Public Health Reports* 113, no. 1 (1998): 83–86.

Parthasarathy, Shobita. "Regulating Risk: Defining Genetic Privacy in the United States and Britain." *Science, Technology, and Human Values* 29, no. 3 (2004): 332–352.

Patterson, Scott. *The Quants: How a New Breed of Math Whizzes Conquered Wall Street and Nearly Destroyed It.* New York: Crown Business, 2010.

Pearson, Karl. *Chances of Death and Other Studies of Evolution.* London: Edward Arnold, 1897.

Peeler, David P. *Hope among Us Yet: Social Criticism and Social Solace in Depression America.* Athens: University of Georgia Press, 1987.

Pietruska, Jamie L. "US Weather Bureau Chief Willis Moore and the Reimagination of Uncertainty in Long-Range Forecasting." *Environment and History* 17, no. 1 (2011): 79–105.

Pinch, Trevor and Richard Swedberg, eds. *Living in a Material World: Economic Sociology Meets Science and Technology Studies.* Cambridge, MA: MIT Press, 2008.

Poovey, Mary. *A History of the Modern Fact: Problems of Knowledge in the Sciences of Wealth and Society.* Chicago: University of Chicago Press, 1998.

Porter, Theodore M. "Funny Numbers." *Culture Unbound* 4 (2012): 585–598.

———. "How Science Became Technical." *Isis* 100, no. 2 (2009): 292–309.

———. *Karl Pearson: The Scientific Life in a Statistical Age.* Princeton, NJ: Princeton University Press, 2004.

———. "Life Insurance, Medical Testing, and the Management of Mortality." In *Biographies of Scientific Objects.* Edited by Lorraine Daston, 226–246. Chicago: University of Chicago Press, 2000.

———. "Precision and Trust: Early Victorian Insurance and the Politics of Calculation." In *The Values of Precision.* Edited by M. Norton Wise, 173–197. Princeton, NJ: Princeton University Press, 1995.

———. *The Rise of Statistical Thinking 1820–1900.* Princeton, NJ: Princeton University Press, 1986.

———. "Thin Description: Surface and Depth in Science and Science Studies." *Osiris* 27, no. 1 (2012): 209–226.

———. *Trust in Numbers: The Pursuit of Objectivity in Science and Public Life.* Princeton, NJ: Princeton University Press, 1995.

Power, Michael. *Organized Uncertainty: Designing a World of Risk Management.* New York: Oxford University Press, 2007.

"Practical Object Lesson." *Medical Examiner* 5, no. 12 (1895): 255.

The Problems of a Changing Population: Report of the Committee on Population Problems to the National Resources Committee May 1938. Washington, DC: Government Printing Office, 1938.

"The Progress of Fourteen Years." *Our Society Journal* 12, no. 70 (July 1889): 1–16.

"Proper Price for Life Insurance Examination." *Medical Examiner* 5, no. 46 (1895): 137.

Provost Marshal General. *Final Report of the Provost Marshal General to the Secretary of War on the Operations of the Selective Service System to July 15, 1919.* Washington, DC: Government Printing Office, 1920.

Puckett, Carolyn. "The Story of the Social Security Number." *Social Security Bulletin* 69, no. 2 (2009): 55–74.

Pusey, Merlo J. *Charles Evans Hughes.* 2 vols. New York: Macmillan, 1951.

Quadagno, Jill. "Welfare Capitalism and the Social Security Act of 1935." *American Sociological Review* 49, no. 5 (1984): 632–647.

Ralph, Michael. " 'Life . . . in the midst of death': Notes on the Relationship between Slave Insurance, Life Insurance, and Disability." *Disability Studies Quarterly* 32, no. 3 (2012). http://dsq-sds.org/article/view/3267/3100.

Ramsden, Edmund. "Carving Up Population Science: Eugenics, Demography and the Controversy over the 'Biological Law' of Population Growth." *Social Studies of Science* 32, no. 5/6 (2002): 857–899.

Ransom, Roger L., and Richard Sutch. "Tontine Insurance and the Armstrong Investigation: A Case of Stifled Innovation, 1868–1905." *Journal of Economic History* 47, no. 2 (1987): 379–390.

"Reduced Fees." *Medical Examiner* 5, no. 12 (1895): 253.

Reiser, Stanley Joel. "The Emergence of the Concept of Screening for Disease." *Milbank Memorial Fund Quarterly. Health and Society* 56, no. 4 (1978): 403–425.

Review of *The Conduct of Life* by Ralph W. Emerson. *Ladies' Companion* 19 (1861): 105–108.

Rice, Dorothy P., and Thomas A. Hodgson. "The Value of Human Life Revisited." *American Journal of Public Health* 72, no. 6 (1982): 536–538.

Richter, Otto C. "Actuarial Basis of Cost Estimates of Federal Old-Age Insurance." *Law and Contemporary Problems* 3, no. 2 (April 1936): 212–220.

Riley, James C. *Rising Life Expectancy: A Global History.* New York: Cambridge University Press, 2001.

Robertson, Thomas. *Malthusian Moment: Global Population Growth and the Birth of American Environmentalism.* New Brunswick, NJ: Rutgers University Press, 2012.

Rockoff, Hugh. "Banking and Finance, 1789–1914." In *Cambridge Economic History of the United States.* Edited by Stanley L. Engerman and Robert E. Gallman. Vol. 2. New York: Cambridge University Press, 2000.

Rodgers, Daniel T. *Atlantic Crossings: Social Politics in a Progressive Age.* Cambridge, MA: Belknap Press of Harvard University Press, 1998.

Rogers, Oscar H. "Build as a Factor Influencing Longevity." *Abstract of the Proceedings of the Association of Life Insurance Medical Directors of America* 12 (1901): 280–288.

———. "Medical Selection and Substandard Business." *Abstract of the Proceedings of the Association of Life Insurance Medical Directors of America* 18 (1907): 81–95.

Rogers, Oscar H., and Arthur Hunter. "The Need in Medical Selection of Standards by Which to Measure Border-Line Risks." *Transactions of the Actuarial Society of America* 17, no. 56 (1916): 281–289; and 18, no. 57 (1917): 164–169.

———. "The Numerical Method of Determining the Value of Risks for Insurance." *Transactions of the Actuarial Society of America* 20, no. 62 (1919): 273–332.

Rosenthal, Caitlin. "Storybook-keepers: Narratives and Numbers in Nineteenth Century America." *Common-place* 12, no. 3 (2012). http://www.common-place.org/vol-12/no-03/rosenthal.

Rothstein, Mark A. ed. *Genetics and Life Insurance: Medical Underwriting and Social Policy.* Cambridge, MA: MIT Press, 2004.

Rothstein, William G. *American Medical Schools and the Practice of Medicine: A History.* New York: Oxford University Press, 1987.

———. *Public Health and the Risk Factor: A History of an Uneven Medical Revolution.* Rochester, NY: University of Rochester Press, 2003.

Roubini, Nouriel, and Stephen Mihm. *Crisis Economics: A Crash Course in the Future of Finance.* New York: Penguin, 2011.

Rubinow, Isaac. *The Quest for Security.* New York: Henry Holt, 1934.

Sandage, Scott A. *Born Losers: A History of Failure in America.* Cambridge, MA: Harvard University Press, 2005.

Sanford, Jon. "Fourteenth Annual Report of the Insurance Commissioner of the Commonwealth of Massachusetts." Reprinted in *Insurance Times* 2 (September 1869): 652–665.

Schweber, Libby. *Disciplining Statistics: Demography and Vital Statistics in France and England, 1830–1885.* Durham, NC: Duke University Press, 2006.

Schiltz, Michael E. *Public Attitudes toward Social Security 1935–1965.* Washington, DC: Government Printing Office, 1970.

Schwartz, Hillel. *Never Satisfied: A Cultural History of Diets, Fantasies and Fat.* New York: Free Press, 1986.

Scott, James C. *Seeing Like a State: How Certain Schemes to Improve the Human Condition Have Failed.* New Haven, CT: Yale University Press, 1999.

Sellers, Christopher C. *Hazards of the Job: From Industrial Disease to Environmental Health Science.* Chapel Hill: University of North Carolina Press, 1997.

Shepherd, George R. "The Relation of Build (*I.E.* Height and Weight) to Longevity." *Abstract of the Proceedings of the Association of Life Insurance Medical Directors of America* 17 (1906): 46–66.

Shepherd, Pearce. "Principles and Problems of Selection and Underwriting." In *Life Insurance Trends at Mid-Century.* Edited by David McCahan, 47–66. Philadelphia: University of Pennsylvania Press, 1950.

Silver, Nate. *Signal and the Noise: Why So Many Predictions Fail—but Some Don't.* New York: Penguin Press, 2012.

Sklansky, Jeffery. "The Elusive Sovereign: New Intellectual and Social Histories of Capitalism." *Modern Intellectual History* 9, no. 1 (2012): 233–248.

Skocpol, Theda. *Protecting Soldiers and Mothers: The Political Origins of Social Policy in the United States.* Cambridge, MA: Harvard University Press, 1995.

Skocpol, Theda, and Edwin Amenta. "Did Capitalists Shape Social Security?" *American Sociological Review* 50, no. 4 (1985): 572–575.

Smith, J. V. C. "Physical Indications of Longevity in Man." In American Popular Life Insurance Company. *Sources of Longevity, Its Indications and Practical Applications.* New York: Wm. Wood, 1869.

Social Security Board. "Why Social Security?" Washington, DC: Government Printing Office, 1937.

Soll, Jacob. "From Note-Taking to Data Banks: Personal and Institutional Information Management in Early Modern Europe." *Intellectual History Review* 20, no. 3 (2010): 355–375.

Solove, Daniel J. *Nothing to Hide: The False Tradeoff between Privacy and Security*. New Haven, CT: Yale University Press, 2011.

Spirn, Anne Whiston. *Daring to Look: Dorothea Lange's Photographs and Reports from the Field*. Chicago: University of Chicago Press, 2008.

Stalson, J. Owen. *Marketing Life Insurance: Its History in America*. Cambridge, MA: Harvard University Press, 1942.

Stapleford, Thomas A. *The Cost of Living in America: A Political History of Economic Statistics, 1880–2000*. New York: Cambridge University Press, 2009.

Starr, Paul. *The Social Transformation of American Medicine: The Rise of a Sovereign Profession and the Making of a Vast Industry*. New York: Basic Books, 1982.

Starr, Paul, and Ross Corson. "Who Will Have the Numbers? The Rise of the Statistical Services Industry and the Politics of Public Data." In *The Politics of Numbers*. Edited by William Alonso and Paul Starr, 415–447. New York: Russell Sage Foundation, 1987.

Stebbins, G. S. "Practical Suggestions concerning Life Insurance." *Medical Examiner* 5, no. 46 (1895): 131–133.

Stern, Alexandra Minna. *Eugenic Nation: Faults and Frontiers of Better Breeding in Modern America*. Berkeley: University of California Press, 2005.

Stone, Deborah A. "Beyond Moral Hazard: Insurance as Moral Opportunity." In *Embracing Risk: The Changing Culture of Insurance and Responsibility*. Edited by Tom Baker and Jonathan Simon, 287–317. Chicago: University of Chicago Press, 2002.

———. "The Struggle for the Soul of Health Insurance." *Journal of Health Politics, Policy and Law* 18, no. 2 (1993): 287–317.

Stuart, M. S. *An Economic Detour: A History of Insurance in the Lives of American Negroes*. College Park, MD: McGrath, 1969.

Sutter, Paul S. "Nature's Agents or Agents of Empire? Entomological Workers and Environmental Change during the Construction of the Panama Canal." *Isis* 98, no. 4 (2007): 724–754.

Swan, Robert Joseph. "Thomas McCants Stewart and the Failure of the Mission of the Talented Tenth in Black America, 1880–1923." PhD diss., New York University, 1990.

Sydenstricker, Edgar, and Rollo H. Britten. "The Physical Impairments of Adult Life: General Results of a Statistical Study of Medical Examinations by the Life Extension Institute of 100,924 White Male Life Insurance Policy Holders since 1921." *American Journal of Hygiene* 11, no. 1 (1930): 73–94.

———. "The Physical Impairments of Adult Life: Prevalence at Different Ages, Based on Medical Examinations by the Life Extension Institute of 100,924 White Male Life Insurance Policy Holders since 1921," *American Journal of Hygiene* 11, no. 1 (1930): 95–135.

Symonds, Brandreth. "The Influence of Overweight and Underweight on Vitality." *Journal of the Medical Society of New Jersey* 5, no. 4 (1908): 159–167. Reprinted in *International Journal of Epidemiology* 39, no. 4 (2010): 951–957.

———. *Life Insurance Examinations: A Manual for the Medical Examiner and for All Interested in Life Insurance*. New York: G. P. Putnam's Sons, 1905.

———. *A Manual of Chemistry for the Use of Medical Students*. Philadelphia: P. Blakiston, Son, 1891.

———. "The Medical Jurisprudence of Life Insurance." In *A System of Legal Medicine.* Edited by Allan McLane Hamilton and Lawrence Godkin, 1:493–582. New York: E. B. Treat, 1895.

———. "The Mortality of Overweights and Underweights." *McClure's Magazine* 32, no. 3 (January 1909): 319–327.

———. "A Plea for Under Graduate Instruction in Making Life Insurance Examinations." *Medical Examiner* 9, no. 7 (July 1899): 207–208.

Sypher, F. J., ed. *Frederick L. Hoffman: His Life and Works.* Philadelphia: Xlibris, 2002.

Szreter, Simon. "The Importance of Social Intervention in Britain's Mortality Decline c. 1850–1914: A Re-interpretation of the Role of Public Health." *Social History of Medicine* 1, no. 1 (1988): 1–38.

Thomas, Owen. "The Guy Who Might Sell His Company to Microsoft for $1B Says, 'Let Him Eat Cake.' " *Business Insider,* 14 June 2012.

Thompson, Warren S., and P. K. Whelpton. "The Population Forecasts of the Scripps Foundation." *Population Index* 14, no. 3 (July 1948): 188–195.

Thornton, Tamara Plakins. "'A Great Machine' or a 'Beast of Prey,': A Boston Corporation and Its Rural Debtors in an Age of Capitalist Transformation." *Journal of the Early Republic* 27, no. 4 (2007): 567–597.

"Three Conceptions of Death." *American Health* 2, no. 2 (1909): 20–21.

Tillman, Benjamin R. *The Race Problem: Speech of Hon. Benjamin R. Tillman of South Carolina: in the Senate of the United States: February 23–24, 1903.* Washington, DC, 1903.

Tobey, J. A. "The Health Examination Movement." *Nation's Health* 5, no. 9 (1923): 610–611.

Tomes, Nancy. *The Gospel of Germs: Men, Women, and the Microbe in American Life.* Cambridge, MA: Harvard University Press, 1998.

Tone, Andrea. *The Business of Benevolence: Industrial Paternalism in Progressive America.* Ithaca, NY: Cornell University Press, 1997.

Toon, Elizabeth. "Managing the Conduct of the Individual Life: Public Health Education and American Public Health, 1910 to 1940." PhD diss., University of Pennsylvania, 1998.

Toulmin, Harry. "The American Negro as an Insurance Risk." *Abstract of the Proceedings of the Association of Life Insurance Medical Directors of America* 24 (1913): 152–179.

———. "Some Details of Office Methods." *Abstract of the Proceedings of the Association of Life Insurance Medical Directors of America* 18 (1907): 98–109.

Turow, Joseph. *Breaking up America: Advertisers and the New Media World.* Chicago: University of Chicago Press, 1997.

Ukers, William H. *All about Coffee.* New York: Tea and Coffee Trade Journal Company, 1922.

Ulrich, Laurel Thatcher. *The Life of Martha Ballard, Based on Her Diary, 1785–1812.* New York: Vintage, 1991.

United States Bureau of the Census. *Birth Statistics for the Birth Registration Area of the United States 1919.* Washington, DC: Government Printing Office, 1921.

United States Department of Commerce and Labor. *Statistical Abstract of the United States,* no. 33. Washington, DC: Government Printing Office, 1911.

Vaughan, Victor C. "The Importance of Frequent and Thorough Medical Examinations of the Well." In *Proceedings of the First National Conference on Race Betterment,* 90–96. Battle Creek, MI: Race Betterment Foundation, 1914.

Veit, Helen Zoe. " 'Why Do People Die?' Rising Life Expectancy, Aging, and Personal Responsibility." *Journal of Social History* 45, no. 4 (2012): 1026–1048.

Vogel, Lauren. "The Skinny on BMI Report Cards." *Canadian Medical Association Journal* 183, no. 12 (2011): E787–E788.

Vose, Edward. *Seventy-Five Years of the Mercantile Agency R. G. Dun & Co. 1841–1916.* Brooklyn, NY: R.G. Dun, 1916.

Washington, Booker T. *Negro Education Not a Failure.* Tuskegee, GA: Tuskegee Institute Steam print, 1904.

Weare, Walter B. *Black Business in the New South: A Social History of the North Carolina Mutual Life Insurance Company.* Durham, NC: Duke University Press, 1993.

Webb, Constance. *Richard Wright: A Biography.* New York: G. P. Putnam's Sons, 1968.

Weems, Robert E. *Black Business in the Black Metropolis: The Chicago Metropolitan Assurance Company, 1925–1985.* Bloomington: Indiana University Press, 1996.

Weigley, Emma Seifrit. "Average? Ideal? Desirable? A Brief Overview of Height-Weight Tables in the United States." *Journal of the American Dietetic Association* 84, no. 4 (1984): 417–423.

Welke, Barbara Young. *Recasting American Liberty: Gender, Race, Law, and the Railroad Revolution, 1865–1920.* Cambridge: Cambridge University Press, 2001.

———. "When All the Women Were White, and All the Blacks Were Men: Gender, Class, Race, and the Road to *Plessy*, 1855–1914." *Law and History Review* 13, no. 2 (1995): 261–316.

White, Richard. *Railroaded: The Transcontinentals and the Making of Modern America.* New York: Norton, 2011.

Wiggins, Benjamin Alan. "Managing Risk, Managing Race: Racialized Actuarial Science in the United States, 1881–1948." PhD diss., University of Minnesota, 2013.

Williamson, W. R. "Social Budgeting." *Proceedings of the Casualty Actuarial Society of America* 24, no. 49 (1937): 17–34.

———. "Social Insurance." *Journal of the American Association of University Teachers of Insurance* 14, no. 1 (1947): 84–93.

———. "Some Backgrounds to American Social Security." *Proceedings of the Casualty Actuarial Society* 30, no. 60 (November 1943): 5–30.

———. "Statement of W. R. Williamson." Economic Security Act 1009–1020. http://www .ssa.gov/history/pdf/hr35williamson.pdf.

Wilson, James Grant. *The Memorial History of the City of New-York.* New York: New York History Company, 1893.

Winner, Langdon. "Do Artifacts Have Politics?" *Daedalus* 109, no. 1 (1980): 121–136.

Witt, John Fabian. *The Accidental Republic: Crippled Workingmen, Destitute Widows, and the Remaking of American Law.* Cambridge, MA: Harvard University Press, 2004.

Wolff, Megan Joy. "The Money Value of Risk: Life Insurance and the Transformation of American Public Health, 1896–1930." PhD diss., Columbia University, 2011.

———. "The Myth of the Actuary: Life Insurance and Frederick L. Hoffman's *Race Traits and Tendencies of the American Negro*." *Public Health Chronicles* 121 (January–February 2006): 84–91.

Woods, Edward A., and Clarence B. Metzger. *America's Human Wealth: The Money Value of Human Life.* New York: F.S. Crofts, 1927.

Woodson, C. G. "Insurance Business among Negroes." *Journal of Negro History* 14, no. 2 (April 1929): 202–226.

Woodward, C. Vann. *Strange Career of Jim Crow.* New York: Oxford University Press, 2002.

Wright, Elizur. *Massachusetts Reports on Life Insurance: 1859–1865.* Boston: Wright & Potter, 1865.

———. "National vs. State Supervision." *Insurance Times* 1 (July 1868): 263–265.

———. *Politics and Mysteries of Life Insurance.* Boston: Lee & Shepard, 1873.

Wright, Richard. *American Hunger.* New York: Harper & Row, 1977.

———. *Black Boy: A Record of Childhood and Youth.* New York: Harper & Row, 1969.

———. *Native Son.* New York: Harper's, 1940.

"W. R. Williamson." *Proceedings of the Casualty Actuarial Society* 67, no. 128 (1980): 274.

Wyatt, Birchard E. *Private Group Retirement Plans.* Washington, DC: Graphic Arts Press, 1936.

Wyatt, Birchard E., and William H. Wandel. *The Social Security Act in Operation: A Practical Guide to the Federal and Federal-State Social Security Programs.* Washington, DC: Graphic Arts Press, 1937.

"Wyco" The Building of a Professional Actuarial and Consulting Organization: The Wyatt Company. Wyco, 1984.

Wynes, Charles E. "T. McCants Stewart: Peripatetic Black South Carolinian." *South Carolina Historical Magazine* 80, no. 4 (1979): 311–317.

Wynne, James. "Louis Agassiz." *Harper's New Monthly Magazine* 25, no. 146 (July 1862): 199–200.

———. *Report on the Vital Statistics of the United States, Made to the Mutual Life Insurance Company of New York.* New York: H. Bailliere, 1857.

Yates, JoAnne. *Structuring the Information Age: Life Insurance and Technology in the Twentieth Century.* Baltimore, MD: Johns Hopkins University Press, 2005.

Zakim, Michael. "Bookkeeping as Ideology: Capitalist Knowledge in Nineteenth-Century America." *Common-place* 6, no. 3 (2006). http://www.common-place.org/vol-06/no-03/zakim.

———. "Inventing Industrial Statistics." *Theoretical Inquiries in Law* 11 (January 2010): 283–318.

Zartman, Lester. *The Investments of Life Insurance Companies.* New York: Henry Holt, 1906.

Zelizer, Julian E. "Where Is the Money Coming From?" *Journal of Policy History* 9, no. 4 (1997): 399–424.

Zelizer, Viviana A. Rotman. *Morals and Markets: The Development of Life Insurance in the United States.* New York: Columbia University Press, 1979.

———. *Pricing the Priceless Child: The Changing Social Value of Children.* Princeton, NJ: Princeton University Press, 1994.

Zipp, Samuel. *Manhattan Projects: The Rise and Fall of Urban Renewal in Cold War New York.* New York: Oxford University Press, 2010.

Zunz, Olivier. *Making America Corporate 1870–1920.* Chicago: University of Chicago Press, 1990.